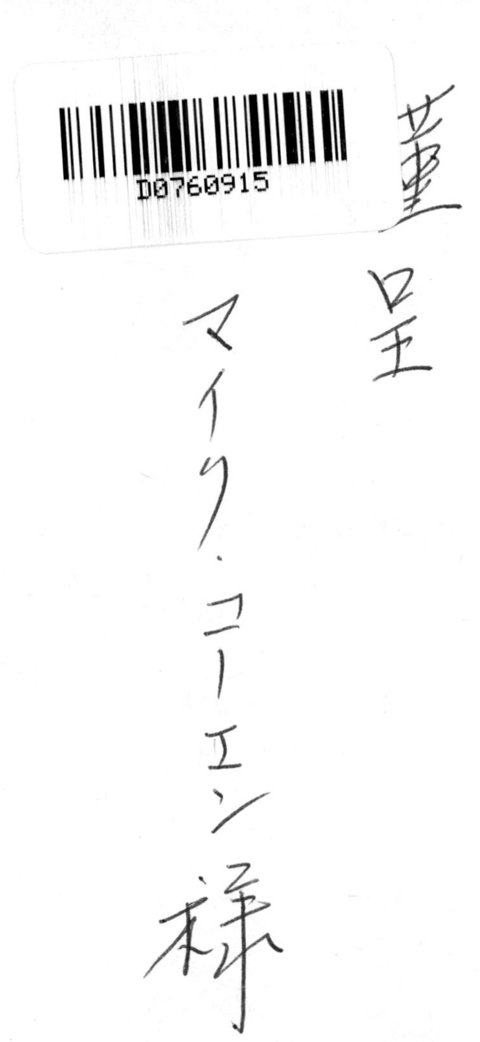

謹呈

マイリ・コーエン様

2011.10.5.

貝沼圭二

Handbook of Amylases and Related Enzymes

Their Sources, Isolation Methods, Properties and Applications

Titles of Related Interest

ALEXANDER *et al.*
Bioreduction in the Activation of Drugs

FRIEDRICH
Supramolecular Enzyme Organization

LINDROS *et al.*
Advances in Biomedical Alcohol Research

WEBER
Enzyme Pattern — Targetted Chemotherapy

Journals of Related Interest *(sample copy gladly sent on request)*

Advances in Enzyme Regulation

Cellular and Molecular Biology

International Journal of Biochemistry

Handbook of Amylases and Related Enzymes

Their Sources, Isolation Methods, Properties and Applications

Edited by

THE AMYLASE RESEARCH SOCIETY OF JAPAN

Committee Office, Osaka Municipal Technical Research Institute, 1–6–50 Morinomiya, Jyoto-ku, Osaka 536, Japan

PERGAMON PRESS

OXFORD · NEW YORK · BEIJING · FRANKFURT
SÃO PAULO · SYDNEY · TOKYO · TORONTO

U.K.	Pergamon Press plc, Headington Hill Hall, Oxford OX3 0BW, England
U.S.A.	Pergamon Press, Inc., Maxwell House, Fairview Park, Elmsford, New York 10523, U.S.A.
PEOPLE'S REPUBLIC OF CHINA	Pergamon Press, Room 4037, Qianmen Hotel, Beijing, People's Republic of China
FEDERAL REPUBLIC OF GERMANY	Pergamon Press GmbH, Hammerweg 6, D-6242 Kronberg, Federal Republic of Germany
BRAZIL	Pergamon Editora Ltda, Rua Eça de Queiros, 346, CEP 04011, Paraiso, São Paulo, Brazil
AUSTRALIA	Pergamon Press Australia Pty Ltd., P.O. Box 544, Potts Point, N.S.W. 2011, Australia
JAPAN	Pergamon Press, 5th Floor, Matsuoka Central Building, 1-7-1 Nishishinjuku, Shinjuku-ku, Tokyo 160, Japan
CANADA	Pergamon Press Canada Ltd., Suite No. 271, 253 College Street, Toronto, Ontario, Canada M5T 1R5

First edition 1988

Library of Congress Cataloging-in-Publication Data

Handbook of amylases and related enzymes: their sources, isolation methods, properties and applications/ edited by the Amylase Research Society of Japan.
— 1st ed.
p. cm.
Includes index.
1. Amylases—Handbooks, manuals, etc. I. Amylase Research Society of Japan.
QP609.A45H36 1988 574.19'254—dc19 88-25465

British Library Cataloguing in Publication Data

Handbook of amylases and related enzymes.
1. Amylases
I. Amylase Research Society of Japan
547.7'58

ISBN 0-08-036141-2

Printed in Great Britain by A. Wheaton & Co. Ltd., Exeter

Editorial Committee

List of Contributors

Akio Kuroda	Yakult Central Institute for Microbiological Research
Bunzo Mikami	The Research Institute for Food Science, Kyoto University
Fumitoshi Hirayama	Faculty of Pharmaceutical Sciences, Kumamoto University
Hajime Takaku	Showa Sangyo Co., Ltd.
Hidemasa Hidaka	Bio Science Laboratories, Meiji Seika Kaisha, Ltd.
Hidetsugu Fuwa	The Institute of Industrial Science Research, Fukuyama University
Hiroko Toda	Institute for Protein Research, Osaka University
Hitoshi Hashimoto	Ensuiko Sugar Refining Co., Ltd.
Iwao Maeda	Hayashi Gakuen Women's Junior College
Kahee Fujita	Faculty of Pharmaceutical Sciences, Fukuyama University
Kaneto Uekama	Faculty of Pharmaceutical Sciences, Kumanoto University
Kaoru Omichi	College of Science, Osaka University
Keiichi Naka	Osaka City University Medical School
Keiji Kainuma	National Food Research Institute
Keisuke Matsumoto	Yakult Central Institute for Microbiological Research
Keitaro Hiromi	College of Agriculture, Kyoto University
Kenzaburo Yoritomi	Towa Kasei Co., Ltd.
Kiyoshi Okuda	Osaka City University Medical School
Koji Yokobayashi	Hayashibara Biochemical Laboratories, Inc.
Kunio Yamane	Institute of Biological Sciences, University of Tsukuba
Kyohei Mizokami	Faculty of Engineering, Fukuyama University
Masakazu Mitsuhashi	Hayashibara Biochemical Laboratories, Inc.
Masami Kusunoki	Institute for Protein Research, Osaka University
Masatake Ohnishi	College of Agriculture, Kyoto University
Noboru Taji	Nagase Biochemicals Ltd.
Nobuo Shimojo	Osaka City University Medical School
Nobuya Matsumoto	Suntory Limited
Noshi Minamiura	Faculty of Science, Osaka City University
Ryu Shinke	Faculty of Agriculture, Kobe University
Seinosuke Ueda	Kamamoto Institute of Technology
Seiya Chiba	Faculty of Agriculture, Hokkaido University
Shigemitsu Osaki	Sanwa Denpun Kogyo Co., Ltd.
Shigetaka Okada	Biochemical Research Laboratories, Ezaki Glico Co.
Shigetoshi Endo	Daiwa Kasei K.K.
Shinsaku Hayashida	Faculty of Agriculture, Kyushu University
Shizuo Mihara	Amano Pharmaceutical Co., Ltd.
Shuzo Sakai	Hayashibara Biochemical Laboratories, Inc.
Sumio Kitahata	Osaka Municipal Technical Research Institute
Takashi Nanmori	Faculty of Agriculture, Kobe University

Takehiko Yamamoto	Faculty of Engineering, Fukuyama University
Tokuji Ikenaka	College of Science, Osaka University
Tomoe Kanno	Showa Sangyo Co., Ltd
Toshiaki Komaki	The Institute of Industrial Science Research, Fukuyama University
Tsukasa Yoshida	Sanmatsu Kogyo Co., Ltd.
Yasuhito Takeda	Faculty of Agriculture, Kagoshima University
Yoshikazu Nakajima	Mitsui Sugar Co., Ltd.
Yoshiki Matsuura	Institute for Protein Research, Osaka University
Yoshikuni Ito	Kato Kagaku Co., Ltd.
Yoshio Hanno	Matsutani Chemical Industry Co., Ltd.
Yoshio Tsujisaka	Hayashibara Institute Corp.
Yoshiyuki Sakano	Faculty of Agriculture, Tokyo Noko University
Yoshiyuki Takasaki	Fermentation Research Institute, Agency of Industrial Science and Technology
Yuhei Morita	The Research Institute for Food Science, Kyoto University
Zenichi Yoshino	Sanwa Denpun Kogyo Co., Ltd.

Preface to the First Edition

The "*Amylase Research Society of Japan*" was organized in 1967 by the proposal of the late Dr. Juichiro Fukumoto, with generous cooperation of his sympathizers. At that time he was Professor of Applied Microbiology at the Faculty of Science of Osaka City University. Several years before the society's organization, he and his collaborators reported a novel enzymic method of glucose production from starch. This method was generally accepted and took the place of the glucose production method so far employed. This epochal event led to the foundation of the Amylase Research Society.

Since then the society has organized symposia on amylases and issued the proceedings every year. Many interesting and useful findings especially of several amylases of new type, studies for mechanisms of amylase catalyzed reactions, applications of amylases and related enzymes to industry have been dealt with in the symposia. This year on the occasion of the twentieth anniversary of the society, the issue of "*Handbook of Amylases and Related Enzymes*" has been planned as the consensus of members of the research society. The "related enzymes" refers hereupon to enzymes such as pullulanase, which at present are not in the category of amylases but are capable of catalyzing certain reactions with starch and glycogen. The "related enzymes" also contain those which are useful for further conversion of amylase reaction products from the point of view of industrial application.

The handbook is designed to present, in concise form, the most reliable and factual information about the enzymic properties and application of amylases and related enzymes. Suggestions and opinions are eliminated as far as possible. The handbook is also designed for the most part to serve as a source of information and material for researchers and engineers working in academic and industrial fields using amylases and related enzymes.

The "*Handbook of Amylases and Related Enzymes*" is divided into seven major parts; (I) enzyme kinetics of amylases and related enzymes, (II) and (III) data on individual amylolytic enzymes, (IV) clinical assay of human blood and urine amylase, (V) analytical and (VI) industrial application of amylases and related enzymes, and (VII) amylolytic enzyme preparations as digestive agents. In this handbook the enzymes are designated generally by their traditional names, but the recommended names are also noted according to "*Enzyme Nomenclature, Recommendations (1984) of the Nomenclature Committee of the International Union of Biochemistry.*"

In part I, the method of kinetic treatment of reactions specific to amylases and related enzymes and the definitions of kinetic parameters are given in text. This part also includes description of certain items to be noticed to assay the enzymes. Many data in parts II and III are quantitative. Some, however, qualitative, especially those concerning the methods of production and purification of enzymes, where the facts at present understood are given in text. The items in parts IV and V are also presented as text. The data in part VI are given mainly in form of diagrams or flow sheets with some comments as quantitative as possible. In part VII, amylolytic enzyme preparations, their quality and standardization according to the Japan Pharmacopoeia are discussed. Commercial suppliers of amylases and related enzymes in Japan are listed in an appendix.

The contributors to this handbook all are the members of the Amylase Research Society of Japan. The committee for issue of this handbook is grateful to the contributors who have generously cooperated in preparation of this edition. We would like to dedicate this handbook to the late Prof. Juichiro Fukumoto. We hope that this handbook will serve well for researchers and engineers who are engaged in work concerned with amylases and related enzymes. We are deeply indebted to the publishers for their unfailing cooperation at every stage of preparation of this handbook. Also, as the president of the "Amylase Research Society of Japan", I would like to express my sincerest thanks to several Foundations and industrial corps. in Japan for their kind supply of the grant in aid of holding the symposium of amylases and publishing the proceedings yearly extending over two decades, as well as that in issuing this handbook.

Takehiko Yamamoto

June 1988

Contents

I. Enzyme Kinetics of Amylases and Related Enzymes

I.1.a. Reactions Catalyzed by Amylases and Related Enzymes

Amylases are classified into "Hydrolases" (EC 3.2.1) that catalyze the hydrolysis of O-glycosyl compounds. However, since 'hydrolysis' is a kind of 'transfer' in which the acceptor is water, confusion could occur in some cases ; for example, cyclomaltodextrin glucanotransferase (EC 2.4.1.19), which was formerly called *Bacillus macerans* amylase, actually 'hydrolyzes' starch, but has been classified into "Transferases", owing to its characteristic transglycosylation including cyclization. Not a few amylases have transferring activity as well as hydrolyzing activity, and some of them have been utilized for producing useful saccharides (e. g., *Streptomyces praecox* α-amylase, etc.).
In contrast to transfer, 'condensation' is merely the reverse reaction of 'hydrolysis'. Phenomenologically, however, condensation and transfer are similar in that products of degrees of polymerization (DP) higher than the starting substrate are produced. In spite of the similarity, condensation and transfer are different in mechanisms.
Since enzymes are catalysts, they should enhance the velocities of the forward and the reverse (backward) reactions by the same factor. Any amylase catalyzing the hydrolysis of an α-glucosidic linkage must catalyze the formation (the condensation) of the same linkage in the same ratio. For example, consider the following reaction :

$$\text{G-G} + H_2O \underset{k_b}{\overset{k_f}{\rightleftharpoons}} \text{G} + \text{G} \tag{1}$$

where G designates a glucose residue or a glucose molecule on the right or the left hand side of the equation, respectively, G–G is maltose, and k_f and k_b are the rate constants of the forward and the backward reactions, respectively. The equilibrium constant of the hydrolysis, K_h, and that of the condensation, K_c, are given by :

$$K_c = 1/K_h = [\text{G-G}][H_2O]/[\text{G}]^2 = k_b/k_f \tag{2}$$

$$\therefore [\text{G-G}] = K_c[\text{G}]^2/[H_2O] = [\text{G}]^2/(K_h[H_2O]) \tag{3}$$

where [G] and [G–G] are the equilibrium concentrations of glucose and maltose, respectively. Thus the condensation is favored by the higher concentration of G and by the lower water concentration. The equilibrium constant for hydrolysis K_h can be expressed without taking water concentration $[H_2O]$ into consideration, as $K_h' = K_h[H_2O]$, and similarly, the rate constant $k_f' = k_f[H_2O]$.
For glycosidic linkages in general, the equilibrium (Eq. 1) is largely inclined towards the hydrolysis side ($k_f' \gg k_b$).
But, it should be noticed that the K_h or K_c value depends on the nature of the

glycosidic linkage to be hydrolyzed. Thus α-1,4-glucosidic linkage is more readily hydrolyzed than α-1,6-glucosidic linkage; conversely, α-1,6-linkage is more easily formed from free glucose at high concentrations, as suggested by the findings that the values of the enthalpy of hydrolysis, ΔH_h, for α-1,4- and α-1,6-glucosidic linkages are about -1 and $+1$ kcal mol^{-1}, respectively (1). In fact, the glucoamylase-catalyzed formation of isomaltose, but not maltose, from glucose of higher concentrations has been confirmed (2).

On the other hand, the transfer reaction may be written as

$$\text{G-G} + \text{A} \underset{k_{-t}}{\overset{k_t}{\rightleftharpoons}} \text{G-A} + \text{G} \tag{4}$$

where A is the acceptor. The equilibrium constant of transfer K_t is given by

$$K_t = [\text{G-A}][\text{G}]/[\text{G-G}][\text{A}] = k_t/k_{-t} \tag{5}$$

$$\therefore [\text{G-A}] = K_t\,[\text{G-G}][\text{A}]/[\text{G}] \tag{6}$$

where the rate constants, k_t and k_{-t}, are those indicated in Eq. 4, and the concentrations here refer to the equilibrium concentrations. The acceptor A could be the substrate G-G itself, and in this case the transfer product G-A would be maltotriose, G-G-G. In hydrolytic reactions catalyzed by certain kinds of amylases, such as *B. subtilis* saccharifying α-amylase, the transient formation of G-G-G from G-G has been observed (3).

The tendency of occurrence of transfer or condensation in an amylase-catalyzed reaction is determined by the equilibrium constant itself (K_t or K_c) of the relevant reaction on one side, and the characteristics of the amylase concerned, on the other. At lower substrate concentrations, we can observe the hydrolysis preferentially. Since the main reaction of amylase is the hydrolysis, only the hydrolytic reaction will be the object of the following kinetic treatment, although kinetic studies including transfer and/or condensation would be possible when quantitative chromatographic analyses are performed.

I.1.b. Analysis of Action Patterns of Amylases Based on the Subsite Theory

The term 'action pattern' here will be used in the following sense : 1) Dependence of the rate parameters (the Michaelis constant K_m and the molar activity k_O (the maximum velocity V divided by the molar concentration of enzyme $[E]_O$)) on DP of linear substrates, which is simply denoted by n ; and 2) Probabilities of splitting of each α-glucosidic linkage in a maltooligosaccharide of a definite DP.

The above defined action pattern of an amylase can be measured quantitatively : K_m and k_O are determined from the concentration dependence of the initial rate of hydrolysis by reducing end measurement (see I.2. below), and the frequency of splitting of each linkage of a maltooligosaccharide whose reducing end glucose residue is labeled with ^{14}C can be measured quantitatively by chromatographic techniques (4). These data are used to evaluate the subsite affinity, A_i, of each subsite constituting the active site of amylase, which is defined as the unitary part of the free energy decrement caused by the binding of glucose residue of a linear substrate with a subsite specified by the subsite number i ($i = 1 \sim m$), where m is the total number of subsites in the active site (see Fig. 1). These treatments are based on "the subsite theory", the essence of which will be outlined below.

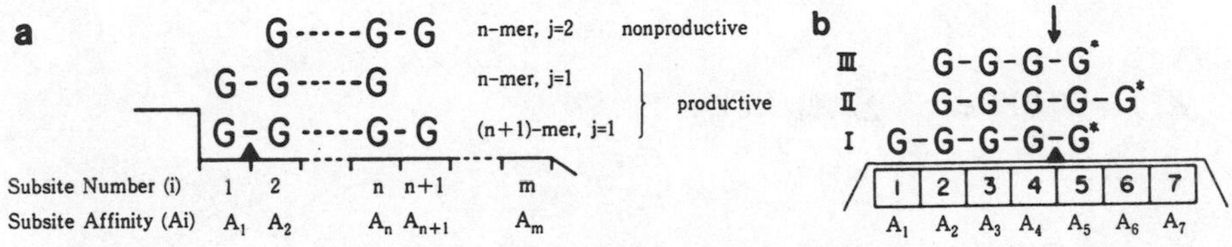

Fig. 1. Schematic Representation in Terms of Subsites for the Active Site of Glucoamylase (a) and that of α-Amylase (b).
Subsites are numbered counting from the terminal one on which the non-reducing end glucose residue, which is generally represented by G, of substrates is situated in the productive binding modes. Each subsite ($i = 1 \sim m$) has its own subsite affinity (A_i) toward a glucose residue. The wedge indicates the catalytic site of the amylase.
Productive and non-productive binding modes are shown in (a), together with how to number the binding modes. In (b) the principle of evaluating some of the subsite affinities by using reducing end-labeled maltooligosaccharides (G* refers to ^{14}C-labeled glucose residue) is shown (see the text). (from *J. Jpn. Soc. Starch Sci.* (in Japanese) *32*, 84-93 (1985).)

The Essence of the Subsite Theory : In amylase-catalyzed hydrolytic reactions of linear substrates, two types of modes of binding of the substrate can occur; the productive mode which can yield products and the non-productive mode leading to no product formation. The number of these binding modes can be more than one: for example, glucoamylase, an exo-amylase, has only one productive binding and $(m-1)$ non-productive binding modes, whereas an α-amylase has $(n-1)$ productive binding modes and m non-productive binding modes, where m and n are the number of subsites of the amylase and DP of the substrate, respectively (see Fig. 1). To specify each binding mode, either productive or non-productive, the subscript j is used, which is taken equal to the subsite number (i) of the subsite on which the non-reducing end glucose residue is situated in the binding mode (if necessary, extra subsite numbers, $0, -1, -2, \cdots$ may be used). For convenience of discriminating productive and non-productive binding modes or complexes, the subscript p and q are used; both of them are included in j.

The probability of occurrence of an arbitrary binding mode j is proportional to, and can conveniently be expressed in terms of, the association constant, $K_{n,j}$, for the binding of an n-mer linear substrate S_n in a binding mode j, which is defined as

$$K_{n,j}=[ES_{n,j}]/([E][S_n]) \quad (7)$$

If we assume that the unitary binding affinity (the unitary free energy decrement) for the substrate S_n to the enzyme active site is simply the sum of the subsite affinities (A_i's) of the subsites actually occupied by that binding mode, j, the association constant $K_{n,j}$ is given by

$$K_{n,j} = 0.018 \exp \left(\sum_i^{\text{cov.}} A_i/RT\right)_{n,j} \quad (8)$$

where R and T are the gas constant and the absolute temperature, respectively. Thus $K_{n,j}$ can be expressed in terms of the subsite affinities.

On the other hand, when there are a number of binding modes of a substrate to an enzyme, the two-step Michaelis-Menten mechanism should be written as follows:

$$E + S_n \underset{}{\overset{K_{n,p}}{\rightleftharpoons}} ES_{n,p} \xrightarrow{k_{\text{int}}} E + P \quad (9)$$
$$K_{n,q} \rightleftharpoons ES_{n,q}$$

where $ES_{n,p}$ and $ES_{n,q}$ are the productive and the non-productive enzyme-substrate complexes, respectively, and $K_{n,p}$ and $K_{n,q}$ the association constants for the formation of $ES_{n,p}$ and $ES_{n,q}$, respectively; k_{int} is the intrinsic rate constant of glucosyl bond cleavage in the productive complexes, which is assumed to be constant regardless of DP of substrates.

The rate equation for the initial velocity of hydrolysis v derived from Eq. 9 becomes:

$$v = k_O[E]_O[S_n]/(K_m+[S_n]) \quad (10)$$

which is the same in form as the familiar Michaelis-Menten equation. However, the meanings of the Michaelis constant K_m and the molar activity k_O are complicated, which are given as follows:

$$1/K_m = \sum_j K_{n,j} = \sum_p K_{n,p} + \sum_q K_{n,q} \quad (11)$$

$$k_O = k_{\text{int}}/[1+(\sum_q K_{n,q}/\sum_p K_{n,p})] \quad (12)$$

and therefore, k_O/K_m becomes

$$k_O/K_m = k_{int} \sum_p K_{n,p} \tag{13}$$

It is to be noted that k_O/K_m, which is proportional to the initial velocity at sufficiently low substrate concentrations ($[S_n] \ll K_m$), includes only the parameters of productive binding modes, although K_m and k_O include those of both productive and non-productive binding modes.

Since all the association constants, $K_{n,p}$ and $K_{n,q}$ (generally expressed by $K_{n,j}$), have been obtained in terms of A_i's, substitution of Eq. 8 into Eqs. 11-13 gives the equations for the rate parameters (K_m, k_O, and k_O/K_m) expressed with the subsite parameters (k_{int} and A_i's). Thus the rate parameters of the linear substrates of various DP's can be used to determine the subsite parameters, and *vice versa*.

Another useful way to determine the subsite affinities, A_i's, is to measure the rate of formation of reducing-end labeled products from the reducing-end labeled initial substrates of different DP's. The principle can be understood from Fig.1b : for example, the rate of production of G-G* from G-G-G-G-G* (v_{II}) divided by that of G* from G-G-G-G* (v_{III}) gives the subsite affinity A_6 in the figure, since v_{II} and v_{III} are proportional to the relevant productive association constants $K_{5,2}$ and $K_{4,2}$, respectively.

These association constants can in turn be expressed by the sum of A_i's of the subsites occupied in the relevant binding modes (see Fig. 1b, and Eq. 8).

The arrangement of subsite affinities, termed 'subsite structure', of various amylases and an α-glucosidase from buckwheat are shown in histograms in Fig. 2. Of these, the subsite structures of glucoamylase and the α-glucosidase were determined by using the rate parameters of a series of linear substrates of different DP's alone, without using the product analysis method with labeled substrates stated above. The latter method is useful especially for endo-amylases. When the both methods are combined, all the subsite parameters (k_{int} and A_i's) can be determined.

The value of subsite affinity, A_i, could be negative or zero, if some kind of strain is brought about at the subsite either by the distortion of substrate pyranose ring or that of enzyme conformation at that subsite. The negative subsite affinity can usually appear at the subsite adjacent to the cleavage point (on the non-reducing end side) of the glucosidic linkage, as found in the case of lysozyme. This negative A_i accounts for the low rate of hydrolysis of maltose, since the probability of occurrence of the productive binding mode for this substrate ($K_{n,p}$) is extremely low, as seen from Eq. 8 (see, for example, β-amylase and Taka-amylase A in Fig. 2).

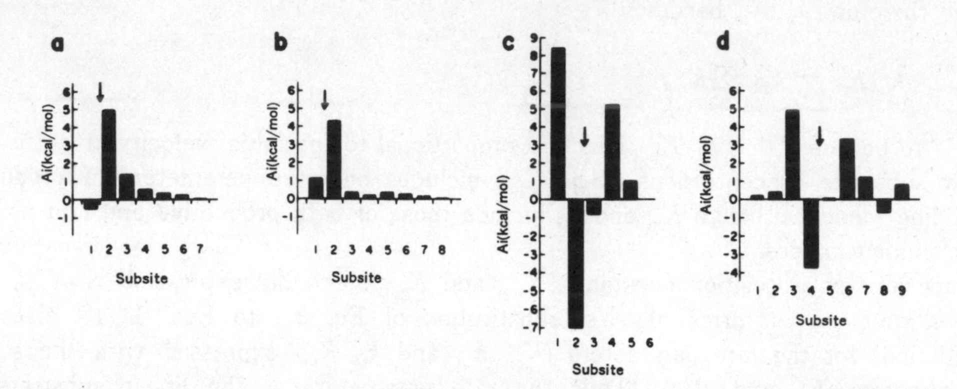

Fig. 2. Histograms Showing the Characteristic Arrangement of Subsite Affinities of Various Amylases and an α-Glucosidase.
(a) Glucoamylase from *Rhi. niveus*. (b) α-Glucosidase from buckwheat.
(c) β-Amylase from soybean. (d) Taka-amylase A from *Asp. oryzae*. (from *Mol. Cell. Biochem. 51*, 79-95 (1983) with modification.)

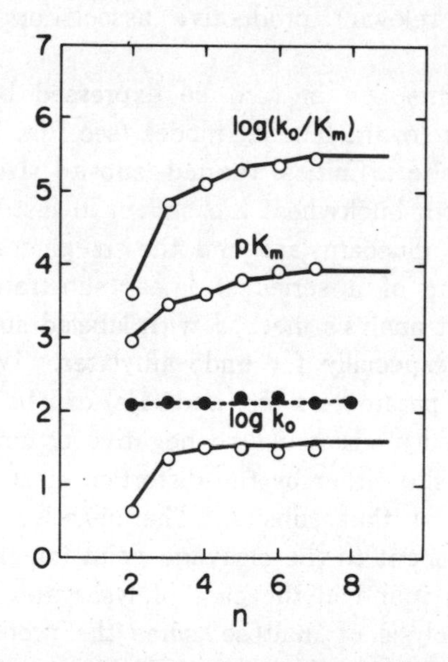

Fig. 3. DP-Dependences of K_m, k_O and (k_O/K_m) for Glucoamylase (open circles) and That of k_O for Buckwheat α-Glucosidase (closed circles). (n=DP).
The solid lines are the theoretical curves drawn according to the subsite theory (see the text). (from *Mol. Cell. Biochem. 51*, 79-95 (1983).)

The comparison of k_O values between glucoamylase and α-glucosidase from buckwheat is interesting. As seen from Fig. 3, k_O increases with DP for glucoamylase, whereas it is constant for α-glucosidase. This difference comes from the fact that $A_1 < A_3$ for the former, while $A_1 > A_3$ for the latter (see Fig. 2). In the case of glucoamylase with maltose as the substrate, $(A_1 + A_2) < (A_2 + A_3)$, then $K_{n,q}/K_{n,p}$ becomes appreciably larger than unity, leading to $k_O < k_{int}$ (cf. Eqs. 8 and 12). However, the contribution of the productive term, $K_{n,p}$, increases with increasing DP of substrate, since the involvement of A_3, A_4 and A_5 which become occupied in the longer substrates increases the $K_{n,p}$ values relative to $K_{n,q}$. This is the reason why k_O tends to increase with DP. In contrast, for buckwheat α-glucosidase, $K_{n,p}$ is always larger than $K_{n,q}$ irrespective of DP of substrates ($K_{n,q}/K_{n,p} \ll 1$), and hence $k_O \fallingdotseq k_{int}$ holds independently of DP. This is one of the examples of useful application of the subsite theory for the explanation of substrate specificity of amylases and related enzymes.

Application of the Subsite Theory : Some of the useful applications of the subsite theory to the analysis of amylase-catalyzed reactions will be stated below.

1. Interpretation of substrate specificity. Besides the example of different DP-dependence of molar activities of glucoamylase and α-glucosidase described above, the specificity of amylases is reasonably accounted for by the theory. For example, very low rate of hydrolysis of phenyl α-glucoside by glucoamylase has successfully been explained in terms of the low subsite affinity of the second subsite of the enzyme towards phenyl residue (5).
2. Characterization of the nature of subsites. Since the binding modes of substrates or analogues (e.g., glucose or gluconolactone) can be predicted based on the known subsite structure, spectroscopic (difference absorption or fluorescence) methods as well as chemical modification technique can be effectively utilized for characterizing the nature of some subsites. The existence of a tryptophan residue at the first subsite of glucoamylase has been predicted from the combined analysis with the above-mentioned methods (6).
3. Prediction of the time-dependent product distribution. As was described earlier, the rate parameters (K_m and k_O) for every linear substrate can be calculated from the subsite parameters (k_{int} and A_i's). Thus the time courses of product distribution starting from an N-mer substrate can be calculated by solving simultaneous differential rate equations with N unknowns by using a computer. This computer simulation enables us to predict, for example, at what percentage of reaction an intermediate substrate (or product) of a given DP can be obtained at a maximum yield, which may be useful in manufacturing oligosaccharides. Moreover, by comparing the theoretical distribution curves with those obtained experimentally using high performance liquid chromatography, the occurrence of multiple attack can be known (7).
4. Chemical modification for changing substrate specificity. If the subsite affinity of a certain subsite is changed by some chemical modification, the action pattern including the substrate specificity will be altered. It is possible, therefore, that chemical modification of an amino acid in a particular subsite may lead to enhanced rate for smaller substrates or synthetic substrates with concomitant decrease in rate

for longer substrates (8). The possibility of artificial alteration of enzyme activity and specificity may be achieved chemically, apart from the protein engineering by using genetic modification.

For more details of the subsite theory and its application, readers may refer to the review article (6) and the literatures cited therein.

I.1.c. Fast Reaction Kinetics

Fast reaction techniques, especially the stopped-flow method which can measure fast reactions in solution down to a little less than a millisecond, is a powerful tool to elucidate more detailed mechanisms of enzyme reactions than the conventionally used steady-state enzyme kinetics (9). The first successful application of the stopped-flow kinetic study was done on the binding of substrates to glucoamylase by monitoring the UV absorption and fluorescence change caused by the ligand binding. The analysis of the results showed that the ligand binding occurs in a two-step mechanism, in which a rapid bimolecular association to form a loosely bound intermediate complex (ES) precedes a slower unimolecular isomerization of ES into a more tightly bound complex (ES*) in the time range of milliseconds. Several lines of evidence strongly suggest that the complexes, ES and ES*, are the non-productive and the productive complexes, respectively, in the latter of which the cleavage of the terminal glucosidic linkage is brought about by the action of catalytic residues of glucoamylase (6). The mechanisms of glucoamylase-catalyzed hydrolysis of maltooligosaccharides are schematically shown in Fig. 4.

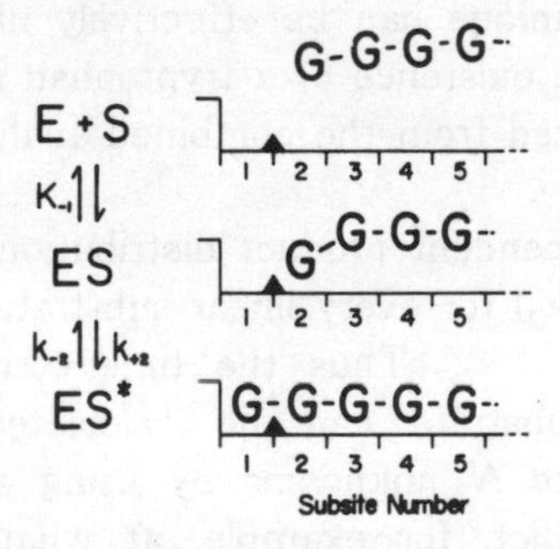

Fig. 4. Schematic Representation of the Proposed Binding Mechanism of Maltooligosaccharides with Glucoamylase.

ES* represents the productive complex, in which the terminal glucosidic linkage at the wedge is cleaved to form products. (from *Mol. Cell. Biochem.* *51*, 79-95 (1983).)

I.1.d. Additional Comments

In addition to the evaluation and the interpretation of the rate parameters (K_m and k_O) obtainable from classical Michaelis-Menten kinetics for a series of substrates of similar structures, the study of pH effect is important and useful to investigate the essential ionizable groups involved in the catalysis. In this case, a simplified procedure for the determining pK_e values of the ionizable groups in the free enzyme molecule, employing sufficiently lower substrate concentrations compared with K_m where $v \propto (k_O/K_m)[S]$ holds, is convenient and useful. If the temperature dependence of the pK_e's is studied to evaluate the heats of ionization, ΔH_e's, those values together with pK_e's are useful to characterize and identify the sorts of the essential ionizable groups involved.

The use of inhibitors of various kinds, substrate analogues and transition state analogues such as gluconolactone, to investigate the type of inhibition, the inhibitor constants, K_i's, is often helpful to get information about the binding site, the complemental structure of the enzyme binding site to fit the ligand structure, etc. In parallel with the kinetic measurements of this kind, spectrophotometric or fluorometric measurements on the binding of substrates or inhibitors are powerful tools for determining the dissociation (or association) constants of the ligands directly, which are to be compared with those values obtained kinetically.

When the transfer and/or condensation reactions are involved in addition to hydrolysis, the kinetic studies may not be so easy as in the case where the hydrolysis is the sole reaction. Generally, these additional reactions may become significant at higher concentrations of substrates and/or acceptors of lower DP's. Chromatographic analysis, using terminally labeled substrates if necessary, is useful for quantitative studies of transfer and condensation reactions.

Finally, it should be emphasized that for reliable kinetic studies most reliable rate measurements with suitable substrates must be made. This problem will be described in the following section, I.2.

(Keitaro Hiromi)

References

1. Ono, S. & Takahashi, K. (1969) in *Biochemical Microcalorimetry* (Brown, H. D., ed.) pp. 99–116, Academic Press, New York
2. Fukumoto, J. & Tsujisaka, Y. (1961) *J. Starch Sweetener Technol. Res. Soc. Jpn.* (in Japanese) *24*, 133–141
3. Fujimori, H., Ohnishi, M., Sakoda, M., Matsuno, R. & Hiromi, K. (1977) *J. Biochem. 82*, 417–427
4. Suganuma, T., Matsuno, R., Ohnishi, M. & Hiromi, K. (1978) *J. Biochem. 84*, 293–316
5. Suetsugu, N., Hirooka, E., Yasui, H., Hiromi, K. & Ono, S. (1973) *J. Biochem. 73*, 1223–1232
6. Hiromi, K., Ohnishi, M. & Tanaka, A. (1983) *Mol. Cell. Biochem. 51*, 79–95
7. Kondo, H., Nakatani, H., Matsuno, R. & Hiromi, K. (1980) *J. Biochem. 87*, 1053–1070
8. Hiromi, K., Ohnishi, M., & Shibata, S. (1973) *J. Biochem. 74*, 397–400
9. Hiromi, K. (1979) *Kinetics of Fast Enzyme Reactions : Theory and Practice*, Kodansha Ltd., Halsted Press, Tokyo, New York

I.2. General Considerations for Conditions and Methods of Amylase Assay

In this section, general aspects of procedures for kinetic studies on amylase-catalyzed reactions, including the determination of the Michaelis parameters (the Michaelis constant, the maximum velocity and the molar (molecular) activity) and that of the anomeric form of the product formed, will be described. Here we shall focus our attention solely on the hydrolytic reaction, although transfer and/or condensation could occur in certain cases, as was mentioned in the preceding section. A general survey on the measurements of the rates of amylase-catalyzed reactions (activities) will briefly be made. For details of measuring activities of individual amylases, readers may refer to relevant sections in this book.

I.2.a. General Survey on the Measurements of Amylase Activities

The most fundamental quantity to be known in the kinetic studies of amylase-catalyzed reactions is the number (or molar concentration) of glucosidic linkages of a substrate hydrolyzed by an amylase at a given time. Generally, this quantity can be measured through the increase in the number (or concentration) of the reducing ends, which is determined, for example, by the Nelson-Somogyi method (see IV.). In the case of glucoamylase, the unique product, glucose, can be specifically and unequivocally measured by using glucose oxidase.

Homogeneous substrates such as maltooligosaccharides of various degrees of polymerization (DP) are more suitable for quantitative kinetic analysis than soluble starch, which is most widely used; it is essentially heterogeneous and comes in a great variety, depending on the manufacturer and on the lot of the product. Amyloses with fairly sharp DP-distribution (e.g., amyloses with average DP ($\overline{DP}$) of 17 or 100 are now commercially available) would be much better substrates to work with.

Maltooligosaccharides of definite DP conjugated with chromophore group(s) are commercially available, and are good substrates in the sense that they facilitate the measurement of reducing ends produced by the reaction because of the lack of initial reducing power, and that continuous absorbance or fluorescence measurements could be made either directly or by using coupled-enzyme assay procedures (see IV.).

For example, Taka-amylase A-catalyzed hydrolysis of *p*-nitrophenyl-α-maltoside can be followed continuously through the absorbance change caused by the liberation of *p*-nitrophenol.

Cycloamyloses (DP=6~9) are often convenient substrates for α-amylases, although they cannot be attacked by glucoamylase and β-amylase. The rate of hydrolysis of the glucosidic linkage of cycloamyloses to form maltooligosaccharides largely increases with increasing ring size, and also dependent on the kind of α-amylase. When a sufficient amount of glucoamylase is added to the α-amylase-cycloamylose system, the maltooligosaccharides produced are quickly hydrolyzed by glucoamylase to form glucose. Thus the sensitivity of the measurement is largely magnified, either in the reducing end measurement or the specific glucose determination by glucose oxidase. Fluorescent dyes such as 2-*p*-toluidinylnaphthalene-6-sulfonate (TNS) are included in cycloamylose (DP=7) with large increase in the fluorescence intensity. Upon the ring opening by α-amylases, the dye is liberated with concomitant decrease in fluorescence intensity. This provides a continuous monitoring method of α-amylase-catalyzed reactions (1).

Optical rotation measurement is one of the useful tools for following amylase-catalyzed reactions continuously. A polarimeter of high sensitivity ($\pm$ 0.0002°) is now commercially available (Otsuka Electronics Co., Ltd. (Photal) PM-101), which has been proved to be very useful for kinetic studies on the hydrolyses of maltooligosaccharides and their derivatives. With this method, it should be remembered that the measurement must be finished within a few minutes, since the mutarotation of the products proceeds (the half-time is about 30 min at 25°C in neutral pH range) which otherwise could obscure the true optical rotation change due to the enzymic hydrolysis.

Continuous monitoring of fluorescence intensity change of a synthetic substrate, a fluorogenic derivative of *p*-nitrophenyl α-maltopentaoside catalyzed by α-amylase in human serum has successfully been made (2).

As was mentioned above, continuous following of the time course of amylase-catalyzed reactions is possible in certain cases, and it is actually more convenient and accurate in determining the initial velocity of the reaction. However, in most cases, a discontinuous method to follow the reaction must be employed: a definite volume of aliquots is withdrawn from the reaction mixture at certain time intervals and immediately poured into a stopping solution (usually, acid or alkali) to terminate the enzyme reaction, and the samples are subjected to suitable analytical measurements. For kinetic studies, sufficient numbers of points should be taken to obtain a good progress curve, which is necessary to determine the initial velocity with good accuracy (see the following section). Examples of two sets of progress curves are shown in Fig. 1.

Finally, the so-called "blue value" method in which I_2-KI solution is added to the reaction mixture in the course of hydrolysis with soluble starch or amylose as the initial substrate, is easy and convenient for assay of relative activities of amylases. However, it is not adequate for quantitative kinetic studies on amylase-catalyzed reactions, since it is difficult to correlate the blue value with the number (or percentage) of hydrolyzed bonds precisely, especially for α-amylases. Nevertheless, the figure showing the blue value versus the extent of hydrolysis is useful for comparing the action patterns of various amylases.

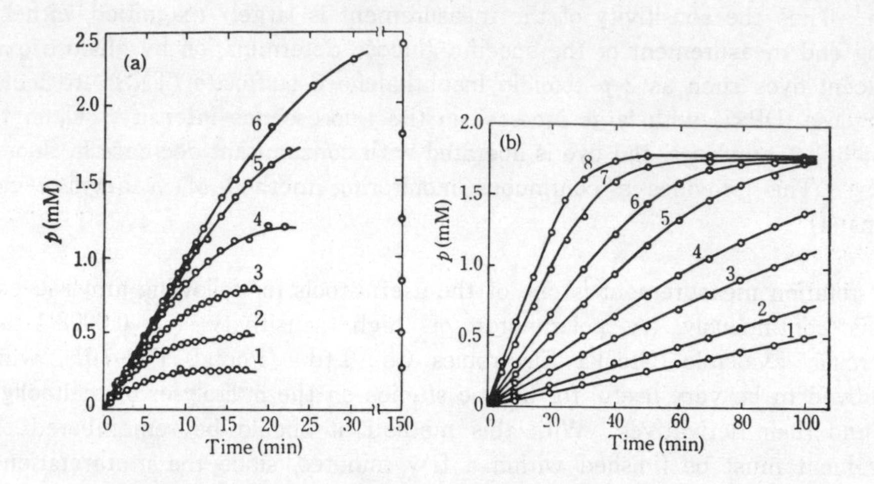

Fig. 1. Examples of Progress Curves of Amylase-Catalyzed Reactions.
Molar concentration of glucose (p) produced from phenyl α-maltoside catalyzed by *Rhizopus* glucoamylase was determined by the Nelson-Somogyi method. (a) $[E]_O = 0.58\ \mu M$ (fixed), and [S] varied. The number of the curves (1~6) refers to [S](mM), respectively, 0.25, 0.50, 0.82, 1.25, 1.88, 2.50. (b) [S] = 1.78 mM (fixed), and $[E]_O$ varied. The number of the curves (1~8) refers to $[E]_O(10^{-8}$ M units), respectively, 3.7, 5.1, 7.5, 10.3, 15.0, 20.5, 29.8, 41.0.

Practical analytical procedures, substrates and products involved in amylase-catalyzed reactions are described in detail in the recent monograph (3).

I.2.b. Determination of Michaelis Parameters

The Michaelis parameters (or rate parameters), i.e., the Michaelis constant, K_m, the maximum velocity, V, and the molar (or molecular) activity, k_O (or k_{cat}) are defined as follows: According to the familiar Michaelis-Menten equation,

$$v = V[S]/(K_m + [S]) = k_O[E]_O[S]/(K_m + [S]) \quad (1)$$

the initial velocity (v) increases hyperbolically with increasing substrate concentration ([S]) to reach the maximum velocity (V), i.e., the asymptote. The substrate concentration that gives v which is equal to $V/2$ is defined as the Michaelis constant, K_m. The molar activity, k_O, is obtained from $V/[E]_O$, where $[E]_O$ is the molar concentration of the enzyme, that can be determined from the molecular weight of the enzyme. If the molecular weight is not known, V must be expressed in the molar concentration of substrate converted to the products per unit time (second or minute) at the "fixed concentration of the enzyme" in the reaction mixture (e.g., mg protein per liter).

To obtain reliable values of the Michaelis parameters, the following should be remembered: first, the initial velocity at every substrate concentration must be determined as accurately as possible, and second, the range of [S] must be chosen appropriately (the range from 0.2 K_m to 2 K_m is recommendable). Fig. 1 (a) shows the example of a set of progress curves for the determination of the values of K_m and V. Obviously, the curve tends to concave upwards with decreasing [S], for which the error involved in obtaining the initial velocity (v) (the slope of the tangent at zero time) becomes larger. In Fig. 1(b), the effect of $[E]_0$ on the progress curve at a fixed [S] is shown. Referring to these figures, suitable conditions for performing and measuring the reactions may be chosen. It is important to check the proportionality between v and $[E]_0$, to confirm the validity of the procedure for evaluation of v's.

Linear plots : The rate equations Eq. 1 are transformed into linear equations as follows:

$$1/v = (K_m/V)(1/[S]) + 1/V \quad : (1/v \text{ vs. } 1/[S] \text{ plot}) \qquad (2)$$

$$[S]/v = [S]/V + K_m/V \quad : ([S]/v \text{ vs. } [S] \text{ plot}) \qquad (3)$$

$$v = -K_m (v/[S]) + V \quad : (v \text{ vs. } v/[S] \text{ plot}) \qquad (4)$$

Any of these equations can be used to evaluate K_m and V; Eq. 2 is most frequently used, known as the "Lineweaver-Burk plot" or the "double reciprocal plot". The least-squares method for straight lines ($y = ax + b$) can be applied to obtain K_m and V values together with their standard deviations. From the statistical viewpoint, Eq. 3 is the best, and Eq. 2 is the worst, since the "weight" is higher for smaller v values at lower [S] where the error becomes bigger. Cornish-Bowden proposed a novel form of plot, called the "direct linear plot" ($V = v + (v/[S])K_m$), which is reasonable in the statistical sense. It is desirable to describe the standard deviations as well as K_m and V values, to show the reliability of the data. The procedure for these least-squares methods to evaluate the Michaelis parameters with their standard deviations are written in detail in his book (4). A non-linear least squares method applicable directly to the hyperbolic form, Eq. 1, is also available (5).

If the [S]-dependence of v does not fit the hyperbolic form (Eq. 1) but shows a sigmoidal curve, it suggests possible involvement of a transfer reaction besides the hydrolysis (cf. Ref. 3 cited in I.1.).

I.2.c. Determination of the Anomeric Form of the Product

Whether α-anomer or β-anomer of the product saccharide is formed, which reflects the retention or inversion of configuration of the glucosidic linkage upon the enzymic hydrolysis, respectively, is a very important problem closely related to the mechanism of amylase-catalyzed reactions. Moreover, the anomeric form of the product is one of the important criteria for discriminating and classifying glucoamylase and α-

glucosidase.

The usual way to distinguish the anomeric forms of the amylase-catalyzed reaction products is to observe the direction of optical rotation change of the reaction mixture caused by adding alkali within a few minutes after the commencement of the reaction. Since the anomeric equilibrium between α- and β-forms of the products is reached immediately at alkaline pH, the decrease or increase in the optical rotation indicates, qualitatively or semi-quantitatively, that the product is of α-form or β-form, respectively.

To determine precisely the percentage of the anomeric forms of the products, more sophisticated quantitative methods should be employed, by using homogeneous and well-characterzyed substrates. It is desirable to use substrates in which only one glucosidic linkage is cleaved preferentially by the enzyme to be studied.

Two methods for quantitative determination of the anomeric forms have been proposed. One is based on the simultaneous measurements of the time course of the reaction with respect to the optical rotation change and the number of linkages hydrolyzed. This method was successfully applied to glucoamylase- and α-glucosidase-catalyzed reactions, by using synthetic substrates. Simple calculations could afford to show that β-glucose and α-glucose are produced exclusively by the action of glucoamylase and α-glucosidase, respectively. For the principle and procedure of this method, the original paper (6) may be referred to. Another more direct and decisive method has been developed employing gas-liquid chromatographic separation of the products after trimethylsilylation (7). The anomer forms of products from a number of glucosides, maltosides, and maltooligosaccharides by the action of glucoamylases, β-amylases and α-glucosidases have been determined unequivocally by this method.

(Masatake Ohnishi and Keitaro Hiromi)

References

1. Kondo, H., Nakatani, H. & Hiromi, K. (1976) *J. Biochem. 79*, 393-405
2. Omichi, K. & Ikenaka, T. (1986) *J. Biochem. 99*, 291-294
3. Chaplin, M. F. & Kennedy, J. F. eds. (1986) *Carbohydrate Analysis : A Practical Approach*, IRL Press, Oxford
4. Cornish-Bowden, A. (1976) *Principles of Enzyme Kinetics*, Butterworths, London
5. Sakoda, M. & Hiromi, K. (1976) *J. Biochem. 80*, 547-555
6. Ono, S., Hiromi, K. & Hamauzu, Z. (1965) *J. Biochem. 57*, 34-38
7. Chiba, S., Kimura, A. & Matsui, H. (1983) *Agric. Biol. Chem. 47*, 1741-1746

I.3. Catalytic Flexibility of Amylases and Glycosylases

— For Comprehension of Various Catalytic Reaction Abilities of These Enzymes

Following development of the glycosyl transfer concept (1), transferases have been classified apart from glycosidases and glycanases (2). However, all hydrolytic and glycosyl transfer reactions may be seen to bring about the same, simple chemical change (3),

$$\text{Glycosyl-X} + \text{H-X'} \rightleftharpoons \text{Glycosyl-X'} + \text{H-X}$$

This equation predicts that nonglycosidic compounds which can bind and yield a glycosyl residue on protonation (reducing sugars, glycosyl fluorides and glycals) may serve as substrates for glycosylases.

Reversibility of Hydrolytic Reaction : The hydrolysis catalyzed by α-, β- and gluco-amylases is reversible. Hehre and his colleagues reported in 1969 that glucoamylase synthesized maltose specifically from β-D-glucose (4). This synthetic reaction is especially clear immediately after incubation of the reaction mixture. Isomaltose is also synthesized by the enzyme from the substrate, but at a slower rate than maltose (4).

Transfer Reaction of Amylolytic Enzymes : The catalytic reactions brought about by amylolytic enzymes are not necessarily restricted to hydrolytic reactions. Hehre and his colleagues demonstrated that α-amylases dominantly catalyze transfer reactions from α-D-glucosyl fluoride and especially from maltosyl fluoride (5,6,7).

Functional Flexibility : Even the enzymes which had long been considered to be strictly specific for hydrolyzing α-glucosidic substrates with configurational inversion and to be without glycosyl transfer ability, were found to show functional flexibility of catalytic groups. β-Amylase, glucoamylase, glucodextranase, and trehalase utilize both α- and β-anomers of an appropriate glycosyl fluoride, and catalyze stereocomplementary hydrolytic/nonhydrolytic reactions with the α- and β-fluoride, respectively (8,9,10).

$$\alpha\text{-Glucosyl-F} + H_2O \xrightarrow{\text{glucoamylase}} \beta\text{-glucose} + \text{HF}$$

$$\beta\text{-Glucosyl-F} + \beta\text{-glucosyl-F} \xrightarrow{\text{glucoamylase}} (\text{glucosyl-}\alpha\text{-1,4-}\beta\text{-glucosyl-F} + \text{HF})$$

$$\downarrow \text{glucoamylase}$$

$$\beta\text{-glucose} + \beta\text{-glucosyl-F}$$

Comparable findings have also been obtained in studies of reactions catalyzed with onolic glycosyl donors as substrate, lacking α- or β-configuration and glycosidic linkage. D-Glucal, for example, is shown to be hydrated both by α- and β-glucosidases. In each case, the D-glucal is protonated by the enzyme from a direction opposite of what assumed for protonating glucosidic substrates. NMR spectroscopy of reactions conducted in D_2O revealed that the glucal hydration reactions resulted a *de novo* creation of product configuration by enzymes which are considered to be limited to retaining substrate configuration (11).

CH_2OD
O
DO OD
D-Glucal

α-Glucosidase in D_2O

CH_2OD
O
DO OD D OD

β-Glucosidase in D_2O

CH_2OD
O
OD
DO OD
D

A synthetic enolic glycosyl donor, D-glucoheptenitol, is hydrated by α- and β-glucosidases, and by glucodextranase, with formation of α-D-glucoheptulose (α-glucosidases) or β-D-glucoheptulose (β-glucosidases and glucodextranases)(12). α-D-glucoheptulosyl transfer products are abundantly formed by α-glucosidases and glucodextranase (which releases β-D-glucose from dextran). The formation of a β-product on hydration, but α-transfer products with carbohydrate acceptors by glucodextranase, reveals the influence of different acceptors (different channels of approach to the reaction center) in determining product configuration (13).

Cellobial is hydrated by cellulases with protonation from below the double bond and formation of β-2-deoxycellobiose (14). On the other hand, maltal is hydrated by β-amylase with protonation from above the double bond (15), producing β-anomer of 2-deoxymaltose (16). The protonation by each enzyme is opposite of what has generally been assumed for glucosidic substrates, yet the product configuration matches that obtained with glycosidic substrates.

The frequent occurrence of the difference in protonating enolic and glycosidic substrate, but correspondence in the configuration of products from both types of substrates, was recently demonstrated. Several α-glucosidases which rapidly catalyzes D-glucal hydration, show this characteristic (17) ; likewise two α-glucosidases and a mold trehalase hydrate D-gluco-octenitol from a direction opposite of what assumed for α-glucosidic substrates, yet yields products with a configuration matching that from α-D-glucosidic substrates (18).

These findings indicate that, for many glycosylases the catalytic process consists of two separate parts ; a variable substrate-dependent phase concerned with how the enzyme protonates the substrate, and a conserved phase concerned with substrate-independent control of product configuration by the enzyme.

Recent measurements of the secondary alpha deuterium and tritium isotope effects on the hydrolysis of α-D-glucosyl fluoride have proved the first experimental evidence that exo-α-glucanases such as glucoamylase catalyze the hydrolysis via a transition state with considerable carbonium ion character (19).

(Takehiko Yamamoto)

References

1. Hehre, E. J. (1951) *Advances in Enzymol. 11*, 297-337
2. IUPAC Enzyme Classification (1961)
3. Hehre, E. J. (1960) *Bull. Soc. Chem. Biol. 42*, 1713-1714
4. Hehre, E. J., Okada, G. & Genghof, D. S. (1969) *Arch. Biochem. Biophys. 135*, 75-89
5. Hehre, E. J., Genghof, D. S. & Okada, G. (1971) *Arch. Biochem. Biophys. 142*, 382-393
6. Hehre, E. J., Okada, G. & Genghof, D. S. (1973) *Advances in Chem. Ser. 117*, 309-333
7. Okada, G., Genghof, D. S. & Hehre, E. J. (1979) *Carbohydr. Res. 71*, 287-298
8. Hehre, E. J., Brewer, C. F. & Genghof, D. S. (1979) *J. Biol. Chem. 254*, 5942-5950
9. Kitahata, S., Brewer, C. F., Genghof, D. S., Sawai, T. & Hehre, E. J. (1981) *J. Biol. Chem. 256*, 6017-6026
10. Hehre, E. J., Sawai, T., Brewer, C. F., Nakano, M. & Kanda, T. (1982) *Biochemistry 21*, 3090-3097
11. Hehre, E. J., Genghof, D. S., Sternlicht, H. & Brewer, C. F. (1977) *Biochemistry 16*, 1780-1787
12. Hehre, E. J., Brewer, C. F., Uchiyama, T., Schlesselman, P. & Lehmann, J. (1980) *Biochemistry 19*, 3557-3564
13. Schlesselman, P., Fritz, H., Lehmann, J., Uchiyama, T., Brewer, C. F. & Hehre, E. J. (1982) *Biochemistry 24*, 6606-6614
14. Kanda, T., Brewer, C. F., Okada, G. & Hehre, E. J. (1986) *Biochemistry 25*, 1159-1165
15. Hehre, E. J., Kitahata, S. & Brewer, C. F. (1986) *J. Biol. Chem. 261*, 2147-2153
16. Kitahata, S., Chiba, S., Brewer, C. F. & Hehre, E. J. Unpublished findings
17. Chiba, S., Brewer, C. F., Okada, G., Matsui, H. & Hehre, E. J. (1988) *Biochemistry 27*, 1564-1569
18. Weiser, W., Lehmann, J., Chiba, S., Matsui, H. & Hehre, E. J. (1988) *Biochemistry 27*, 2294-2300
19. Matsui, H., Blanchard, J., Brewer, C. F. & Hehre, E. J., Unpublished manuscript

II. Data on Individual Amylases

II.1. α-Amylase

The amylase generally known as the enzyme which catalyzes the hydrolysis of starch and glycogen, is among the most important industrial enzymes. Also, the enzyme has long been studied and it may be said that the history of study of amylase is as long as that of enzymes. The amylase was probably the first enzyme to be discovered. It was observed to degrade starch by Kirchhoff in 1811. In 1831, Leuchs discovered the digestive action of saliva on starch (1). Payen and Persoz discovered a starch digestion active substance in malt and they isolated this active substance as an alcohol precipitate (2). Märker pointed out in 1878 that malt amylase consisted of two different enzymes (3). Far later, Ohlsson suggested the classification of starch digestive enzymes in malt into α- and β-amylases according to the anomeric type of sugars produced by the enzyme reaction (4). In 1950, Myrbäck and Neümuler proposed another classification for the amylases, namely, a) exoamylases and b) endoamylases. Saccharifying amylase or β-amylase was included into the former and starch liquefying or α-amylase, into the latter amylases. Their criterion for the amylase classification was based on the action mode of amylases. However, it was demonstrated recently that an exoamylase is not always a β-amylase, as several α-amylases which produce maltotriose, -tetraose, -pentaose or -hexaose have been reported to attack starch exowise. In this part for enzymic and proteochemical properties, and production and isolation, therefore, various α-amylases are grouped depending on their reaction products and then, further subgrouped depending on their sources. Application of α-amylase is described in another part.

(Takehiko Yamamoto)

References

1. Leuchs, E. F. quoted from Sumner, J. B. & Somes, G. F. (1947) *Chemistry and Methods of Enzymes* p 103
2. Payen, A. & Persoz, J. (1833) *Ann. Chem. Phys. 33,* 73
3. Märker, M. (1878) *Chem. Zentr.* s 559.
4. Ohlsson, E. (1930) *Z. Physiol. Chem. 189,* 17
5. Myrbäck, K. & Neumüler, G. (1950) *The Enzymes* vol. 1, p 668

II.1.a. α-Amylases Which Produce Various Molecular Size Dextrins

II.1.a.1-1 Human Salivary and Pancreatic α-Amylases

Human α-amylase contains two different forms secreted from salivary gland and pancreas. These two α-amylases are known to be very similar but not identical.
The two α-amylases are also found in blood serum and in urine as normal components, but their activities are small in normal adults. α-Amylases in urine

originate to blood serum (1). No papers have been published on the detailed properties of α-amylases in blood, but it is certain that the enzymes hydrolyze glycogen as well as starch, as described below.

Reaction :

Glycogen (starch) +n H_2O ⟶ G1 + G2 + G3 + various branched dextrins

The two α-amylases have nearly the same amino acid composition and are very similar in the optimum pH, mode of action on various substrates and activation by chlorine ion (2,3). Additionally, the nucleotide sequences of the cDNAs coding for the two α-amylases are 96% homologous and the deduced amino acid sequences 94% homologous (4). Also, the two enzymes are immunologically identical in their reactions with polyclonal antibodies (5). However, the two α-amylases are different in molecular weight, isoelectric point, carbohydrate contents, and transglycosylation activity (6). The crystalline preparation of human salivary α-amylase consists of three components which show different migrations on electrophoresis using a cellulose acetate membrane. The crystalline pancreatic α-amylase obtained from the tissue also consists of at least three components (7). However, α-amylase isolated from fresh pancreatic juice consists of one major and one minor component (8). The minor component and certain two components of the pancreatic tissue are considered to be newly formed by enzymic modification from the major component with pancreatic peptidoglutaminase during storage of the crude preparation or during purification of the enzyme (9).

Specificity : No significant difference is observed for the action mode of human salivary and human pancreatic α-amylases on several maltooligosaccharides, glycogen and starch. Human salivary α-amylase hydrolyzes G4 to produce G2 and a minor portion of G4 into G and G3, G5 into G2 and G3, and G6 into G2 and G4 or partially into two moles of G3. The enzyme does not hydrolyze G3. Human pancreatic α-amylase shows almost the same action patterns in hydrolysis of the above maltooligosaccharides as human salivary α-amylase (10,11). The attack sites of several maltooligosaccharides by salivary and pancreatic α-amylases are described below (10) :

Human salivary α-amylase hydrolyzes β-limit dextrin to produce G1, G2, G3 and several branched dextrins. The hydrolysis products are the same as those by human pancreatic α-amylase, as shown in Fig. 1 (7). Both enzymes seldom attack G3, as described above. However, when a relatively large amount of the enzyme is used for the digestion, no production of G3 is observed in the digests, as shown in the right paper chromatograms in Fig. 1. This fact indicates that both enzymes have the ability to attack G3, if they are applied in large amount.

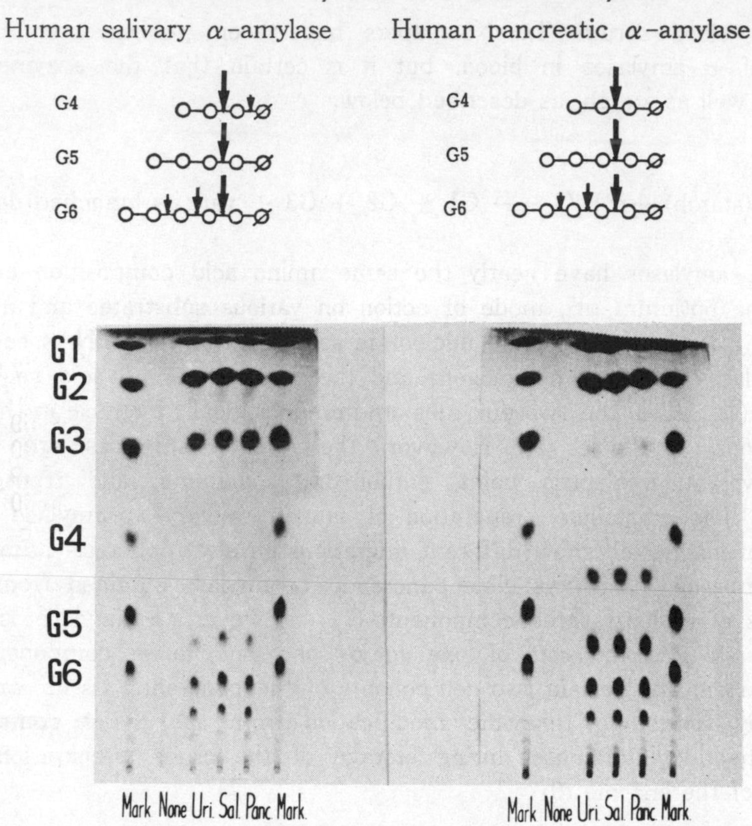

Fig. 1. Paperchromatograms of Hydrolyzates of β-Limit Dextrin by Human α-Amylases

Substrate, 8% β-limit dextrin in a solution containing M/100 each of sodium chloride and calcium-acetate, pH 7.0 ; 40°C; 24 h. Crystalline α-amylase : Uri, urinary ; Sal, salivary ; Panc ; pancreatic. Amylase added per g of β-limit dextrin : left, 100 units ; right, 680 units.

Purification : Purification of human salivary α-amylase can be readily done by the method of using raw corn starch as a certain affinity adsorbent (3). The saliva collected in the presence of a small amount of sodium azide is stored overnight or longer in a freezer to remove viscous materials which become insoluble by freezing. The frozen saliva is thawed and centrifuged to remove the insoluble substance. To this supernatant, sodium chloride and calcium acetate are added in a concentration of M/100, respectively, for stabilizing the enzyme. Then solid ammonium sulfate is added at a concentration of 0.25 saturation with adjusting the pH at 6.0. After removing the resulting precipitate by centrifugation, the corn starch which has previously been heated to about 70°C at pH 8.0 for 30 min in a 0.25 saturated

Table 1. Some Enzymic and Proteochemical Properties of Human Salivary and Pancreatic α-Amylases

		Salivary (Family A)[a]	Pancreatic
Molecular weight (2)		62,000	56,000
E (1%, 280nm)		26 (12)	21.5 (8)
Carbohydrate (2) (Neutral, mol/mol protein)		8	1
SH (mol/mol protein) (2)		1	1
pH Stability (2)		4–10	4–10
Thermal stability (℃) (3)		50	50
Optimum pH (3)		6.8–7.0	6.8–7.0
Activation by Cl^- (3)		+	+
*K*m (mM) (10)	G4	3.14	3.59
	G5	3.41	2.30
	G6	4.62	2.49
	G7	10.6	5.90
Adsorption on raw starch (3)		+	+

a) Besides Family A, Family B which has no carbohydrate is reported to exist in human saliva. Family A and Family B are quite similar in their enzymic and proteochemical properties except for the carbohydrate content (2). Number in parentheses shows reference number.

solution of ammonium sulfate and then cooled (13), is added to the supernatant in a ratio of one g corn starch per 1 x 10^4 units of α-amylase activity. After gentle agitation at 4℃ for several hours, the starch is collected on a suction filter and washed with 0.25 saturated ammonium sulfate solution containing M/ 100 each of sodium chloride and calcium acetate of pH 6.0. The washed starch is then suspended in three to four volumes of a solution consisting M/100 each of sodium chloride and calcium acetate of pH 6.0. Then the starch suspended in the same sodium chloride and calcium acetate solution as above is incubated at 40℃ for 2 h with gentle agitation to elute the enzyme. This procedure is repeated several times. The eluates are combined and concentrated on a ultrafilter membrane. The concentrate is chromatographed on a column of Sephadex G-100 in the solution used for elution. This procedure serves as a kind of affinity chromatography since the enzyme has a weak affinity with Sephadex G-100. The enzyme is eluted as a symmetrical peak and also separated from a small amount of impurity. The human salivary α-amylase is obtained in a highly purified state with 50 to 60% of activity recovery by the above purification procedures and the specific activity of the enzyme obtained is more than 700 units per mg protein. Human pancreatic α-amylase is also purified from the extract of pancreatic homogenate with a mixture of M/100 each of sodium chloride and calcium acetate of pH 6.0 or from pancreatic juice by a similar purification procedures (6).

Assay Method : Several methods have been reported for assay of human salivary and pancreatic α-amylases for diagnosis. However, the assay method by starch saccharifying activity is described here, since this is the most accurate among various methods, though slightly troublesome. One ml of enzyme is added to 5 ml of 0.5% soluble starch in M/50 Tris-HCl buffer, pH 7.2, containing M/100 each of sodium chloride and calcium acetate. After 5 min incubation at 37°C, the reducing sugars formed in the mixture are determined as glucose by the Shaffer-Somogyi method. One unit of α-amylase activity is defined as the amount of enzyme which produces reducing sugar equivalent to one μmol of glucose per min under the conditions (14).

(Noshi Minamiura)

References

1. Ogawa, M. (1981) *J. Jap. Soc. Starch Sci.* (in Japanese) *28*, 87-91
2. Stiefel, D. J. & Keller, P. J. (1973) *Biochim. Biophys. Acta 302*, 345-361
3. Minamiura, N., Umeki, K., Tsujino, K. & Yamamoto, T. (1971) *Proceedings of the Symposium on Amylase* (in Japanese) *6*, 69-77
4. Nakamura, Y., Ogawa, M., Nishide, T., Emi, M., Kosaki, G., Himeno, S. & Matsubara, K. (1984) *Gene 28*, 263-270
5. Takeuchi, T., Nakagawa, Y., Ogawa, M., Kawachi, T. & Sugimura, T. (1977) *Clin. Chim. Acta 77*, 203-206
6. Omichi, K. & Ikenaka, T. (1983) *J. Biochem. 94*, 1797-1802
7. Minamiura, N., Umeki, K., Tsujino, K. & Yamamoto, T, (1972) *J. Biochem. 72*, 1295-1298
8. Matsuura, K., Ogawa, M., Kosaki, G., Minamiura, N. & Yamamoto, T. (1978) *J. Biochem. 83*, 329-332
9. Ogawa, M., Kosaki, G., Matsuura, K., Fujimoto, K., Minamiura, N., Yamamoto, T. & Kikuchi, M. (1978) *Clin. Chim. Acta 87*, 17-21
10. Saito, N. (1982) *J. Jap. Soc. Starch Sci.* (in Japanese) *29*, 153-160
11. Nakagiri, Y., Kanda, T., Otaki, M., Inamoto, K., Asai, T., Okada, S. & Kitahata, S. (1982) *J. Jap. Soc. Starch Sci.* (in Japanese) *29*, 161-166
12. Fischer, E. D. & Stein, E. (1960) *The Enzymes* (Boyer *et al.* eds.) 2nd ed., Vol. 4, p319
13. Yamamoto, T. (1955) *Bull. Agr. Chem. Soc. 19*, 121-128
14. Minamiura, N., Kimura, Y., Tsujino, K. & Yamamoto, T. (1975) *J. Biochem. 77*, 163-169

II.1.a.1-2. Porcine Pancreatic α-Amylase

Porcine pancreatic α-amylase (PPA) hydrolyzes the α-1,4-glucosidic linkages of starch, amylose, amylopectin and glycogen. This α-amylase was obtained in a crystalline form in 1943 (1). The crystalline preparation of this α-amylase consists of two isozymes, PPA I and PPA II. The amino acid sequences of PPA I (2) and II (3), and the three-dimensional X-ray structure of PPA I (4) have recently been made clear. The amino acid sequences of the two α-amylases are 92% homologous. The

N-terminal and C-terminal sequences are the same. The enzymic properties of PPA I and II are the same except for their isoelectric points and mobilities on polyacrylamide gel electrophoresis. The action of this enzyme is endowise and of the multiple attack type.

Reaction :

Starch + n H_2O ⟶ G1 + G2 + G3 + G4 + various branched dextrins

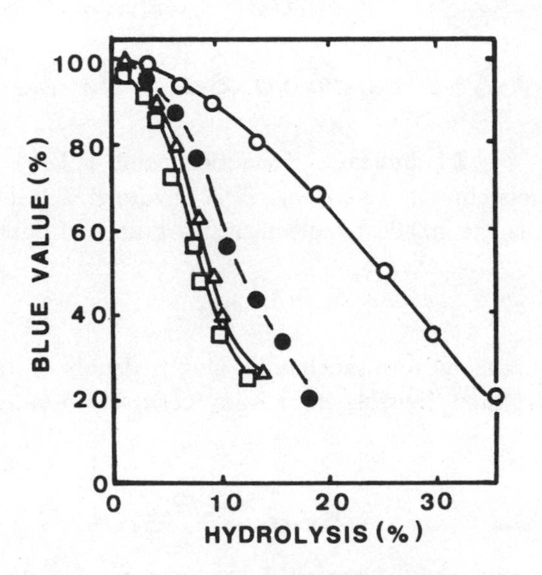

Fig. 1. A Comparison of Decrease in Blue Value and Progress of Hydrolysis Degree for Amylose Reacting with Several α-Amylases and with Sulfuric Acid (5).

○, PPA; ●, HSA (human salivary α-amylase ; △, TAA (*Aspergillus oryzae* α-amylase) ; □, acid (1.0 M sulfuric acid).

PPA hydrolyzes starch, glycogen, amylose and amylopectin to produce mainly maltose and maltotriose. The action of PPA on amylose is shown in Fig. 1. The curve showing the relationship of the decrease in iodine color reaction (blue value) with the degree of hydrolysis of amylose by PPA is different from those for other α-amylases. This fact may be attributed to the difference in the frequency of multiple attack. The frequency of attack in every productive binding is about 6, 2, 2 and 1 for PPA, human salivary α-amylase (HSA), *Aspergillus oryzae* α-amylase (TAA) and 1 M sulfuric acid, respectively.

Specificity : PPA hydrolyzes amylose to produce mainly maltose and maltotriose. The action patterns of PPA on maltooligosaccharides (G3-G7) are similar except for that on G3. PPA cleaves the α-1,4-glucosidic linkage penultimate to the reducing end more preferably than other α-1,4-linkages in G4-G7, but in G3 it hydrolyzes the α-1,4-linkage at the reducing end more readily (6). PPA hydrolyzes exclusively α-1,4-glucosidic linkages in starch, glycogen and amylopectin except for the reducing and α-1,4-linkages adjacent or penultimate to the glucosyl residue, in which the C-6 position is connected to another glucosyl residue. The action of PPA on glucosyl branched dextrins (from hexasaccharide to nonasaccharide) is shown below (7,8) :

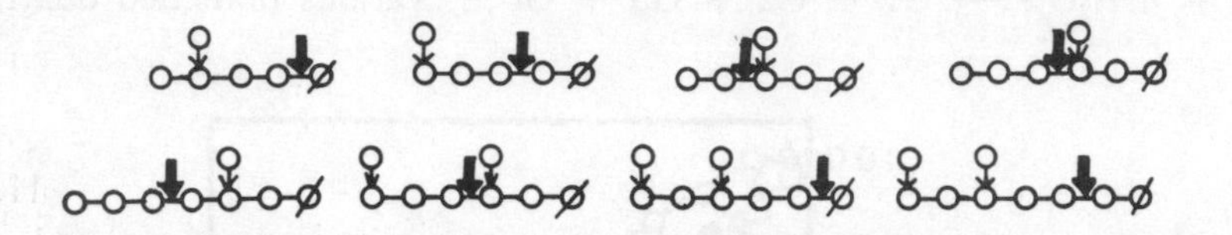

↓, Indicates the action point of PPA.

Acting upon amylopectin or glycogen, PPA hydrolyzes them to produce 6^3-α-glucosylmaltotriose as the smallest molecular size branched dextrin :

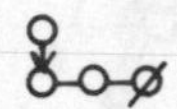

In addition to this branched tetrasaccharide, singly, doubly or triply branched dextrins are obtained as PPA-limit dextrins from waxy corn starch (amylopectin) (7,8,9). These dextrins are :

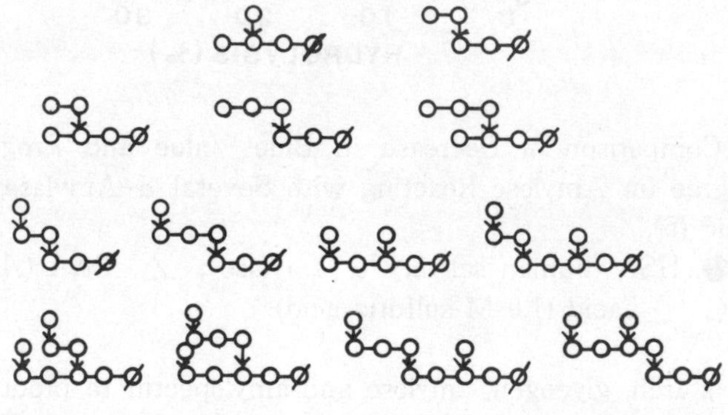

Isolation and Purification : Porcine pancreatic α-amylase (PPA) can be purified by slightly modifying the method of Hatfaludi *et al.* (11). Porcine pancreas stored at −80°C is partially defrosted at room temperature and cut into pieces with scissors and homogenized in 0.8 volumes of a solution containing 3% 1-butanol and 10 mM $CaCl_2$ with a Polytron homogenizer for 3 min. After the homogenized mixture is centrifuged at 10,000 rpm for 15 min, the supernatant is filtered through adsorbent cotton to remove floating fat. Cold acetone (−20°C) is added to the filtrate to bring the acetone concentration to 40%, and the mixture is centrifuged at 10,000 rpm for 15 min. The enzyme is precipitated by further addition of acetone (final concentration, 67%) to the supernatant. The precipitated enzyme is dissolved in a small amount of

Table 1. Enzymic and Proteochemical Properties of Porcine Pancreatic α-Amylase (10)

Isozyme	PPA 1	PPA 2
Molecular weight	56,000	56,000
E (1%, 280nm)	24*1	24*1
Optimum pH	6.9	6.9
Km (DP=17) mM	0.13	0.12
k_o (short chain amylose, DP=17) sec^{-1}	5.6×10^2	5.5×10^2
pH Stability		
($-Ca^{2+}$)	6.0–8.0	6.0–8.0
($+Ca^{2+}$)	5.0–11.0	5.0–11.0
Thermal stability (°C)	35	35
Inactivation by EDTA	+*2	+*2
Isoelectric point	6.5	6.1

*1 Elodi, P. (1972) *Acta Biochim. Biophys. Acad. Sci. Hung.* *7*, 24.
Hsiu, J., Fischer, E. H. & Stein, E. A. (1964) *Biochemistry*, *3*, 61–66
*2 Watanabe, I. (1988) *Dissertation for the Master's Degree*, Tokyo University of Agriculture and Technology.

1 mM $CaCl_2$ solution, and salted-out with ammonium sulfate (40% saturation) followed by dialysis against 1 mM $CaCl_2$ solution overnight. The dialyzed solution is centrifuged at 10,000 rpm for 15 min, and PPA is crystallized by addition of $CaCl_2$ (final concentration, 10 mM) to the enzyme solution. This enzyme crystallization is carried out for about 4–6 days in the cold. The crystals are centrifuged, and washed twice with cold 1 mM $CaCl_2$. The first crystals are dissolved in a least amount of 1 mM $CaCl_2$, centrifuged at 10,000 rpm for 15 min, and then $CaCl_2$ is added (final concentration, 10 mM) to the supernatant to recrystallize the enzyme. The second crystalline preparation is washed twice with cold 1 mM $CaCl_2$, and stored in 20 mM phosphate buffer (pH 6.9) containing 7 mM NaCl, 1 mM $CaCl_2$ and 0.02% NaN_3 at 4°C. The crystalline preparation (20 mg) is dialyzed against 20 mM Tris-HCl buffer (pH 8.3) containing 1 mM $CaCl_2$ and 0.02 % NaN_3, and the dialyzed enzyme solution is applied on a TSK-Gel DEAE-Toyopearl column (1.5 × 25 cm) buffered with the same buffer. Elution is performed with a linear gradient of NaCl from 0 to 0.3 M in the same buffer at a flow rate of 28 ml/cm·h, and 3 ml fractions are collected. PPA is separated by this procedure to PPA I (first peak) and PPA II (second peak) showing a single band on PAGE and SDS-PAGE, respectively (10). The specific activities of PPA I and II are more than 900 units/mg protein (10).

Assay Method : As far as the enzyme is pure, the saccharifying activity method is the most accurate among several assay methods of α-amylase.
Starch saccharifying activity assay —— Fifty μl of the enzyme solution is incubated with 200 μl of 0.5 % soluble starch buffered with 20 mM glycerophosphate (pH 6.9) containing 7 mM NaCl, 1 mM $CaCl_2$ and 0.02% NaN_3 at 30°C for 5 min. The reducing sugar produced is determined by the Somogyi-Nelson method (12). One unit of the

saccharifying activity is defined as the amount of the enzyme that catalyzes the hydrolysis of 1 μmol of glycosidic bonds per min under the assay conditions.

Starch dextrinizing activity assay —— One hundred μl of the enzyme solution is incubated with 300 μl of 0.12% soluble starch buffered with 20 mM glycerophosphate (pH 6.9) containing 7 mM NaCl, 1 mM $CaCl_2$ and 0.02% NaN_3 at 30°C. After 5 min reaction, the enzyme reaction is stopped by adding 1.0 ml of 0.5 N acetic acid and 3.0 ml of 0.015% iodine-0.15% potassium iodide solution. The color developed is measured at 700 nm. One unit of the dextrinizing activity is defined as the amount of enzyme that catalyzes 10% decrease of the absorbance per min under the conditions.

(Yoshiyuki Sakano)

References.

1. Fischer, E.D.H. & Bernfeld, P. (1948) *Helv. Chim. Acta 31*, 1831-1839
2. Pasero, L., Maggie-Pierron, Y. Abadie, B., Chicheportiche, Y. & Marchis-Mouren, G. (1986) *Biochim. Biophys. Acta 869*, 147-157
3. Kluh, I. (1981) *FEBS Letters 16*, 231-234
4. Buisson, G., Duee, E., Haser, R. & Payan, F. (1987) *The EMBO Journal 6*, 3906-3916
5. Robyt, J. F. & French, D. (1967) *Arch. Biochem. Biophys. 122*, 8-16
6. Robyt, J. F. & French, D. (1970) *J. Biol. Chem. 245*, 3917-3927
7. Kainuma, K. & French, D. (1969) *FEBS Letters 5*, 257-261
8. Kainuma, K. & French, D. (1970) *FEBS Letters 6*, 182-186
9. Kainuma, K. & French, D. (1970) *Proceedings of the Symposium on Amylase* (in Japanese) *5*, 35-37
10. Sakano, Y., Takahashi, S. & Kobayashi, T. (1983) *J. Jpn. Soc. Starch Sci. 30*, 30-37
11. Hatfaluti, F., Strashilov, T. & Straub. F. B. (1956) *Acta Biochim. Biophys. Acad. Sci. Hung. 1*, 39-44
12. Somogyi, M. (1952). *J. Biol. Chem. 195*, 19-23

II.1.a.2 Plant α-Amylases (Malt)

Malt enzyme was once called "diastase" and the first enzyme which was separated by scientific processing from organisms by Payen and Persoz (1833). Later the enzyme preparation was found to consist of two components and they were named α-amylase and β-amylase by Kuhn (1924) and Ohlsson (1930) (1). The former is characterized by decrease in viscosity or liquefaction of starch and the latter, by formation of fermentable sugars or saccharification. Ripe cereal grains generally contain little or no α-amylase, but on germination, they produce substantial amounts of the enzyme. A typical time course of development of α-amylase during germination was first studied by Hesse (1908). The same study done relatively recently was summarized by Luers *et al.* (1). As shown in Fig. 1, the difference in activities between the saline extracts (I) and the papain extracts (II) suggested the existence of a latent form of α-amylase as in the case of β-amylase. However, recent works have shown that no latent form of α-amylase is present (1). Schwimmer (1951) also studied the fate of amylases during growth, malting and brewing of barley (Fig. 2) (2). These results clearly showed the occurrence of *de novo* synthesis of α-amylase during germination.

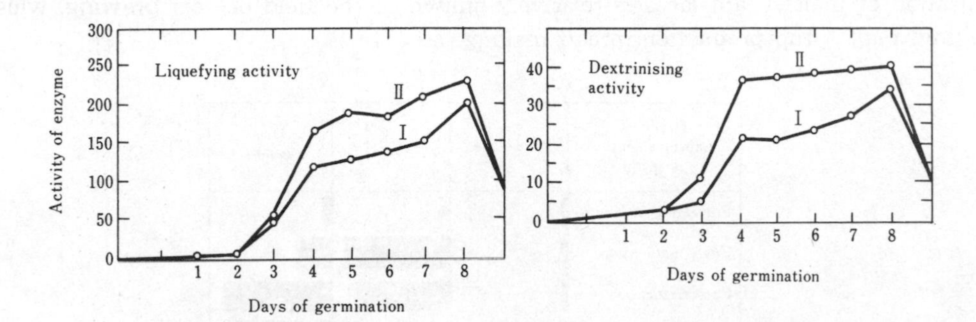

Fig. 1. Development of α-Amylase during Germination of Barley (Luers *et al.*, 1933). I , Saline extracts ; II , papain extracts.

Yomo (1960) (3) and Paleg (1960) (4), using barley half seeds with and without embryo or embryoless half seeds and gibberellic acid, found independently the fact that α-amylase synthesis is initiated by gibberellic acid. Varner *et al.* (1965) (5) reported that the aleurone layer is the site of α-amylase synthesis. This is also confirmed by many investigators using isolated aleurone layers. For example, Filner *et al.* (1967) (6) observed that *de novo* synthesis of ^{18}O containing α-amylase occurred when aleurone layer was incubated with $H_2{}^{18}O$ and gibberellic acid. Higgins *et al.* (1976) (7) reported that gibberellic acid in the aleurone layers stimulated transcription of DNA into mRNA for α-amylase. McGregor *et al.* (1982) (8), basing on patterns of starch degradation in barley and wheat grains, suggested that the initial site of α-amylase synthesis was the embryo and not the aleurone layer, because the degradation of starch started at the ventral edge of endosperm-embryo junction and moved along this junction to the dorsal edge (Fig. 3).
Gibbons *et al.* (1983) (9) studied the relative roles of scutellum and the layer during germination of barley. They conclusively reported that even when the function of aleurone is inhibited by abscisic acid, the scutellar system can produce hydrolytic enzymes and modify at least 50% of the endosperm at the normal rate. Chandler *et al.* (1984) also reported that abscisic acid had an inhibitory effect at the same stage of transcription of hydrolytic enzymes as gibberellic acid (10).

On the other hand, Gibbons *et al.* (1983) summarized the major findings obtained during the period from 1886 to 1983 (Fig. 4) (9). A theory derived from the summary is that enzymes are produced from the scutellum during an early stage of germination followed by a gibberellic acid induced production of enzymes from the aleurone layer in the later stage in the process of production of barley malt. Similar phenomena have been reported on α-amylase synthesis in other cereal grains such as wheat (Boston, *et al*, 1982) (11), rice (Miyata, *et al*, 1982) (12), etc.

α-Amylases are widely used in various fields such as food and fermentation industries, pharmaceutical industry, dyeing industry, etc. However, recently the utilization of malt α-amylases is relatively limited in the field of beer brewing, whisky making, malt syrup production, bread making, etc. (13,14).

Stage of development of grain	α-amylase	β-amylase	Dry weight
Formation of Kernel			
Early maturation			
Late maturation			
Ripening			
Steeping			
Germination			
Kilning			
Mashing			
"Lautering" *			
Boiled wort			

Fig. 2. Development and Fate of Amylases during Growth, Malting and Brewing of Barley (Schwimmer, 1951) (2).
* This applies only in decoction mashing.

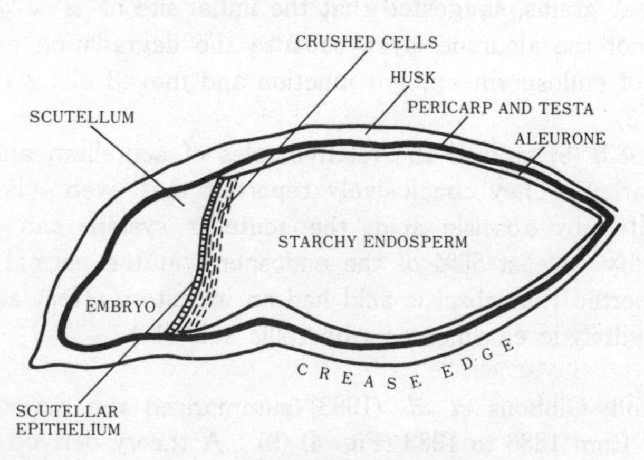

Fig. 3. Longitudinal Section of Barley Grain (McGregor *et al.*, 1982) (8).

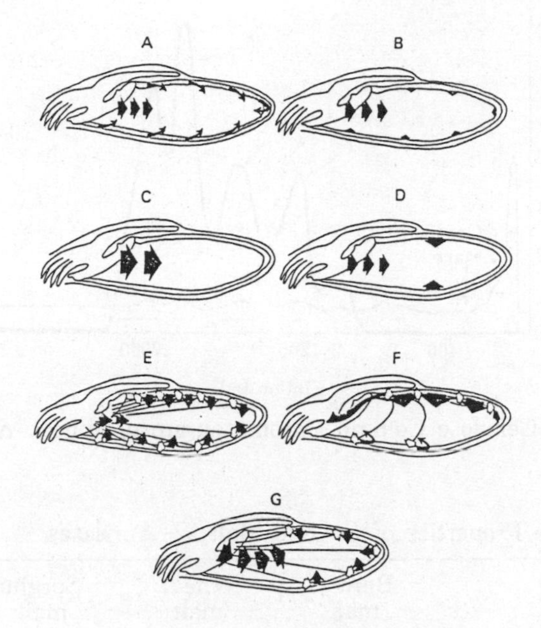

Fig. 4. Theories of Germination.

The solid arrows represent the movement of hydrolytic enzymes and the open arrows sites of action of gibberellic acid. The size of the arrows indicate the relative importance of each tissue in the process of endosperm modification suggested in the different theories. The patterns are superimposed on a diagram of barley seedling, germinated at 15°C for 60 h. The embryo is shown on the left-hand side.

A, Tangl's theory (1886); B, Haberlant's theory (1890); C, Brown & Morris's theory (1890); D, Liderstrøm-Lang & Engel's conclusion (1938) ; E, the generally accepted view *anno* (1965) ; F, Parmer's theory (1966,1982) ; G, the present theory (1982).

Isolation and Purification : In 1949 Schwimmer *et al.* reported on isolation and crystallization of α-amylase from germinated barley (15). Crystalline α-amylase was also obtained by Fischer *et al.* (1951) from fresh green malt (16).
Tkachuk *et al.* (1974) purified wheat α-amylase by heat-treatment, salting out and DEAE-Cellulose column chromatography, and obtained four active fractions as shown in Fig. 5 (17).

Enzymic and Proteochemical Properties : Malt α-amylase randomly hydrolyzes the α-1,4-glucosidic linkages in starch. The enzyme also attacks raw starch granules of various sources to certain extents (Ueda, S., 1974) (18). Table 1 shows some properties of cereal α-amylases (19).

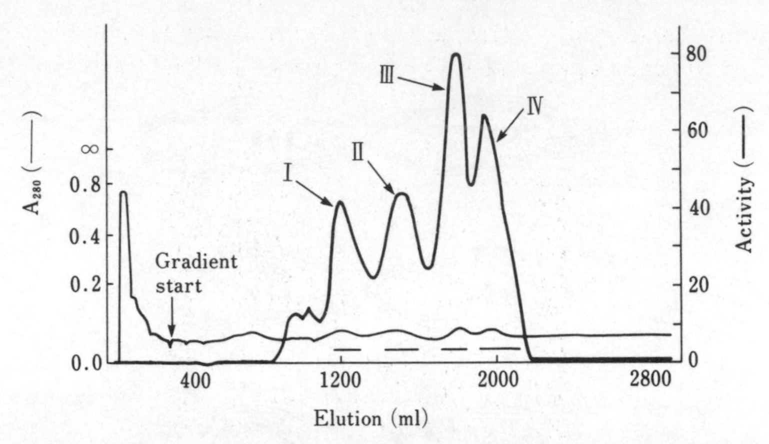

Fig. 5. DEAE-Cellulose Column Chromatogram of Wheat α-Amylase (1).

Table 1. Some Properties of Several Malt α-Amylases

	Barley malt	Wheat malt	Sorghum malt	Rice malt
Optimum pH	5.5	5.5	—	6.0
Molecular weight	45,000	41,500	48,000	44,000
Optimum temp.	51-60℃	60-66℃	—	45-50℃

Plant α-amylases, as shown in Table 1, are similar to each other in their properties. The molecular weight ranges from 40,000 to 50,000, and the optimum pH's are from 5.5 to 6.0. Plant α-amylases are sensitive to acid and lose their activities by incubation at pH 3.7. Most of them have an activation energy of 8-14 kcal per mole and need calcium ion for their activity and stability. The removal of calcium with EDTA or dialysis results in loss of activity.

Table 2. Properties of Four Fractions of Wheat α-Amylase

	Ⅰ	Ⅱ	Ⅲ	Ⅳ
Activity/mg protein (μg maltose/min)	1,480	1,300	1,510	1,570
Molecular weight	42,500	42,200	42,000	41,500
Isoelectric point	6.16	6.20	6.05	6.17
Optimum pH	5.5	5.7	5.5	5.5
E (1% in 0.1N NaOH, 275nm)	23.8	25.6	—	23.4
SH group	0	0	0	0

Table 2 shows some properties of the four fractions obtained by Tkachuk *et al.*

(17). There are some minor differences in isoelectric points and molecular weights, etc., but no enzymically significant difference among the four fractions.

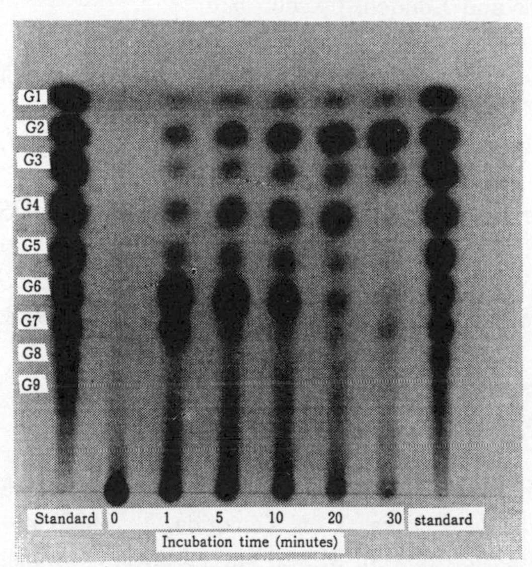

Fig. 6. Digestion of Soluble Starch by Malt α-Amylase.

Specificity : Plant α-amylases hydrolyze amylose to produce maltose and various oligosaccharides in the initial reaction. Then, the intermediary relatively high molecular weight dextrins larger than G8 disappear and finally only oligosaccharides from G2 to G7 remain, as shown in Fig. 6 (14).

Polymorphism of plant α-amylase is one of the characteristics and several kinds of isozymes were reported on barley and rice (14). Nishikawa *et al.* observed fifteen bands of wheat α-amylases by electrofocusing. The enzymically active bands were found to change during ripening and germination (20). Jacobsen *et al.* (1982) also fractionated barley α-amylases into four fractions by a DEAE-Cellulose column. The fractions were different electrophoretically and immunologically (21). These results show that several genes are responsible to α-amylase in cereal grains (Gale *et al.*, 1983) (22).

Assay Methods : Assay methods of α-amylase are as follows : 1) measurement of reducing sugars by Somogyi-Nelson method or 3,5-dinitrosalicylic acid method, 2) colorimetric method by iodine-starch color reaction, 3) measurement of colored substances released from insoluble colored starch as substrate, 4) conjugated enzyme systems on maltopentaose as substrate using α-glucosidase, hexokinase, glucose-6-P dehydrogenase and NADP (14).

(Ryu Shinke)

References

1. Harris, G. (1962) *α-Amylases* in *Barley and Malt* (Cook, A. H., ed.) Academic Press, New York and London, pp. 606-620
2. Schwimmer, S. (1951) *Brew. Digests 26*, 29T, 47T
3. Yomo, H. (1960) *Hakko-Kyokaishi* (in Japanese) *18*, 600-602
4. Paleg, L. G. (1960) *Plant Physiol.* 35, 902-906
5. Varner, J. E., Chandra, G. R. & Chrispeels, M. J. (1965) *J. Cell. Comp. Physiol. 66*, 55-68
6. Filner, P. & Varner, J. E. (1967) *Proc. Natl. Acad. Sci. USA 58*, 1502-1526
7. Higgins, T. J. V., Zwar, J. A. & Jacobsen, J. V. (1976) *Nature 260*, 166-169
8. MacGregor, A. W. & Matsuo, R. R. (1982) *Cereal Chem. 59*, 210-216
9. Gibbons, G. C. & Nielsen, E. B. (1983) *J. Inst. Brew. 89*, 8-14
10. Chandler, P. M., Zwar, J. A., Jacobsen, J. V., Higgins, T. J. V. & Inglis, A. S. (1984) *Plant Molec. Biol. 3*, 407-418
11. Boston, R. S., Miller, T. J., Mertz, J. E. & Burgess, R. R. (1982) *Plant Physiol. 69*, 150-154
12. Miyata, S. & Akazawa, T. (1982) *Plant Physiol. 70*, 147-153
13. Aizawa, T., Ono, M., Tezuka, T. & Yanagida, T. (1980) *α-Amylases* in *Handbook of Enzyme Utilization* (Ozaki, M., ed.)(in Japanese) Chijinshokan, pp. 3-54
14. Taniguchi, H. (1986) *α-Amylases* in *Amylases* (Nakamura, M., ed.)(in Japanese) Gakkai-Shuppan Center, pp. 141-186
15. Schwimmer, S. & Balls, A. K. (1949) *J. Biol. Chem. 179*, 1063-1074
16. Fischer, E. H. & Haselbach, C. H. (1951) *Helv. Chim. Acta 34*, 325-329
17. Tkachuk, R. & Kruger, J. E. (1974) *Cereal Chem. 51*, 508-529
18. Ueda, S. (1974) *J. Jap. Soc. Starch Sci.*(in Japanese) *21*, 210-221
19. Thoma, J. A., Spradlin, J. E. & Dygert, S. (1971) *Plant and Animal Amylases* in *The Enzymes* Vol. 5 (Boyer, P. D., ed.) Academic Press, pp. 115-189
20. Nishikawa, K., Furuya, Y., Hina, Y. & Fuji, S. (1978) *Wheat Inf. Serv.* 47-48
21. Jacobsen, J. V. & Higgins, T. J. V. (1982) *Plant Physiol. 70*, 1647-1653
22. Gale, M. D., Law, C. N., Chojecki, A. J. & Kempton, R. A. (1983) *Theor. Appl. Genet. 64*, 309-316

II.1.a.3. Fungal α-Amylases

II.1.a.3-1. Taka-amylase A (*Aspergillus oryzae* α-Amylase)

Taka-amylase A (TAA) is an α-amylase, catalyzing the hydrolysis of α-1,4-glucosidic linkage of α-1,4-glucans. It was purified and crystallized from Taka-diastase (a commercial preparation of a culture medium of *Aspergillus oryzae*) by Akabori *et al.* (1). The enzyme is a single polypeptide containing a high-mannose type carbohydrate moiety linked to an asparagine residue (2,3). Toda *et al.* determined the amino acid sequence of the enzyme (4). Matsuura *et al.* showed the three-dimensional structure of the enzyme from their X-ray crystallographic analysis and proposed a three-dimensional model of the interaction between the active site of the enzyme and a substrate (5). The model provides much information about the enzyme action on the substrate.

Reaction and Specificity : It is conceivable that the active site of TAA is composed of tandem subsites geometrically complementary to several glucose residues S1, S2, etc. (Fig. 1).

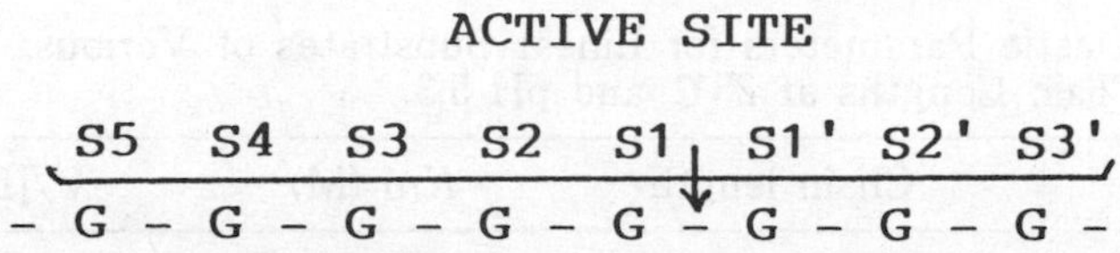

ACTIVE SITE

S5 S4 S3 S2 S1 ↓ S1' S2' S3'

- G - G - G - G - G - G - G - G -

Fig. 1. Schematic Representation of a Substrate Binding to the Active Site of TAA.

The active site is composed of subsites S1, S2, etc. and the glycosidic bonds of the substrate are split between subsites S1 and S1'.

The features of three-dimensional distribution of ionic charge and hydrophobicity or hydrophilicity at each subsite are to be the determinants of the substrate specificity of the enzyme. Studies on the substrate specificity of TAA have been carried out using amylose or maltooligosaccharides or their derivatives to examine the interaction of substrate and subsites and to draw the mechanism of the enzyme action.

1. Action on Starch, Glycogen and β-Amylase Limit Dextrins.

TAA hydrolyzes amylose completely to maltose and glucose. The branched substrates including the branched fractions from corn starch, waxy maize starch, glycogen and β-amylase limit dextrins are hydrolyzed less readily and less extensively than the linear fraction from corn starch (6).

2. Action on Maltooligosaccharides.

The action of TAA on maltooligosaccharides has been extensively studied to examine the active site and the mechanism of action of the enzyme. The bond-cleavage frequencies for maltooligosaccharides were qualitatively determined using oligosaccharides labelled at the reducing end with radio-active glucose (7).

Nitta *et al.* (8) determined the values of kinetic parameters for the hydrolyses of oligosaccharides by TAA to evaluate subsite affinities between each subsite and the glucose residues of the oligosaccharides (Table 1). Allen and Thoma (9) and Suganuma *et al.* (10) independently investigated the bond-cleavage frequencies of oligosaccharides to quantitatively estimate the subsite affinity of the active site, based on the different principle from that of Nitta *et al.* (8). Table 2 shows the data by Suganuma *et al.* These kinetic parameters and bond-cleavage frequencies for maltooligosaccharides indicate that the number of subsites is seven to nine. The bond-cleavage frequencies depend on the substrate concentration because transglycosylation and condensation are not negligible at high substrate concentrations. Allen and Thoma (11) estimated that repetitive attacks occur on the degradation of maltooligosaccharides by TAA from the data of bond-cleavage frequencies. The extents of repetitive attack calculated for maltotriose and maltotetraose are not significant. However, for maltopentaose through maltodecaose, a gradual increase of repetitive attack is observed.

Table 1. Kinetic Parameters for Linear Substrates of Various Chain Lengths at 25℃ and pH 5.3.

Substrate	Chain length	*K*m (M)	$V/[E]_0$ (min^{-1})
Maltose	2	9.6×10^{-2}	1.2
Maltotriose	3	6.9×10^{-2}	1.3×10^{2}
Maltotetraose	4	8.0×10^{-3}	3.3×10^{3}
Maltopentaose	5	2.0×10^{-3}	7.0×10^{3}
Maltohexaose	6	1.6×10^{-3}	9.6×10^{3}
Maltoheptaose	7	9.0×10^{-4}	1.2×10^{4}
Maltodextrin I	15.5	1.5×10^{-4}	1.2×10^{4}
Maltodextrin II	28.4	6.5×10^{-5}	8.9×10^{3}
Maltodextrin III	38.4	4.6×10^{-5}	1.1×10^{4}
Maltodextrin IV	50.5	2.6×10^{-5}	8.8×10^{3}
Maltodextrin V	71.5	1.4×10^{-5}	7.0×10^{3}
Maltodextrin VI	117	9.5×10^{-6}	9.9×10^{3}

Table 2. Cleavage Distribution of Maltooligosaccharides at Low and High Substrate Concentrations by TAA

Substrate	1	2	3	4	5	6	[S]
G—G—G	.058	.942					9 μM
	.759	.241					83.2 mM
G—G—G—G	.019	.971	.010				20 μM
	.359	.629	.011				87.1 mM
G—G—G—G—G	.001	.715	.278	.006			10 μM
	.136	.589	.269	.006			19.3 mM
G—G—G—G—G—G	.010	.133	.677	.165	.016		5 μM
	.132	.164	.511	.189	.005		17.6 mM
G—G—G—G—G—G—G	—	.282	.123	.504	.091	.000	6 μM
	.070	.155	.168	.441	.163	.002	14.0 mM

The numerals indicate the fractional cleavage distribution.

3. Action on Cyclodextrins.

TAA catalyzes the ring opening hydrolysis of cyclodextrins. The linear maltooligosaccharides produced are immediately degraded to smaller oligosaccharides. Suetsugu *et al.* observed multiple attack on the hydrolyses of cyclodextrins (12). Table 3 summarizes the values of the kinetic parameters for the ring opening hydrolysis (12).

Table 3 Kinetic Parameters for the Hydrolyses of Cyclodextrins by TAA at 25℃ and pH 5.3

Substrate	$Km(\times 10^3 M)$	$Ki\ (\times 10^3 M)$	$V/[E]_O\ (min^{-1})$
α-CD	4.7	5.2	3.3
β-CD	10.2	9.3	270
γ-CD	2.4	—	3,270

4. Action on Modified Amylose and Starch.

Huges found 6^3-glucosylmaltotriose in TAA digest of starch (13). Fujinaga-Isemura *et al.* prepared partially O-methylated amyloses and analyzed their TAA digests (14). They identified 6-O-methylglucosyl-glucose, but did not find oligosaccharides whose reducing-end glucose residues were methylated in the digests. The data indicates that any O-methylated glucose residue can not fit on subsite S1, and that 6-O-glucosyl-glucose and 6-O-methylglucose residues can fit on subsite S1' to be hydrolyzed. Takeda and Hizukuri identified 6^3-phosphorylmaltotriose and 3^3-phosphorylmaltotetraose in TAA digest of potato starch (15). Their data indicate that 6-phosphorylglucose residue can bind to subsites S3, S1' and S2' and 3-phosphorylglucose residue to S3 and S2'.

5. Action on Phenyl α-Maltoside and Its Derivatives.

TAA hydrolyzes a synthetic substrate, phenyl α-maltoside, to give maltose and phenol, though the rate of hydrolysis is rather low compared to that of starch or amylose (16). The synthetic substrate is very suitable for examining subsites S1 and S2, because its hydrolysis can be easily detected using phenol reagent (17). Ikenaka reported that TAA can not hydrolyze phenyl 6-O-acetyl-α-maltoside and that its C6-OH is important for formation of the enzyme-substrate complex (18). Isemura *et al.* examined the action of the enzyme on partially O-methylated phenyl α-maltoside (19). Phenyl 2-O-methyl-α-maltoside, phenyl 3-O-methyl-α-maltoside, phenyl 6-O-methyl-α-maltoside and phenyl 2'-O-methyl-α-maltoside were resistant to TAA. On the other hand, phenyl 3'-O-methyl-α-maltoside, phenyl 4'-O-methyl-α-maltoside and phenyl 6'-O-methyl-α-maltoside were hydrolyzed by the enzyme. This result is consistent with that of the partially O-methylated amylose (14).

Arita *et al.* synthesized various derivatives of phenyl α-maltoside and examined their hydrolyses by TAA quantitatively (20-25). Table 4 shows the relative rates of hydrolyses of various derivatives of phenyl α-maltoside by TAA. It is likely that C2-OH, C3-OH, C6-OH and C2'-OH of phenyl α-maltoside are essential for hydrolysis by the enzyme, while C3'-OH, C4'-OH and C6'-OH are not so important.

Table 4. Action of TAA on Phenyl α-Maltoside and Its Derivatives.

Substrate	Rate	Substrate	Rate
Phenyl α-maltoside	1.00	Phenyl 6'-deoxy-α-maltoside	0.05
Phenyl 6-deoxy-α-maltoside	0.02	Phenyl 6'-chloro-6'-deoxy-α-maltoside	0.24
Phenyl 6-chloro-6-deoxy-α-maltoside	0.00	Phenyl 6'-deoxy-6'-iodo-α-maltoside	0.29
Phenyl 6-deoxy-6-iodo-α-maltoside	0.00	Phenyl 6'-O-methyl-α-maltoside	2.10
Phenyl 6-deoxy-6-fluoro-α-maltoside	trace	Phenyl 4'-deoxy-α-maltoside	0.75
Phenyl 6-O-methyl-α-maltoside	0.00	α-D-Gal-(1→4)-α-D-Glc-P[2]	0.25
Phenyl 2-deoxy-α-maltoside	0.00	Phenyl 4'-O-methyl-α-maltoside	2.08
α-D-Glc-(1→4)-α-D-Man-P[1]	0.00	Phenyl 4'-O-t-butyl-α-maltoside	0.50
Phenyl 2-O-methyl-α-maltoside	0.00	Phenyl 2'-O-methyl-α-maltoside	0.00
Phenyl 3-O-methyl-α-maltoside	0.00	Phenyl 3'-O-methyl-α-maltoside	+

The numerals indicate the relative rates of hydrolyses to phenyl α-maltoside

[1] Phenyl O-α-D-glucopyranosyl-(1→4)-α-D-mannopyranoside
[2] Phenyl O-α-D-galactopyranosyl-(1→4)-α-D-glucopyranoside.

6. Action on Phenyl α-Maltotetraoside, Phenyl α-Maltopentaoside and Their Derivatives.

Substrate	Cleavage	Products	Relative rates of hydrolysis
G-G-G-G-P (G4P)	2.4% →	G-G-G + G-P	1.0
	96.0% →	G-G + G-G-P	
	1.6% →	G + G-G-G-P	
AG-G-G-G-P (AG4P)	29.9% →	AG-G-G-G + P	0.0081
	63.0% →	AG-G-G + G-P	
	7.1% →	AG-G + G-G-P	
CG-G-G-G-P (CG4P)	1.0% →	CG-G-G + G-P	0.073
	99.0% →	CG-G + G-G-P	
G-G-G-G-G-P (G5P)	0.8% →	G-G-G-G + G-P	2.0
	72.9% →	G-G-G + G-G-P	
	20.6% →	G-G + G-G-G-P	
	5.7% →	G + G-G-G-G-P	
AG-G-G-G-G-P (AG5P)	2.1% →	AG-G-G-G + G-P	1.6
	97.9% →	AG-G-G + G-G-P	
CG-G-G-G-G-P (CG5P)	2.8% →	CG-G-G-G + G-P	0.15
	75.7% →	CG-G-G + G-G-P	
	21.5% →	CG-G + G-G-G-P	

Fig. 2. Action of TAA on Phenyl α-Maltotetraoside, Phenyl α-Maltopentaoside and their Derivatives.

AG, 6-amino-6-deoxy-glucose residue; CG, glucuronic acid residue; P, phenyl residue. The rates of hydrolysis are presented taking the value for G4P as unity.

Nagamine *et al.* prepared ionic derivatives of phenyl maltooligosides to examine subsites S3 and S4 (26). The action of TAA on the substrates is summarized in Fig. 2. TAA acts on AG4P or AG5P in quite a different manner from that on G4P or G5P, respectively, whereas no remarkable change is seen on the hydrolysis of CG4P or CG5P. This strongly suggests the presence of acidic amino acid residues at both subsites S3 and S4, which interact with the positive charges on non-reducing end glucose residues of AG4P and AG5P.

7. Acceptor Specificity on Transglycosylation.
TAA has transferase activity in the presence of an appropriate acceptor as well as hydrolytic activity (27,7). The acceptor should interact with subsite(s) S1' (and S2') at transglycosylation. Omichi and Matsushima examined the interaction between subsite S1' and a substrate taking advantage of the transfer reaction (28). When amylose is used as the substrate, its glycon moiety is transferred to maltitol, phenyl α-glucoside, phenyl β-glucoside, phenyl 2-O-methyl-α-glucoside and phenyl 6-O-methyl-α-glucoside. But, the transfers to phenyl α-mannoside, phenyl β-mannoside and phenyl 3-O-methyl-α-glucoside do not occur.

Production and Purification : The purified TAA can be obtained from Taka-diastase, a commercial product of the Sankyo Co., Tokyo, Japan. The outlines of the procedures of making Taka-diastase are : 1. Cultivation of *Aspergillus oryzae* on wheat bran at appropriate conditions. 2. Extraction of enzymes from bran with water. 3. Precipitation of the enzymes by addition of ethanol to final concentration of 70%. 4. Drying under reduced pressure and powdering. The purification of TAA from Taka-diastase includes ammonium sulfate fractionation, precipitation with 2-ethoxy-6,9-diamino-acridinium lactate (Rivanol) and crystallization from aqueous acetone (1). The enzyme preparation crystallized three times is further purified by DEAE-ion exchange column chromatography to homogeneity on polyacrylamide gel electrophoresis (29,31).

Assay Method : TAA is determined spectrophotometically at 280 nm as $A^{1cm}_{1\%}$=22.1(29) and its molecular weight is 53,662 (4). The activity of TAA is usually estimated by measuring the reducing power produced by the enzyme using amylose or soluble starch (30). The reducing power is determined colorimetrically according to the Somogyi-Nelson method. TAA has also activity to hydrolyze *p*-nitrophenyl α-maltoside to maltose and *p*-nitrophenol. The assay method using this synthetic substrate is convenient, because the *p*-nitrophenol liberated is easily determined spectrophotometrically at 410 nm at alkaline conditions (16).

(Kaoru Omichi and Tokuji Ikenaka)

References

1. Akabori, S., Ikenaka, T. & Hagihara, B. (1954) *J. Biochem. 41*, 577-582
2. Yamaguchi, H., Ikenaka, T. & Matsushima, Y. (1971) *J. Biochem. 70*, 587-594

3. Saita, M., Ikenaka, T. & Matsushima, Y. (1971) *J. Biochem. 70,* 827-833
4. Toda, H., Kondo, K. & Narita, K. (1982) *Proc. Jpn. Acad. 58B,* 208-212
5. Matsuura, Y., Kusunoki, M., Harada, W. & Kakudo, M. (1984) *J. Biochem. 95,* 697-702
6. Hanrahan, V. & Caldwell, M. L. (1953) *J. Am. Chem. Soc. 75,* 2191-2197
7. Okada, S., Kitahata, S., Higashihara, M. & Fukumoto, J. (1969) *Agric. Biol. Chem. 33,* 900-906
8. Nitta, Y., Mizushima, M., Hiromi, K. & Ono, S. (1971) *J. Biochem. 69,* 567-576
9. Allen, J. D. & Thoma, J. A. (1976) *Biochem. J. 159,* 121-132
10. Suganuma, T., Matsuno, R., Ohnishi, M. & Hiromi, K. (1978) *J. Biochem. 84,* 293-316
11. Allen, J. D. & Thoma, J. A. (1978) *Carbohydr. Res. 61,* 377-385
12. Suetsugu, N., Koyama, S., Takeo, K. & Kuge, T. (1974) *J. Biochem. 76,* 57-63
13. Hughes, R. C. (1959) *Ph. D. Thesis, University of London* ; cited in Whelan, W. J. (1960) *Stärke, 12,* 358-364
14. Fujinaga-Isemura, M., Ikenaka, T. & Matsushima, Y. (1968) *J. Biochem. 64,* 63-80
15. Takeda, Y. & Hizukuri, S. (1986) *Carbohydr. Res. 153,* 295-307
16. Matsubara, S., Ikenaka, T. & Akabori, S. (1959) *J. Biochem. 46,* 425-431
17. Folin, O. & Ciocalteu, V. (1927) *J. Biol. Chem. 73,* 627-633
18. Ikenaka, T. (1963) *J. Biochem. 54,* 328-333
19. Isemura, M., Ikenaka, T. & Matsushima, Y. (1969) *J. Biochem. 66,* 77-85
20. Arita, H., Isemura, M., Ikenaka, T. & Matsushima, Y. (1970) *Bull. Chem. Soc. Jpn. 43,* 818-823
21. Arita, H., Isemura, M., Ikenaka, T. & Matsushima, Y. (1970) *J. Biochem. 68,* 91-96
22. Arita, H. & Matsushima, Y. (1970) *J. Biochem. 68,* 717-722
23. Arita, H. & Matsushima, Y. (1971) *J. Biochem. 69,* 401-407
24. Arita, H. & Matsushima, Y. (1971) *J. Biochem. 69,* 409-413
25. Arita, H. & Matsushima, Y. (1971) *J. Biochem. 70,* 795-801
26. Nagamine, Y., Sumikawa, M., Omichi, K. & Ikenaka, T. (1987) *J. Biochem. 103,* 767-775
27. Matsubara, S. (1961) *J. Biochem. 49,* 226-231
28. Omichi, K. & Matsushima, Y. (1970) *J. Biochem. 68,* 303-309
29. Toda, H. & Akabori, S. (1963) *J. Biochem. 53,* 102-110
30. Fuwa, H. (1954) *J. Biochem. 54,* 583-603
31. Yamakawa, Y. & Okuyama, T. (1973) *J. Biochem, 73,* 447-454

II.1.a.3-2. α-Amylase of *Rhizopus niveus*

α-Amylase of *Rhizopus delemar* and some properties of the enzyme were at first reported by Tsujisaka, Y. and his collaborators in connection with isolation of glucoamylase of the fungus (1). The commercial preparations of glucoamylase of *Rhizopus* fungi have been applied to industrial production of glucose from starch. Commercial glucoamylase preparations including those from *Aspergillus niger,* however, usually contain α-amylase which effects to accelerate the action of glucoamylase not only on gelatinized starch, but also on raw starch (2). It may be said in this sense that α-amylase in commercial glucoamylase preparations is an industrially important enzyme.

Reactions and specificity : α-Amylase of *Rhizopus niveus* hydrolyzes gelatinized potato starch to the degree of about 35% in hydrolysis. This degree of hydrolysis is slightly higher (by 3.0 to 4.0%) than that achieved by Taka-amylase A under the same conditions. The major products by *Rhizopus* α-amylase are : G2, G3 > branched (G5, G6) > high molecular branched dextrins. *Rhizopus niveus* α-amylase hydrolyzes raw corn starch slightly. But, the addition of this amylase to the incubation mixture of the purified glucoamylase with raw corn starch greatly accelerates the production of glucose, as shown in Fig. 1. Commercial glucoamylase preparations, however, usually contain α-amylase in such amounts that no more addition of α-amylase is necessary.

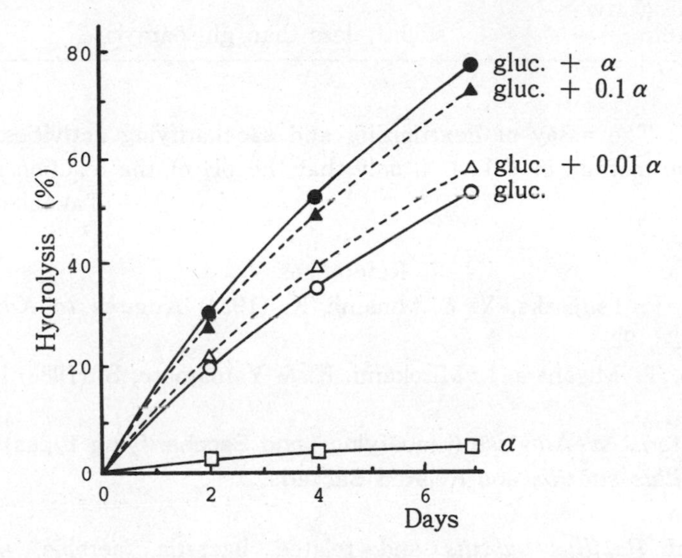

Fig. 1. Effect of α-Amylase on Hydrolysis of Raw Corn Starch by Glucoamylase of *Rhizopus niveus* (8.3% raw corn starch suspension, pH 4.5, 40℃).
Amounts of α-amylase added : α-, with the same ratio of α-amylase to glucoamylase in commercial enzyme ; 0.1- and 0.01-α ; one tenth and one hundredth of α-amylase, respectively.

Isolation and Purification : Commercial *Rhizopus* glucoamylase is a convenient source of *Rhizopus* α-amylase. A solution of this preparation (1.0% in water) is incubated at 40℃ and at pH 7.8 buffered with sodium carbonate for 30 min, in order to destroy proteolytic enzymes. After adjusting the pH at 5.6 with acetic acid, the solution is dialyzed against changing 0.01 M acetic-acetate buffer, pH 6.0, for 24 h in the cold. The dialyzed solution is then applied to a column of DEAE-Sephadex A50 equilibrated with above acetate buffer. Glucoamylase comes out in the effluent while α-amylase is adsorbed. The adsorbed enzyme is eluted with 0.02 M acetate buffer, pH 6.0, separating into three peak fractions. The firstly eluted α-amylase is the major component and the enzymic and proteochemical properties are shown in Table 1.

Table 1. Enzymic and Proteochemical Properties of *Rhizopus* α-Amylase

Optimum pH	4.6-4.8
pH Stability	3.6-8.0 (40℃, 1 h)
Thermal stability	<40℃ (pH 5.0, 1 h)
Molecular weight	51,000
Sugar content	+
Dextrinizing activity	65,000 units/mg protein
Saccharifying activity	150 units/mg protein
Effect of EDTA	not observed
Effect of Hg^{2+}	not observed
Adsorption on raw corn starch	slightly less than glucoamylase

Enzyme assay : The assay of dextrinizing and saccharifying activities is carried out according to the method in Ⅱ.1. a. 4, only that the pH of the reaction mixture is 4.6.

(Takehiko Yamamoto)

References

1. Fukumoto, J., Tsujisaka, Y. & Minamii, K. (1954) *Kagaku to Kogyo*, Osaka (in Japanese) *28*, 92
2. Yamamoto, T., Miyahara, I., Mizokami, K. & Yamamoto, S. (1988) in preparation

Ⅱ.1.a.4. Bacterial α-Amylase (Liquefying- and Saccharifying types) of *Bacillus subtilis* and Related Bacteria

α-Amylase of *Bacillus subtilis* and related bacteria (aerobic, spore forming, mesophile) is divided into two types from the specificity (1) : One is starch liquefying and the other, saccharifying type (2). The two α-amylases are also distinguished from their protein nature (3). The type of α-amylase of those bacteria is bound to the strain which produces the enzyme and a strain belonging to those *bacilli* produces only one amylase of either liquefying or saccharifying type (5). The liquefying α-amylase-secreting bacterial strains have been grouped as *Bacillus amyloliquefaciens* (5,6) whose species name was first given by Fukumoto (7).

The liquefying α-amylase is one of the industrially important enzymes and utilized in various fields (See part Ⅵ). The saccharifying α-amylase, however, has not been so applied yet, but this α-amylase is useful for structual elucidation study of branched dextrins and cyclodextrins, as described below and somewhere of this book.

Reaction :

liquefying α-amylases

Starch + n H_2O ⟶ G1 + G2 + G3 + G4 + G5+ G6 + various branched oligosaccharides

saccharifying α-amylases

Starch + n H_2O ⟶ G1 + G2 + G3 + various branched oligosaccharides

The extents of hydrolysis of starch, glycogen, amylose and amylopectin to be achieved by saccharifying α-amylase is at least twice that achieved by liquefying α-amylase, as shown in Fig. 1 (8).

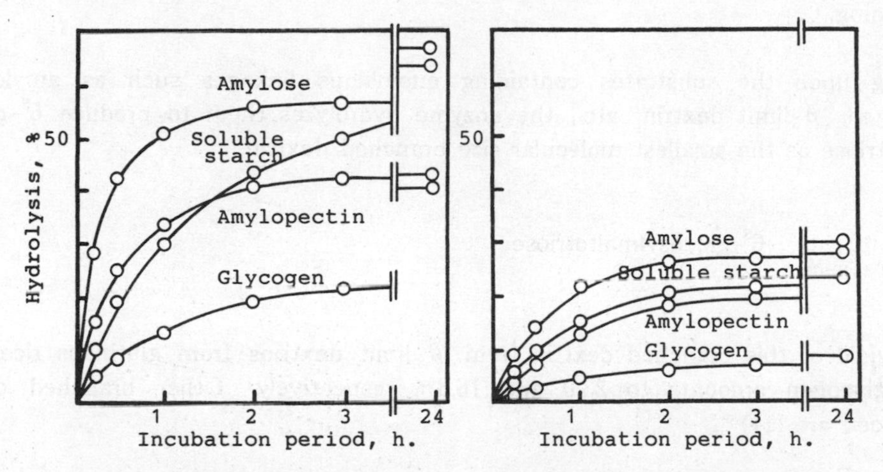

Fig. 1. Progressive Curves of Hydrolysis of Several Substrates by Saccharifying (left) and Liquefying (right) α-Amylase.

The degree of hydrolysis of soluble starch by the former amylase at which the iodine color reaction reaches the achromic stage, is also nearly double that at the achromic stage by the latter enzyme. However, the degrees of hydrolysis of starch by those α-amylases more or less vary depending on the amount of enzyme added.

Specificity : The liquefying α-amylase hydrolyzes amylose to produce glucose, maltose, maltotriose, -tetraose, -pentaose, and -hexaose. The enzyme splits maltoheptaose and -octaose into maltohexaose and glucose, and maltohexaose and maltose, respectively. The glucose to split from maltoheptaose is of the reducing end of the dextrin (9). The liquefying α-amylase never attack α-1,4-glucosidic linkages adjacent to α-1,6-glucosidic linkage in starch and glycogen. Acting upon amylopectin and glycogen, the enzyme produces the following dextrins as the smallest molecular size branched ones (10) :

```
0-0       0-0-0       0-0          0-0
  ↓           ↓         ↓            ↓
0-0-φ,     0-0-φ,    0-0-0-φ,   0-0-0-φ ,
```

where 0 is glucose residue and – and ↓ are α-1,4- and α-1,6-glucosidic linkages, respectively.

In the reaction mixture described above, various other dextrins of doubly, triply or more multiply branched are also produced as liquefying α-amylase-limit dextrins.

The saccharifying α-amylase hydrolyzes amylose to produce glucose, maltose and maltotriose. On a long incubation, the enzyme splits maltotriose into glucose and maltose. Under this condition, isomaltose is formed from maltose by this enzyme action, but the amount of isomaltose formed is much less than that of the maltose remaining.

Acting upon the substrates containing anomalous linkages such as amylopectin, glycogen, β-limit dextrin, etc., the enzyme hydrolyzes them to produce 6^3-glucosyl maltotriose as the smallest molecular size branched dextrin.

```
0
↓          6³-glucosylmaltotriose.
0-0-φ,
```

The yield of this branched dextrin from β-limit dextrins from glutinous rice starch and glycogen amounts to 23.0 and 16.8% respectively. Other branched dextrins produced are (11) :

```
0           0     0        0              0           0  0
↓           ↓     ↓        ↓              ↓           ↓  ↓
0-0         0-0-0-0-0-φ,   0-0-0          0-0    0    0-0-0-0
  ↓                            ↓            ↓    ↓          ↓
  0-0-φ,                       0-0-φ,     0-0-0-0-φ,        0-0-φ, etc.
```

Fig. 2. Several Doubly Branched Dextrins Produced by Saccharifying α-Amylase.

Saccharifying α-amylase hydrolyses α-1,4-linked glucosidic chains existing outside the anomalous linkage of amylopectin, glycogen, etc. and produces glucose, maltose and maltotriose, as described above. The enzyme also hydrolyzes β- and γ-cyclodextrins, but hardly attacks α-cyclodextrin (12). α-Amylase of *Streptococcus bovis* is very similar in the specificity to saccharifying α-amylase described above (13).

Table 1. Enzymic and Proteochemical Properties of Two Bacterial α-Amylases

Amylase	Saccharifying	Liquefying
Molecular weight	4.73 x 10^4	4.89 x 10^4*
N (%)	17.3	16.0
E (1%, 280nm)	19.8	25.0
Optimum pH	5.3	5.9
*K*m (soluble starch, %)	0.068	0.069
Activity/mg protein	416	480
pH Stability	4.5-9.2	5.1-10.4
Thermal stability	up to 55℃	up to 80℃
Stabilization by Ca^{++}	−	+
Inactivation by EDTA	−	+
Ca content (%)	0.076	>0.08
Cystein content (moles)	1.0 (Masked)**	0
Adsorption on raw starch	−	+

*,** See Ref. 14, 15 and 3, respectively.

Production and Purification : Starch liquefying α-amylase is produced by submerged culture of a strain of *Bacillus amyloliquefacines*. An example of the medium composition for the bacterial culture is as follows : 5.0% defatted soybean cake extract (Defatted soybean cake is boiled with twenty weights of 0.2% sodium hydroxide solution for one h, and the filtrate is neutralized with hydrochloric acid to pH 6.6-6.8.), 3.0-5.0% dextrin prepared by applying liquefying α-amylase, 0.5% $(NH_4)_2HPO_4$, and 0.02% each of KCl, $MgSO_4 \cdot 7H_2O$ and calcium acetate, pH 6.8. Secretion of the α-amylase occurs most actively over the stage from the late logarithmic growth to an early stationary phase. The addition of lactose at this culture stage greatly intensifies the enzyme secretion (17). Calcium is an indispensable element for synthesis of the enzyme (18) and phosphate is necessary for the enzyme secretion (19). The amount of enzyme to be secreted on a 36 h culture amounts to 8.0-9.0 mg as enzyme protein per ml medium. In the presence of ammonium sulfate (ca. 18%) or ethanol (ca. 20%), the liquefying α-amylase is readily adsorbed on raw corn starch in the cold (pH 5.6-6.0), and when the enzyme-adsorbed starch is suspended in a slightly alkaline solution (pH 8.0, 30℃), the enzyme is eluted. This method is effective for isolation and purification of the enzyme (20).

The production of saccharifying α-amylase is also done by submerged culture. The medium employed is also composed similarly to that for liquefying α-amylase, but amino acid mixture or peptone (1.0%) is added, and the culture is carried out with maintaining the pH at around 6.6. For purification, saccharifying α-amylase in the culture filtrate is precipitated by adding ammonium sulfate to 0.7 saturation, dialyzed and then, subjected to chromatography using a column of Duolite A2 (active group, amines) buffered with 0.02 M phosphate at pH 5.6. The α-amylase is adsorbed on the resin and eluted with 0.3 M phosphate buffer of pH 6.6.

Assay Method : The starch dextrinizing activity method is relatively convenient among various assay methods of α-amylase.

Starch dextrinization activity assay —— Two ml of 0.5% potato starch solution, pH 5.9 (Fifty ml of 0.05 N sodium hydroxide solution is added to 1.0 g of potato starch and the mixture is gently boiled for 10 min. The solution is then brought up to 200 ml with water, adjusting the pH to the desired value with acetic acid. This starch solution is available as substrate for several h after preparation.) and 1.0 ml of enzyme are mixed at 40℃. After 5 to 15 min incubation, a 0.3 ml aliquot of the reaction mixture is taken into 0.1 ml of 0.01 N iodine solution containing 0.1 N hydrochloric acid. To this mixture is added 10 ml water and the color developed is measured at 660 nm. When the color decreases to a half in 10 min under the conditions, the activity is defined as one unit activity of dextrinization. For determination of dextrinization activity of saccharifying α-amylase, the pH of the starch solution is adjusted to 5.4.

Saccharification activity assay —— One ml enzyme is let react at 40℃ with 2.0 ml of 0.5% starch solution prepared as described above. Ten min later, the reducing sugar formed is determined as glucose by the method of Somogyi-Nelson (21). One unit of saccharification activity is defined as the enzyme amount which produces reducing sugar equivalent to 10 μmol of glucose under the conditions.

(Takehiko Yamamoto)

References

1. Kneen, E. & Beckord, L. D. (1946) *Arch. Biochem.* *10,* 41.
2. Fukumoto, J., Yamamoto, T. & Ichikawa, K. (1951) *Proc. Japan Academy 27,* 352
3. Nishida, A., Fukumoto, J. & Yamamoto T. (1967) *Agric. Biol. Chem.* *31,* 682
4. Fukumoto, J., Yamamoto, T. & Okada, S. (1965) *Proceedings of the Symposium on Amylase* (in Japanese) *1,* 47
5. Welker, N. E. & Campbell L. L. (1967) *J. Bacteriol.* *94,* 1124
6. Welker, N. E. & Campbell, L. L. (1967) *J. Bacteriol.* *94,* 1131
7. Fukumoto, J. (1943) *Nippon Nogeikagaku Kaishi* (in Japanese) *17,* 488
8. Mantani, S., Yamamoto, T. & Fukumoto, J. (1967) *Proceedings of the Symposium on Amylase.* (in Japanese) *2,* 95
9. Okada, S., Kitahata, S., Higashihara, M. & Fukumoto, J. (1969) *Agric. Biol. Chem.* *33,* 900
10. Umeki, K. & Yamamoto, T. (1975) *J. Biochem.* *72,* 101, (1972) *78,* 889
11. Umeki, K. & Yamamoto, T. (1975) *J. Biochem.* *78,* 897
12. Fujita, K., unpublished
13. Mizokami, K., Kozaki, M. & Kitahara, K. (1977) *Nippon Nogeikagaku Kaishi* (in Japanese) *51,* 299
14. Junge, M. J., Stein, E. A., Neurath, H. & Fischer, E. H. (1959) *J. Biol. Chem.* *234,* 556
15. Toda, H. & Narita, K. (1968) *J. Biochem.* *63,* 302

16. Yamamoto, T. (1956) *Bull. Agr. Chem. Sci.* (in Japanese) *20*, 188
17. Fukumoto, J. Yamamoto, T. & Tsuru, D. (1957) *Nature 180*, 438
18. Hamada, N., Fukumoto, J. & Yamamoto, T. (1967) *Agric. Biol. Chem. 31*, 1
19. Hamada, N., Yamamoto, T. & Fukumoto, J. (1971) *Agric. Biol. Chem. 35*, 1052
20. Yamamoto, T. (1955) *Bull. Agr. Chem. Sci. 19*, 121
21. Somogyi, M. (1952) *J. Biol. Chem. 195*, 19

II.1.a.5. Thermostable Bacterial α-Amylases

Stark and Tetrault (1) and Campbell (2) reported *Bacillus stearothermophilus*, a thermophile, to produce a thermostable α-amylase. Ogasahara *et al.* (3) investigated the physicochemical and biochemical properties of this enzyme. Shimamura *et al.* (4) reported that *Bacillus licheniformis*, a mesophile, also produced a thermostable α-amylase. Hattori *et al.* (5) observed that a strain of *Bacillus subtilis* also produced a thermostable α-amylase. Thermophile V-2 (6), *Bacillus acidocaldarius* (7) of acido-thermophile, *Clostridium* sp. RS-0001 (8) of thermophilic anaerobe, and *Dictyo glomus thermophilum* (9) of super-thermophilic anaerobe were also reported to produce thermostable α-amylases. Goto *et al.* (10), on the other hand, treated *B. licheniformis* IFO 12196 (ATCC 9789) with N-methyl-N'-nitro-N-nitrosoguanidine and obtained a mutant strain, Y5, which was capable of producing α-amylase with an improved resistance to heat and acid. Among the above thermostable α-amylases, those of *B. licheniformis, B. subtilis* and partially *B. stearothermophilus* are being produced on an industrial scale. These thermostable α-amylases are stable to temperatures higher by 10°C than the traditional α-amylase derived from *Bacillus amyloliquefaciens*. Thus these thermostable α-amylases are gradually replacing the traditional bacterial amylase in many application fields such as glucose production and desizing in textile industries.

Reaction :

$$\text{Starch} + n\,H_2O \xrightarrow{\text{thermostable bacterial } \alpha\text{-amylase}} G1+G2+G3+G4+G5+\cdots\cdots + \text{branched oligosaccharides}$$

Specificity : α-Amylase of *B. licheniformis* is characteristic in that at early stage of reaction with amylose, the enzyme produces dextrins of G5 to G9 in amounts much larger than G2 and G3. The amounts of G1 and G4 at this stage are trace. In later stage of the reaction, G1 and G5 increase while G6 and G9 decrease. An example of sugar composition of the digest after a long incubation is as follows : 33.3% G5, 15.8% G3, 8.2% G2, 7.7% G4, 6.8% G1, and 37.5% dextrins of G6 to G12. The action patterns of the enzyme on amylopectin, glycogen, and soluble starch are very similar to that on amylose with the exception of producing two unidentified dextrins which come between G2 and G3, and G3 and G4, respectively on paper chromatograms. G3 as well as G2 are not available as substrates for the enzyme (11).

α-Amylase of *B. acidocaldarius,* on the other hand, predominantly produces G2 and G3 in the digestion of soluble starch. The formation of G6 and G1 by the enzyme is significantly small. At 22.8% hydrolysis of soluble starch by the enzyme, G2 and G3 are formed in amounts of 18% and 21% of the substrate, respectively. This enzyme hydrolyzes *p*-nitrophenyl-α-D-maltoside (PNPM) to produce *p*-nitrophenol and maltose. As for the ratio of PNPM-hydrolyzing activity to soluble starch-hydrolyzing activity, *B. acidocaldarius* α-amylase has a smaller one than that of *B. subtilis* saccharifying α-amylase, but larger than that of *B. licheniformis* α-amylase (12).

The main hydrolysis products by α-amylase of a certain *Clostridium sp.* have been reported to be G3 and G4 (8). The action mode and substrate specificity of thermostable α-amylases other than those described above have remained unclarified.

Table 1. Thermostable Bacterial α-Amylases

Source	Opt. temp. (℃)	Opt. pH	pH Stability	Molecular weight	Ref.
B. stearothermophilus[a]	65–73	5–6	6–11	48000*	3
B. licheniformis	90	7–9	7–10	62650**	13
B. subtilis[b]	95–98	6–8	5–11	—	
Thermophile V-2[c]	70	6–7	9.2	50000**	6
B. acidocaldarius[d]	70	3.5	4–5.5	66000**	12
Clostridium sp.	80	4	2–7	—	8
D. thermophilum	85–90	5	—	70000**	9

Data for thermal stability are not listed, because the experimental conditions reported are too different for comparison.

*Sedimentation equilibrium method.

**SDS-PAGE method.

a. One cysteine residue per mole is found.

b. This amylase is more stable than *B. stearothermophilus* amylase at 90℃ and pH 6.0 in the presence of an excess amount of calcium ions (14). The optimum temperature for hydrolysis of thick corn starch slurry (35%) is 95–100℃, but it is 90℃ for *B. stearothermophilus* amylase and 85℃ for *B. amyloliquefaciens* amylase (14).

c. This amylase contains 2 moles of half-cystine residues per mole, but does not contain disulfide crosslinkages.

d. Calcium ions do not improve the thermostability of this amylase. Properties of α-amylase of *B. acidocaldarius* have been reported to differ more or less depending on the bacterial strain which produces the enzyme.

Table 2. Properties of *B. licheniformis* α-Amylases Obtained from Parent and Mutant Strain (10)

	Parent (IFO 12196)	Mutant (Y5)
Molecular weight (SDS-PAGE)	59000	61000
E (1%, 280nm)	21.0	25.5
Specific activity ($L_J$unit*/mg)	578	558
Isoelectric point (chromatofocusing)	8.47	7.17
pH Stability (90℃, 60 min)	6.0–9.0	5.5–9.0
Optimum pH	6.0	6.0
Optimum temperature (℃)	90	95
Remaining activity (%) at 90℃, 30 min	90	100

*See "Assay Method".

Production and Purification : The composition of the medium employed by Maruo and Tojo (15) for examining productivity of thermostable α-amylase of *B. licheniformis* is 1% meat extract, 2% peptone, 0.4% NaCl, 0.4% yeast extract, and 10% soluble starch. For industrial production of the enzyme, media mainly containing starch and defatted soybean (10–15% total solids) are used and the bacterium is submergedly cultured. The enzyme production amounts to 5–10 g per liter medium in 3–5 days culture. An affinity chromatography with a modified potato starch (16) is convenient for purification of α-amylases. To 10 g of potato starch is added 7 g of ammonium sulfate dissolved in 32 ml of water. The starch suspension is heated for 30 min with agitating in a water bath at 75–80℃. To this suspension is additionally added 5.4 g of ammonium sulfate dissolved in 8 ml of water, and mixed well. The mixture is cooled to 5℃, and applied onto a column, and the starch layer is settled to be 5–8 cm in height. A clear filtrate of bacterium-cultured broth is also cooled to 5℃ and applied onto the starch bed. The starch bed is washed with a least amount of 0.25 M sodium chloride solution after the filtrate has passed. The amylase adsorbed on starch is eluted with 0.05 M Na_2CO_3. The maximum amount of α-amylase to be adsorbed on starch under the conditions is about 0.1 g per g of starch on a dry basis.

Assay Method : In the case of liquefying starch by α-amylases of different specificities, the results relate better to the α-amylase activity estimated by the starch liquefying, i.e. viscosity reducing method than by the starch dextrinizing method based on the starch-iodine coloration. This is because of that the former method directly fits for the present purpose, while the later method represents the degradation of only amylose of starch directly. For assay of α-amylase by the activity of starch liquefaction, there is a Japanese Industrial Standard (JIS) K7001-1976, Amylase for Industrial Use. The method is as follows :

The starch suspension as substrate —— A mixture of 10 g of potato starch on a dry basis and 10 ml of 1.0 M sodium acetate-acetic acid buffer, pH 6.0, is filled up to 100 ml with deionized water. The suspension is constantly agitated to protect from precipitation of starch. This suspension should be freshly prepared for every assay. Viscosity standard —— Silicone oil of 500cSt viscosity is used as the viscosity standard. Eleven ml of the oil is placed in a test tube of 18 mm in diameter and the test tube is sealed for use at any time.

Procedure —— An aliquot of 10 ml of the starch suspension is taken into a test tube of 18 mm in diameter, and one ml enzyme solution is added to this starch suspension at room temperature. Then, this test tube is transferred into a boiling water bath to heat the suspension with violent shaking. As soon as the suspended starch is gelatinized, the test tube is transferred to a water bath at 65℃. After 15 min incubation at 65℃, the viscosity is compared with the standard which has been kept at 65℃. When the reaction mixture flows at the same time as the viscosity standard, the starch liquefying activity of the sample amylase is defined as one L_J unit per ml.

(Shigetoshi Endo)

References

1. Stark, E. & Tetrault, P. A. (1951) *J. Bacteriol. 62,* 247–249
2. Campbell, L. L. Jr. (1955) *Arch. Biochem. Biophys. 54, 154–161*
3. Ogasahara, K., Imanishi, A., & Isemura, T. (1970) *J. Biochem. 67,* 65–75
4. Shimamura, M., Amano, H. & Onuma, S. (1971) *Japan patent* 623056, filed 1967
5. Hattori, F., Nakamura, M., Taji, N., Nojiri, M., Nakai, T. & Kusai, K. (1974) *Proceedings on the Symposium of Amylase (in Japanese) 9,* 79–87
6. Hasegawa, A., Miwa, N., Oshima, T. & Imahori, K. (1976) *J. Biochem. 79,* 35–42
7. Buonocore, V., Caporale, C., de Rosa, M. & Gambocorta, A. (1976) *J. Bacteriol. 128,* 515–521
8. Yoshiga, R., Tsuchiya, M. & Ishida, M. (1986) *Abstracts of Papers, Annual Meeting of the Agricultual Chemical Society of Japan (in Japanese)* p.653
9. Shoun, H., Ojima, K., Fukuzumi, M., Uozumi, T. & Beppu, T. (1986) *Abstracts of Papers, Annual Meeting of the Agricultural Chemical Society of Japan* (in Japanese) p.652
10. Goto, K., Yao, M. & Eda, M. (1986) *Abstracts of Papers, Annual Meeting of the Agricultural Chemical Society of Japan* (in Japanese) p.653
11. Saito, N. (1973) *Arch. Biochem. Biophys. 155,* 290–298
12. Kanno, M. (1986) *Agric. Biol. Chem. 51,* 23–31
13. Morgan, F. J. & Priest, F. G. (1981) *J. Appl. Bact. 50,* 107–114
14. Kotaka, T., Sasaibe, M., Miyashita, Y. & Goto, K. (1980) *J. Jap. Soc. Starch Sci.* (in Japanese) *27,* 151–157
15. Maruo, B. & Toji, T. (1985) *J. Gen. Appl. Microbiol. 31,* 323–328
16. Kasabo, T., Kotaka, T. & Ono, M. (1961) *Japan patent* 279671, filed 1958

II.1.a.6 *Streptococcus bovis* α-Amylase

A rumen bacterium, *Streptococcus bovis*, isolated from bovine secretes an amylase whose specificity is very similar to that of the saccharifying α-amylase of *Bacillus subtilis*. However, α-amylase of *Sc. bovis* differs from *Bacillus* enzyme not only in its strong activity of hydrolyzing raw grain starches, but also in the proteochemical properties (1). α-Amylase of *Sc. bovis* is produced when the bacterium is cultured in media containing raw starch as the main carbon source under anaerobic conditions saturated with carbon dioxide. Because of a strong adsorbability on raw starch, *Sc. bovis* α-amylase is readily isolated in a highly purified state by the starch adsorption method (2).

Specificity on Soluble Starch and Raw Starch : The α-amylase of *Sc. bovis* hydrolyzes soluble starch to produce glucose, maltose and several α-limit dextrins. The degrees of hydrolysis of soluble starch at achromic point against iodine and that reached after a long incubation of the reaction mixture are 26 and 62%, respectively. The degrees of hydrolysis of raw rice and corn starch by the α-amylase are also around 66% and the hydrolysis products observed are the same as those found on soluble starch.

Enzymic and Proteochemical Properties : The enzymic and proteochemical properties of *Sc. bovis* α-amylase are shown in Table 1.

Table 1. Enzymic and Proteochemical Properties of *Sc. bovis* α-Amylase

Activity on raw starch (unit/mg protein)※	175 units
pH Stability	5.0~7.0 (2 h at 40°C)
Optimum pH	5.6 (40°C)
Thermal stability	45°C (15 min at pH 5.6)
Optimum temperature	50°C (pH 5.6)
Stabilization by Ca^{2+}	—
Adsorbability on raw starch	+
Molecular weight	7.9×10^4
E (1%, 280nm)	14.7
N content (%)	15.5
Isoelectric point	3.85

※ The specific activity was determined by the method described in this article.

Production and Purification : The α-amylase of *Sc. bovis* is produced by culture in a broth containing 1% peptone, 0.5% yeast extract and 1% corn starch with blowing carbon dioxide and maintaining the pH at around 6.3 by addition of $NaHCO_3$ at 40°C for 7–8 h. The cell growth rate is almost the same on either soluble starch or raw starch as the main carbon source. The amylase production is not observed on glucose, but it is observed on maltose. Furthermore the enzyme production on soluble starch and on raw corn starch, is twice and six times as much as on maltose, respectively. The amylase secreted in the cultured broth is isolated by adsorption on raw corn starch in the presence of 20% $(NH_4)_2SO_4$ and the amylase adsorbed is eluted from the starch with 30 mM phosphate buffer, pH 6.5. The eluted enzyme is precipitated by ammonium sulfate and after dialysis the dialyzed enzyme is subjected to column chromatographies on DEAE-Sephadex A-50 and Bio-Gel P-100. The enzyme purified as above is crystallized from 25% isopropyl alcohol. A rhombic shaped crystalline preparation of the enzyme is homogeneous in the tests of disc electrophoresis and ultracentrifugation.

Assay Method : The enzyme is assayed by estimation of saccharifying activity on raw starch since the α-amylase of *Sc. bovis* is characterized by its strong activity of hydrolyzing raw cereal starch. The assay of the enzyme activity is carried out using raw rice starch as substrate. A reaction mixture consisting of 500mg of rice starch (screened through a 100 mesh sifter) in 19 ml of 0.025 M acetate buffer, pH 5.6 and 1 ml of enzyme is incubated at 40℃ without agitation. The reducing sugar produced on a 10 min incubation is determined as glucose by the Fehling-Lehman-Schoorl method (3). One unit enzyme activity is defined as the enzyme amount that produces 1.0 mg reducing sugar as glucose under the conditions. If this activity assay is done using soluble starch instead of rice starch powder, one unit activity would be 5.7 units under the same conditions as above.

(Kyohei Mizokami)

References:

1. Mizokami, K., Kozaki, M. & Kitahara, K. (1977) *Nippon Nogeikagaku Kaishi* (in Japanese) *51* 299-307
2. Yamamoto, T. (1955) *Bull. Agr. Chem. Soc. 19*, 121-128
3. Van der Harr (1920) *Anleitung zum Nachweis, zur Trennung und Bestimmung der Monosaccharide und Aldehydsauren*, Berlin *8*, 120-124

II.1.b. α-Amylases Which Produce Specific Oligosaccharides

Amylases have long been classified into α-amylases, β-amylases, glucoamylases and isoamylases according to their action on starch and other substrates. Since the first discovery by Robyt and Ackerman (1) of the unusual microbial amylase which produces only maltotetraose from starch by exo-mechanism, several new amylases have been discovered from various microorganisms including maltohexaose forming amylase by Kainuma and Suzuki (2), maltotriose forming amylase by Wako *et al.* (3). Saito (4) reported that a strain of *Bacillus* produced thermostable α-amylase which accumulated a large quantity of maltopentaose. By the discovery of these new amylases, the production of maltooligosaccharides in large quantities became easier. More recently, Nakakuki *et al.* (5) studied the action pattern of these maltosaccharides forming amylases and proposed that some of them belong to the

"Exo-α-amylase", as was proposed by Sakano for the maltotetraose forming amylase. Each of these new amylases is described in detail.

(Keiji Kainuma)

References

1. Robyt, J. F. & Ackerman, R. J. (1971) *Arch. Biochem. Biophys. 145,* 105–114
2. Kainuma, K. & Suzuki, S. (1971) *Proc. Int. Symp. Conversion Manuf. Foodstuffs Microorganisms,* Kyoto, Japan, pp. 95–98
3. Wako, K., Hashimoto, S., Kubomura, S., Yokota, A., Aikawa, K. & Kanaeda, J. (1979) *J. Jap. Soc. Starch Sci. 26,* 175–181
4. Saito, N. (1973) *Arch. Biochem. Biophys. 155,* 290–298
5. Nakakuki, T., Azuma, K. & Kainuma, K. (1984) *Carbohydr. Res. 128,* 297–310

II.1.b.1 Maltotriose Forming Amylase

Maltotriose forming amylase was first discovered by Wako *et al.* (1,2) in 1978 as an extracellular amylase produced by *Streptomyces griseus.* The enzyme was the third exo-α-amylase found after the maltotetraose and maltohexaose forming amylases. The enzyme is extremely useful in production of highly pure maltotriose from starch (Fig. 1).

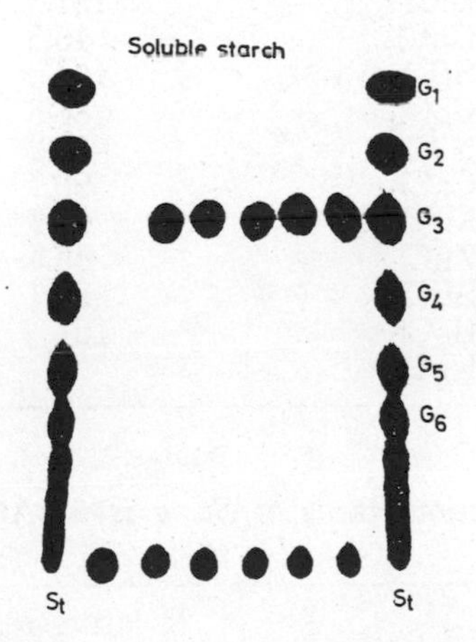

Fig. 1. Action of G_3-Forming Amylase on Soluble Starch (Wako, 1978).

Reaction :

O-O-O-O* ⟶ O-O-O + O*
O-O-O-O-O* ⟶ O-O-O + O-O*
O-O-O-O-O-O* ⟶ O-O-O + O-O-O*

Specificity : The enzyme hydrolyzes its substrates by exo-mechanism and forms α-maltotriose molecules from the nonreducing end of the maltooligosaccharides and α-1,4-glucans. The anomeric configuration of the reaction product is α (3). The enzyme hydrolyzes short chain amylose ($\overline{DP}$ 20) completely. It also hydrolyzes soluble starch, waxy corn starch and glycogen but by the hydrolysis degree of 55%, 51% and 40%, respectively. The action of the enzyme is similar to that of β-amylase whose reaction is blocked by the existence of branched linkages in the substrate molecules.

Enzymic and Proteochemical Properties : The optimum pH and temperature for the enzyme reaction are 5.6–6.0 and 45°C, respectively. The enzyme retains more than 80% of activity between pH 3.5–6.5. The enzyme is stable up to 40°C but its activity decreases rapidly at temperatures higher than 45°C. The molecular weight of the enzyme is estimated at 55,000 by SDS disc electrophoresis. The effect of metal ions on the activity is shown in Table 1. The relative activities on various substrates are in Table 2. As shown in Table 2, short chain amylose is hydrolyzed at a slightly higher rate than soluble starch.

Table 1. Effects of Metal Ions on the Activity of *St. griseus* Amylase

Reagent (2 mM)	Activity (%)
None	100
LiCl	131
$CuCl_2$	15.5
$SrCl_2$	104
$CoCl_2$	86.6
$NiCl_2$	74.2
$CaCl_2$	90.7
$MgCl_2$	88.7
$ZnCl_2$	46.4
$SnCl_2$	89.7
$BaCl_2$	101
$HgCl_2$	0

Table 2. Relative Reaction Rates of *St. griseus* Amylase on Various Substrates

Substrates	Relative reaction rates
Soluble starch	100
Amylose ($\overline{DP}$ 20)	114
Waxy β-LD	0
Waxy starch	91.6
Oyster glycogen	62.6
Phytoglycogen	40.2
Pullulan	0
β-Cyclodextrin	0

Waxy β-LD, waxy starch β-limit dextrin. A mixture of 1 ml of 1% substrate and 1 ml of enzyme (11 units/ml) was incubated at 40°C, pH 5.8 for 20 min.

Table 3. Purification Scheme for *St. griseus* Amylase

Procedure	Total activity (u/ml)	Protein (mg)	Specific activity (u/mg protein)	Recovery (%)
Culture filtrate	87,000	1,560	56	100
Acetone ppt. (0–0.55 vol.)	63,000	334	189	72.4
CM-Sephadex-1	39,400	20.1	1,960	45.3
CM-Sephadex-2	24,700	7.4	3,340	28.4
Sephadex G-100	18,000	3.3	5,450	20.7

Production and Purification : Five hundred ml of precultured *Streptomyces griseus* is added to a 30 ℓ jar fermentor which contained 15 ℓ of a medium consisting of 3% dextrin, 2% soybean flour, 0.1% Polypepton, 0.3% potassium dihydrogen phosphate, 0.5% calcium chloride and 0.1% ammonium sulfate. Incubation is conducted for 5 days at 27℃. The culture broth is then filtered through a celite bed, and the filtrate is purified by the steps mentioned in Table 3. The activity recovered at the last step of purification is 20.7%.

Assay Method : The enzyme activity is assayed by determining the increase in reducing power by the Ferricyanide method with Technicon Autoanalyzer (4) or by the Somogyi-Nelson method (5). One unit of the enzyme activity is defined as the enzyme amount that produces 1 mg of maltotriose in 1 h reaction at pH 6.0 and at 40℃.

(Keiji Kainuma)

References

1. Wako, K., Takahashi, C., Hashimoto, S. & Kanaeda, J. (1978) *J. Jap. Soc. Starch Sci.* (in Japanese) *25, 155–161*
2. Wako, K., Hashimoto, S., Kubomura, S., Yokota, K., Aikawa, K. & Kanaeda, J. (1979) *J. Jap. Soc. Starch Sci.* (in Japanese) *26,* 175–181
3. Nakakuki, T., Azuma, K. & Kainuma, K. (1984) *Carbohydr. Res. 128,* 297–310
4. Kainuma, K., Wako, K., Nogami, A. & Suzuki, S. (1973) *J. Jap. Soc. Starch Sci.* (in Japanese) *20,* 112–119
5. Nelson, N. (1944) *J. Biol. Chem. 153,* 375–380

II.1.b.2. Maltotetraose Forming Amylase (EC 3.2.1.60)

Exo-maltotetraohydrolase was discovered by Robyt and Ackerman (1) from *Pseudomonas stutzeri* in 1971. This enzyme is the first demonstration of an exo-amylase which produces a specific oligosaccharide from starch. Also this enzyme is the third exo-amylase found after β-amylase and glucoamylase. The discovery of Robyt and Ackerman stimulated the search for other amylases that might produce specific oligosaccharides. Schmidt and John (2) purified the enzyme partially and Sakano (3) purified the enzyme into two components by the chromatofocusing technique (3,4). The action mode was studied by Nakakuki *et al.* (5) using reducing end labelled oligosaccharides. The enzyme is being applied for production of maltotetraose in a large scale by a continuous bioreactor system.

Reaction :

O-O-O-O* ⟶ O-O-O + O*
O-O-O-O-O* ⟶ O-O-O-O + O*
O-O-O-O-O-O* ⟶ O-O-O-O + O-O*
O-O-O-O-O-O-O* ⟶ O-O-O-O + O-O-O*
O-O-O-O-O-O-O-O* ⟶ O-O-O-O + O-O-O-O*
O-O-O-O-O-O-O-O-O* ⟶ O-O-O-O + O-O-O-O + O*

Specificity : *Pseudomonas stutzeri* α-amylase releases maltotetraose from the nonreducing end of oligosaccharides. The enzyme also hydrolyzes maltotetraose slowly into G3, G2 and G1. Also, this enzyme hydrolyzes a water insoluble, crosslinked blue starch that has bulky substituents on the chains and is normally considered to be hydrolyzed only by α-amylase, as reported previously by Schmidt and John. The enzyme also hydrolyzed soluble starch and partially periodate-oxidized amylose. The action pattern may change depending on the molecular size and/or structure of the substrate. By this characteristic action on blue starch, *Pseudomonas stutzeri* amylase is distinguished from other exo-amylases.

Enzymic and Proteochemical Properties : Robyt and Ackerman studied the chemical properties of the enzyme. However, the later studies on the enzyme action by other researchers were not consistent with those by Robyt and Ackerman. This discrepancy may be attributed to the presence and absence of a minor contamination of other amylolytic enzymes in the preparations obtained.

Table 1. Characteristics of Maltotetraose Forming Amylase (Sakano, 1983)

	F-1	F-2
Opt. pH	8.0	8.0
Opt. temperature	45℃	45℃
pH Stability	6.0-10.5	6.0-10.5
Thermal stability	35℃	35℃
Molecular weight	56,000	56,000
Isoelectric point	5.6	5.3

Recently, Sakano *et al.* (1983) separated the enzyme into two components naming them F-1 and F-2, respectively. The two components are identical in their enzymatic properties and molecular weight, but differ in their isoelectric point (Table 1). The enzyme is strongly inhibited by the presence of 1 mM Cu^{2+}, Hg^{2+}, Zn^{2+} or

N-Bromosuccinimide (NBS).

Production and Purification : *Pseudomonas stutzeri* NRRL B-3389 is cultured in a liquid medium that consisted of 12% trypticase, 0.5% yeast extract, 0.28% anhydrous potassium monohydrogen phosphate, 0.1% anhydrous potassium dihydrogen phosphate, and 1% soluble potato starch. A loopful of cells of the bacterium from a stock culture slant is inoculated to 2 ml of the liquid medium; after 24 h incubation at 30°C, 1 ml of this culture is transferred to 50 ml of the liquid medium in a 250 ml Erlenmeyer flask and incubated at 30°C with reciprocal shaking for 14-16 h. A large scale culture of the organisms is prepared using a jar fermentor. After addition of 1 ℓ of the inoculum, the culture is incubated for 3 days at 30°C, with an aeration rate of 0.5 ℓ of air/ℓ of culture liquid/min, and agitation rate of 200 rpm.

The following are the purification procedures of the enzyme reported by Sakano (1982). After cultivation of the bacterium at 30°C for 14 h, the culture broth is centrifuged at 12,000 rpm for 20 min to remove the cells. Corn starch (28 g) and cold ethanol (2 ℓ, −20°C) are added to the supernatant (8 ℓ) in an icebox, and the mixture is left to stand overnight with gentle stirring. Then, the enzyme is eluted from the starch and purified through the chromatographic steps on Sephadex G-100 (2.7 × 7.3 cm), DEAE-Cellulose column (1.5 × 23 cm, equilibrated with 50 mM Tris buffer, pH 8.5 containing 5 mM $CaCl_2$). The chromatography by DEAE-Cellulose is repeated twice. The changes in specific activity and purification factors are shown in Table 2.

Table 2. Purification of *Pseudomonas stutzeri* α-Amylase

Purification step	Protein (mg/ml)	Activity (u/ml)	Specific activity (u/mg)	Purification
Culture supernatant	15.0	3.6	0.24	1
Starch adsorption	1.34	78.0	57.8	240
Sephadex G-100 column	1.42	28.8	90.7	378
DEAE-Cellulose column-1				
F-1 Fraction	0.56	58.3	104	434
F-2 Fraction	0.38	38.6	102	425
DEAE-Cellulose column-2	0.24	22.5	90.0	375

Assay Method : Fifty μl of the enzyme solution was incubated with 200 μl of 0.5% soluble starch buffered with 10 mM glycerophosphate(pH 7.0) containing 5 mM $CaCl_2$ at 30°C for 30 min. The reducing sugar produced in the reaction mixture was determined by the Somogyi-Nelson method (6,7). One unit of the amylase activity is defined as the amount of the enzyme that yielded reducing sugar equivalent to 1 μmol of maltotetraose per min under the assay conditions.

(Keiji Kainuma)

References

1 . Robyt, J. F. & Ackerman, R. J. (1971) *Arch. Biochem. Biophys. 145,* 105–114
2 . Schmidt, J. & John, M. (1979) *Biochim. Biophys. Acta 566,* 88–99
3 . Sakano, Y., Kashiyama, E. & Kobayashi, T. (1982) *Agric. Biol. Chem. 46,* 639–646
4 . Sakano, Y., Kashiyama, E. & Kobayashi, T. (1983) *ibid. 47, 1761–1768*
5 . Nakakuki, T., Azuma, K. & Kainuma, K. (1984) *Carbohydr. Res. 128,* 297–310
6 . Nelson, N. (1944) *J. Biol. Chem. 153,* 375–380
7 . Somogyi, M. (1952) *J. Biol. Chem. 195,* 19–23

II.1.b.3 Maltopentaose Forming Amylase

Maltopentaose forming amylase was first reported by Saito (1) in 1973 as a thermostable amylase obtained by *Bacillus licheniformis.* In his screening research of thermostable α-amylase, he found that the product distribution of starch digest by the enzyme gave the highest peak at maltopentaose by endowise attack. From these reasons, maltopentaose forming amylase is classified as an α-amylase which is different from a group of "exo-α-amylases" described in this chapter. The enzyme has been used industrially to liquefy starch at higher temperatures and also to produce maltopentaose from starch. Recently, Okemoto *et al.* (2) discovered a strain of *Pseudomonas* sp. to produce exo-maltopentaohydrolase. This enzyme produces maltopentaose as a sole reaction product from soluble starch. Maltopentaose is used as a substrate for determining the amylase activity of human blood.

Bacillus Licheniformis Amylase

Reaction :

```
G1 *  O*           ┐
G2 *  O-O*         │
G3 *  O-O-O*       │   None
G4 *  O-O-O-O*     │
G5 *  O-O-O-O-O*   ┘
G6 *  O-O-O-O-O-O*
               ↑
G7 *  O-O-O-O-O-O-O*
               ↑ ↑
G8 *  O-O-O-O-O-O-O-O*
               ↑ ↑ ↑
G9 *  O-O-O-O-O-O-O-O-O*
               ↑ ↑ ↑ ↑
G10*  O-O-O-O-O-O-O-O-O-O*
               ↑ ↑ ↑ ↑ ↑
```

Specificity : The amylase hydrolyzes soluble starch, amylose, amylopectin, glycogen and cyclodextrin. At early stage of the reaction, oligosaccharides from G5 to G9 appeared in amount much larger than those of G2 and G3. With progressing of the reaction, G1-G5 increased and G6-G9 decreased. In the final stage of the reaction, the oligosaccharides distribution was 6.8, 8.2, 15.8, 7.7, 33.3 and 37.7%, for G1, G2, G3, G4, G5 and larger dextrins than G6 respectively (1).

Nakakuki *et al.* (3) studied the specificity of the enzyme using reducing end labelled maltooligosaccharides. The autoradiogram indicated that the enzyme hydrolyzed G6* to form G5 and G1*. G7* and G8* are also hydrolyzed to produce G5 and G2*, and G5 and G3*, respectively, by an exo-mechanism. G3* was produced from the reducing end side mainly of maltosaccharides larger than G8. The relative reaction rates of the amylase on G7 and G8 are five times higher than on G6. Similarly to α-amylase, the enzyme hydrolyzes partially oxidized amylose and blue starch. The enzyme also hydrolyzes G6, G7 and G8 to produce maltopentaose by attacking exowise. These characteristics indicate the enzyme action mechanism changes significantly depending on the size of the substrate molecule.

Enzymic and Proteochemical Properties : The optimum temperature of the enzyme reaction is 76˚C. At 40˚C and 90˚C, the activity becomes a half of the maximal activity. The maximal activity is shown at pH values between 5.0 and 8.0. The enzyme is active even on the alkaline side, for example 95 and 76% of activities are shown at pH 9.0 and 10.0. The enzyme is stable at pH 8.0 and at temperatures below 60˚C for 15 min, but loses its activity by 60% at 80˚C. The enzyme is also stable in the pH range between 6 and 11 on incubation for 24 h, but it is unstable at pH values lower than 5 or higher than 11. The molecular weight of the enzyme estimated by Sephadex G-100 gel filtration is 22,500. It is not certain whether the enzyme is of multiple forms or not, because the disc electrophoresis showed four active protein bands.

Table 1. Purification Summary of the Amylase of *B. licheniformis* 584

Purification step[a]	Total protein (mg)	Specific activity (u/mg protein)	Yield (%)	Purification
Culture filtrate	46,800	793	100	1
Starch adsorption	637	46,000	79	58
DEAE-Cellulose	309	78,000	65	98.4
CM-Cellulose	141	100,000	38	126
Sephadex G-100	135	100,000	36	126

a After the step of starch adsorption, all the dilutions of enzyme solutions were made with 0.05 M Tris-acetate buffer (pH 8.0) containing 0.02% bovine serum albumin to prevent the inactivation of the enzyme.

Production and Purification : A liquid medium containing 4% soluble starch, 0.35% $(NH_4)_2HPO_4$, 0.6% yeast extract, 0.05% $MgSO_4 \cdot 7H_2O$, 0.2% sodium citrate and 0.008%

$CaCl_2 \cdot 2H_2O$, is used for enzyme production. The pH of the medium is 7.2 after sterilization at 121°C for 20 min. A loopful of cells of the microorganism from its stock culture slant is inoculated to 500 ml of the liquid medium in a 5 ℓ Erlenmeyer flask, and cultured at 50°C on a rotary shaker for 24 h. Then, 500 ml of this culture is transferred as an inoculum to 100 ℓ of the liquid medium in a 200 ℓ fermentor and cultured at 50°C for 4 days. The purification of the enzyme is performed by starch adsorption, and chromatographies on DEAE-Cellulose, CM-Cellulose and Sephadex G-100, to 126 fold of the original. Changes in the specific activity and yield during the purification are shown in Table 1.

Assay Method : α-Amylase activity is determined by measuring its dextrinizing power using amylose as the substrate, according to a modification of Fuwa's method (4). The reaction mixture containing 0.25 ml of the enzyme solution, 0.25 ml of 0.5 M glycine-NaOH buffer (pH 9.0) and 0.5 ml of 0.2% amylose is incubated at 50°C for 10 min. The reaction is stopped with 1.0 ml of 1.5 N acetic acid and 1.0 ml of iodine reagent (0.2% iodine and 2.0% potassium iodide). Distilled water is added to fill the mixture up to 20 ml and the absorbance at 690 nm is measured on a Klett-Summerson photoelectric colorimeter fitted with No.69 filter. One unit of the amylase activity is defined as the amount of enzyme which produces 10% reduction in the blue color density under the conditions.

Pseudomonas **sp. Amylase (2)**

Pseudomonas sp. produces "exo-maltopentaohydrolase". The isolation and cultivation method of this bacterium are discussed in reference (2), which shows that the enzyme produces only maltopentaose at the initial stage of the reaction.

(Keiji Kainuma)

References

1. Saito, N. (1973) *Arch. Biochem. Biophys. 155,* 290-298
2. Okemoto, H., Kobayashi, S., Momma, M., Hashimoto, H., Hara, K. & Kainuma, K. (1986) *Appl. Microbiol Biotechnol. 25,* 137-142
3. Nakakuki, T., Azuma, K. & Kainuma, K. (1984) *Carbohydr. Res. 128,* 297-310
4. Fuwa, H. (1954) *J. Biochem.* (Tokyo) *41,* 583-603

II.1.b.4. Maltohexaose Forming Amylase (EC 3.2.1.98)

Maltohexaose forming amylase was first discovered accidentally by Kainuma (1) as a contaminating enzyme of partially purified pullulanase obtained from *Aerobacter aerogenes (Klebsiella pneumoniae)*. Kainuma *et al.* (2,3) studied the details of the enzyme, such as purification methods and chemical properties. The enzyme was produced by *Aerobacter aerogenes* as cell bound enzyme and classified as the fourth exo-amylase after β-amylase, glucoamylase, and *Pseudomonas stutzeri* maltotetraose forming amylase, which is classified as "Exo-maltohexaohydrolase". Recently, Nakakuki *et al.* (4.5) obtained an ultraviolet induced mutant which produces the enzyme extracellularly. After the identification of both the enzymes, Nakakuki reported these two enzymes to be identical. Taniguchi (6) found a strain identified as *Bacillus circulans* F-2 which formed maltohexaose-forming amylase. Takasaki (7) also discovered a strain belonging to *Bacillus circulans* to produce a maltohexaose-forming amylase. Kennedy found maltohexaose forming amylase from *B. subtilis*. These three bacterial amylases belong to endo-amylases forming maltohexaose.

Exo-Maltohexaohydrolase Obtained from *Aerobacter aerogenes*

Reaction : The action on reducing end sides of ^{14}C-labelled maltooligosaccharides are summarized as follows.

O-O-O-O-O* ⟶ None
O-O-O-O-O-O* ⟶ O-O-O-O + O-O*
O-O-O-O-O-O-O* ⟶ O-O-O-O-O-O + O*
O-O-O-O-O-O-O-O* ⟶ O-O-O-O-O-O + O-O*
O-O-O-O-O-O-O-O-O* ⟶ O-O-O-O-O-O + O-O-O*
O-O-O-O-O-O-O-O-O-O* ⟶ O-O-O-O-O-O + O-O-O-O*

Specificity : The enzyme cleaves α-1,4-glucan to form α-maltohexaose. The initial velocity of the enzyme reaction determined on various substrates is shown in Table 1. No reaction is observed on α-, β- and γ-cyclodextrin, pullulan and maltose. Maltohexaose is slightly split into maltose and maltotetraose at a rate less than 2% of that on soluble starch. As mentioned in the previous paper (3), the β-amylase limit dextrin of amylopectin was also hydrolyzed by the extracellular enzyme and formed branched oligosaccharides. Though β-limit dextrin of amylopectin is normally considered as a non-substrate of exo-amylase, the exo-maltohexaohydrolase seems to accept α-1,6-linkage substituted to α-1,4-linkage in the structure of β-limit dextrin. Position specificity of α-1,6-linkage in the substrate and structural characteristics of the branched oligosaccharides produced have been discussed in detail elsewhere (3).

Enzymic and Proteochemical Properties : The optimum pH of the enzyme is 7.0 and the enzyme activity at pH 6.0 or 8.0 is more than 80% of the maximum. The enzyme retains about 80% of its activity between pH 5.0 and 10.0. The optimum reaction temperature is 52°C, and the enzyme is stable at temperatures lower than 50°C. The presence of 2 mM calcium ion increases the stability range by about 5°C. The molecular weights of the enzyme determined by SDS disc gel electrophoresis and gel filtration are 65,000 and 48,000, respectively. These enzymic properties are listed in Table 2.

Table 1. Comparison of Relative Reaction Rates of the Extracellular and the Cell-bound Enzymes on Various Substrates

	Extracellular	Cell-bound
Soluble starch	100	100
Amylose*	57	58
S.C.A (DP 23)**	117	115
R.S.C.A.***	108	109
Waxy maize starch	70	65
β-Limit dextrin of amylopectin	16	49
Glycogen	48	46
β-Limit dextrin of glycogen	13	11
Maltohexaose	2	2

No reaction is observed on α-, β- or γ-cyclodextrin, pullulan and maltose.
* Potato amylose. ** Short-chain amylose. *** Reduced short-chain amylose.

Table 2. Comparison of the Extracellular Exo-Maltohexaohydrolase with the Cell-bound Enzyme

	Extracellular	Cell-bound
Opt. temp.(℃)	52	60
Opt. pH	7.0	6.8
pH Stability	5.0–10.0	6.0–9.0
Thermal Stability	50℃	50℃
Molecular weight		
SDS PAGE	65,000	67,000
Sephadex G-100	48,000	48,000

Table 3. Purification Scheme of the Extracellular Exo-Maltohexaohydrolase

Purification step	Total activity (IU)	Specific activity (IU/mg)	Enzyme recovery (%)
Culture supernatant	354	0.017	100
$(NH_4)_2SO_4$ ppt.*	304	0.434	86
DEAE-Cellulose-1	179	1.27	51
DEAE-Cellulose-2	80	15.6	23
Sephadex G-100	46	18.4	13

* 0–0.5 saturation of $(NH_4)_2SO_4$.

Production and Purification : The production of the extracellular enzyme proceeds as follows : A mutant from *A. aerogenes* IFO-3321 is cultured on a rotary shaker (200 rpm) using 1 ℓ liquid medium containing 1% tryptone, 0.28% K_2HPO_4, 0.1% KH_2PO_4, 0.5% CH_3COONH_4, 0.2% $MgSO_4 \cdot 7H_2O$ and 1% soluble starch in 5 ℓ Erlenmeyer flask for 72 h at 30℃. The cells are removed by centrifugation at 15,000 rpm for 30 min at 4℃. The enzyme in this supernatant is purified by ammonium sulfate precipitation, DEAE-Cellulose (repeated) and Sephadex G-100 gel filtration. The changes of the total activity, specific activity and yield during the purification are shown in the Table 3, as an example. The details of the purification of cell-bound exo-maltohexaohydrolase are described in reference (3).

Assay Method : Exo-maltohexaohydrolase activity is assayed using a mixture of 0.5 ml of reduced short-chain amylose solution (0.4%, w/v) and 0.4 ml of 0.1 M sodium phosphate buffer (pH 7.0). After preincubating this mixture at 40℃ for 3 min, 0.1 ml of suitably diluted enzyme is added and incubated at 40℃ for 30 min. The formation of 1 μmol glucosidic bond per min under the conditions is expressed as one unit of enzyme activity.

Maltohexaose Forming Amylase Obtained from *Bacillus circulans*

Bacillus circulans G-6 (7) and *Bacillus circulans* F-2 (6) were independently discovered to produce maltohexaose-forming amylase. Amylase formed by *B. circulans* G-6 is purified from the culture filtrate by ammonium sulfate precipitation, DEAE-Cellulose column, Sephadex G-200 to be electrophoretically homogeneous. The optimum pH and temperature are 8.0 and 60℃, respectively. The enzyme is stable at pH values between 5-10 and at temperatures lower than 50℃. The molecular weight obtained by gel filtration is 76,000. The action pattern at the initial reaction stage seems to be endo-type with production of G1, G2, …… G6.

Bacillus circulans F-2 amylase is purified by ammonium sulfate precipitation, starch adsorption, Bio-Gel P-100, ion exchange chromatography by Whatman DE 32 to a homogeneous state. The molecular weight is estimated to be 93,000. The optimum pH and temperature are at 6.0-6.5 and 60℃. The enzyme has a substantial activity to hydrolyze raw starch. The enzyme forms maltohexaose on gelatinized starch. The action mode of the enzyme is endo-type.

(Keiji Kainuma)

References

1. Kainuma, K. & Suzuki, S. (1971) *Proceedings of the International Symposium on Conversion and Manufacture of Foodstuffs by Microorganisms*, Kyoto, Japan pp. 95-98
2. Kainuma, K., Kobayashi, S., Ito, T. & Suzuki, S. (1972) *FEBS Lett.* 281-285
3. Kainuma, K., Wako, K., Kobayashi, S., Nogami, A. & Suzuki, S. (1975) *Biochim.*

Biophys. Acta 410, 333-346
4. Nakakuki, T., Monma, M., Azuma, K., Kobayashi, S., & Kainuma, K. (1982) *J. Jap. Soc. Starch Sci. 29,* 179-187
5. Nakakuki, T., Azuma, K., Monma, M. & Kainuma, K. (1982) *ibid. 29,* 188-197
6. Taniguchi, H., Chung, M. J., Yoshigi, N. & Maruyama, Y. (1983) *Agric. Biol. Chem. 47,* 511-519
7. Takasaki, Y. (1982) *Agric. Biol. Chem. 46,* 1539-1547
8. Kennedy, J. F. & White, C. A. (1979) *Stärke 31,* 93-99

II.1.c Action Patterns of Taka-amylase A on Cyclodextrin Derivatives

Cyclodextrins are substrates of Taka-amylase A (1,2). Melton and Slessor found that Taka-amylase A hydrolyzed 6-monosubstituted α-cyclodextrins to give 6′-substituted maltose exclusively (3). Since specific chemical modifications on cyclodextrins at the desired position (C2 , C3 , or C6) are much easier than those on linear maltooligosaccharides as shown below, the cyclodextrin derivatives can be easily available as specifically modified substrates. 6^A, 6^X-O-Disulfonylated β- and γ-cyclodextrins are prepared by the reaction of the corresponding cyclodextrins (see Fig. 1) with arenesulfonyl chloride in pyridine and isolated by reverse-phase column chromatography (4,5). Similarly, 6-polysulfonylated α-cyclodextrins are prepared and isolated (6,7). 2-O-Sulfonylated cyclodextrins are prepared by the reaction of cyclodextrins with m-nitrophenylsulfonyl chloride or *m*-nitrophenyl *p*-tosylate in alkaline aqueous solution (8,9). 2^A, 2^X-O-Disulfonylated α-cyclodextrins are similarly obtained and isolated by reverse-phase column chromatography (9). 3-O-Monosubstituted and 3^A, 3^X-O-disulfonylated cyclodextrins (X= C and X= D) are synthesized in an alkaline aqueous acetonitrile by use of β-naphthylsulfonyl chloride as a sulfonylation reagent (10). 3,6-Monoanhydro- and 3^A, 6^A; 3^X, 6^X-dianhydro-cyclodextrins are prepared by treatment of the corresponding 6-O-sulfonates or 3-O-sulfonates with aqueous alkali (5, 11, 12). (2AS)2^A, 3^A-Monoanhydro- and (2AS, 2^XS)2^A, 3^A; 2^X, 3^X-dianhydro-cyclodextrins are synthesized by treatment of the corresponding 2-O-sulfonates with aqueous alkali under conditions milder than those described above (9). (3^AR)2^A, 3^A-Monoanhydro- and (3^AR, 3^XR)2^A, 3^A ; 2^X, 3^X-dianhydro-cyclodextrins are also prepared from the corresponding 3-O-sulfonates by procedures similar to those described above (10).

Since the cyclodextrin derivatives can be regarded as a kind of modified oligosaccharides available as substrates for Taka-amylase, Taka-amylolysis of these cyclodextrin derivatives may provide methods for selective preparation of specifically modified (activated) linear oligosaccharides as well as mechanistic investigation on the interaction between the subsites of the amylase (Fig. 2)(13-15) and the oligosaccharides. The results of Taka-amylolysis are summarized in Fig. 1(5, 9-11, 16). In Fig. 1, the cyclodextrin derivatives are depicted to the left of the reaction schemes and their Taka-amylolysis products to the right. For an example, in the first reaction scheme,

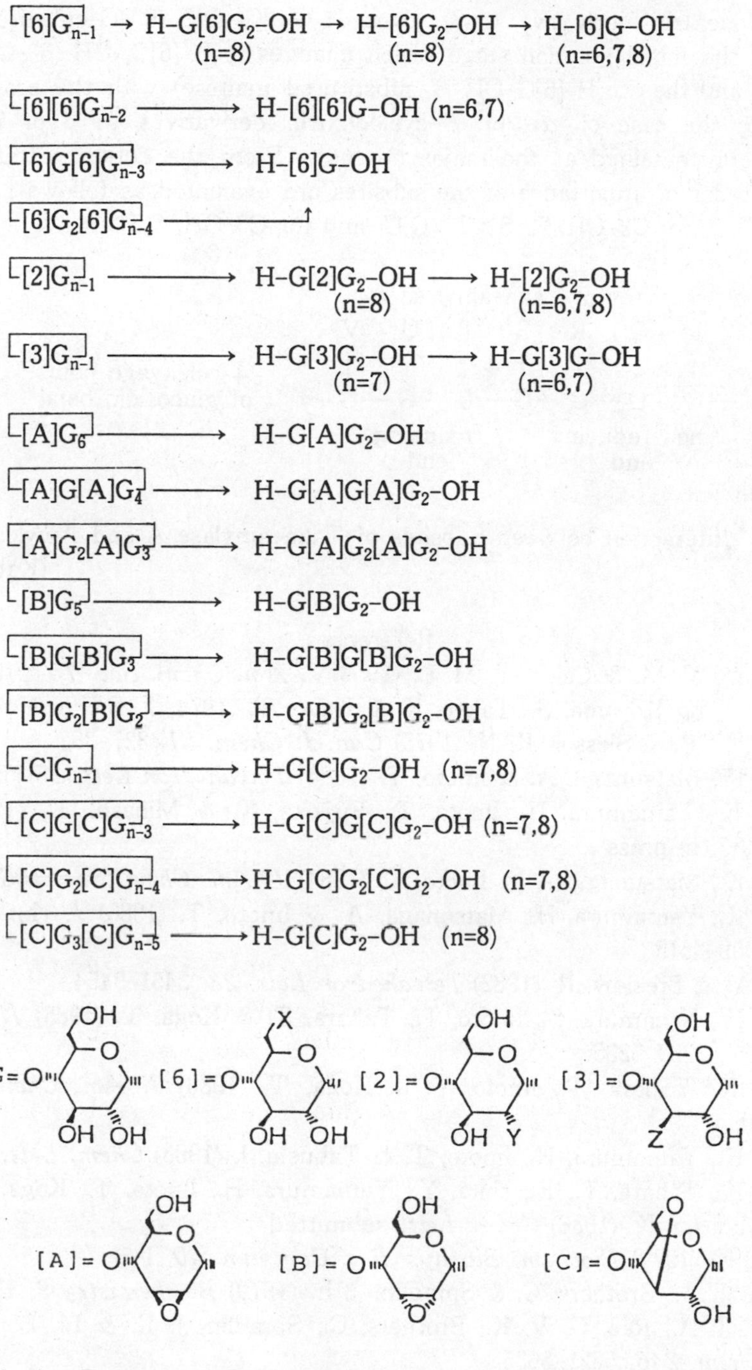

$\overline{[6]G_{n-1}}$ → H-G[6]G_2-OH (n=8) → H-[6]G_2-OH (n=8) → H-[6]G-OH (n=6,7,8)

$\overline{[6][6]G_{n-2}}$ ⟶ H-[6][6]G-OH (n=6,7)

$\overline{[6]G[6]G_{n-3}}$ ⟶ H-[6]G-OH

$\overline{[6]G_2[6]G_{n-4}}$ ⟶ ↑

$\overline{[2]G_{n-1}}$ ⟶ H-G[2]G_2-OH (n=8) ⟶ H-[2]G_2-OH (n=6,7,8)

$\overline{[3]G_{n-1}}$ ⟶ H-G[3]G_2-OH (n=7) ⟶ H-G[3]G-OH (n=6,7)

$\overline{[A]G_6}$ ⟶ H-G[A]G_2-OH

$\overline{[A]G[A]G_4}$ ⟶ H-G[A]G[A]G_2-OH

$\overline{[A]G_2[A]G_3}$ ⟶ H-G[A]G_2[A]G_2-OH

$\overline{[B]G_5}$ ⟶ H-G[B]G_2-OH

$\overline{[B]G[B]G_3}$ ⟶ H-G[B]G[B]G_2-OH

$\overline{[B]G_2[B]G_2}$ ⟶ H-G[B]G_2[B]G_2-OH

$\overline{[C]G_{n-1}}$ ⟶ H-G[C]G_2-OH (n=7,8)

$\overline{[C]G[C]G_{n-3}}$ ⟶ H-G[C]G[C]G_2-OH (n=7,8)

$\overline{[C]G_2[C]G_{n-4}}$ ⟶ H-G[C]G_2[C]G_2-OH (n=7,8)

$\overline{[C]G_3[C]G_{n-5}}$ ⟶ H-G[C]G_2-OH (n=8)

Fig. 1. Action Patterns of Taka-amylase A on Cyclodextrin Derivatives.
X=p-TsO, α- or β-NsO or PhS ; Y=p-TsO or β-NsO ; Z=p-TsO or β-NsO
Ts = toluenesulfonyl ; Ns = naphthylsulfonyl ; Ph = phenyl

the γ-cyclodextrin derivative (n=8) gives H-G[6]G_2-OH (6"-O-substituted maltotetraose) at the initial reaction stage which changes to H-[6]G_2-OH (6"-O-substituted maltotriose) and then to H-[6]G-OH (6'-substituted maltose) with the progress of the reaction. In the case of α- or β-cyclodextrin derivatives (n= 6 or 7), the final product is only obtained as the major product. From the action pattern shown in Fig. 1, the order of importance of the subsites are evaluated as follows : For C6-OH, S>R>T>Q,U, for C2-OH, R, S>T>Q,U, and for C3-OH, S,T>R.

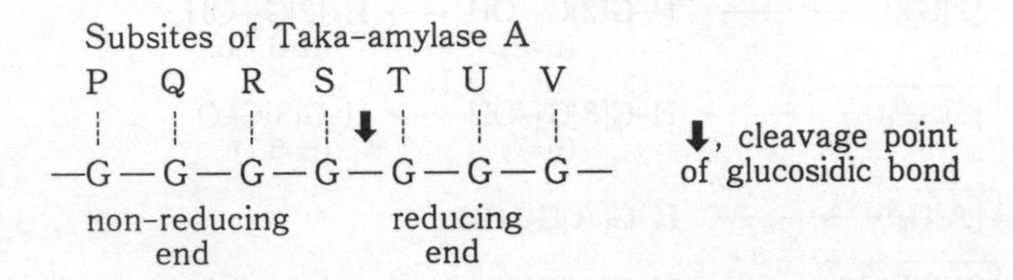

Fig. 2. Interaction between Subsites of Taka-amylase A and Amylose.

(Kahee Fujita)

References

1. Hanrahan, V. M. & Caldwell, M. L. (1953) *J. Am. Chem. Soc. 75*, 2191-2197
2. Suetsugu, K., Koyama, S., Takeo, K. & Kuge, T. (1974) *J. Biochem. 76*, 57-63
3. Melton, L. D. & Slessor, K. N. (1973) *Can. J. Chem. 51*, 327-332
4. Fujita, K., Matsunaga, A. & Imoto, T. (1984) *Tetrahedron Lett. 25*, 5533-5536
5. Fujita, K., Yamamura, H., Imoto, T., Fujioka, T. & Mihashi, K. (1988) *J. Org. Chem. 53*, in press
6. Fujita, K., Matsunaga, A. & Imoto, T. (1984) *J. Am. Chem. Soc. 106*, 5740-5741
7. Fujita, K., Yamamura, H., Matsunaga, A. & Imoto, T. (1986) *J. Am. Chem. Soc. 108*, 4509-4513
8. Ueno, A. & Breslow, R. (1982) *Tetrahedron Lett. 23*, 3451-3454
9. Fujita, K., Nagamura, S., Imoto, T., Tahara, T. & Koga, T. (1985) *J. Am. Chem. Soc. 107*, 3233-3235
10. Fujita, K., Tahara, T., Imoto, T. & Koga, T. (1986) *J. Am. Chem. Soc. 108*, 2030-2034
11. Fujita, K., Yamamura, H., Imoto, T. & Tabushi, I. (1988) *Chem. Lett.* 543-546
12. Fujita, K., Tahara, T., Egashira, Y., Yamamura, H., Imoto, T., Koga, T., Fujioka, T. & Mihashi, K. (1988) *Chem. Lett.* submitted
13. Hiromi, K. (1970) *Biochem. Biophys. Res. Commun. 40*, 1-6
14. Thoma, J. A., Brothers, C. & Spradlin, J. E. (1970) *Biochemistry 9*, 1768-1775
15. Thoma, J. A., Rao, G. V. K., Brothers, C., Spradlin, J. E. & Li, L. H. (1971), *J. Biol. Chem. 246*, 5621-5635
16. Fujita, K., Tahara, T., Nagamura, S., Imoto, T. & Koga, T. (1986) *J. Org. Chem. 52*, 636-640

II.1.d.1. Amino Acid Sequence of Several α-Amylases

Up to date, the amino acid sequences of α-amylases of more than ten kinds from various sources have been determined. Among these, the amino acid sequences of hog pancreatic isozyme I and Taka-amylase A (*Aspergillus oryzae*) were elucidated by sequence analysis of peptides generated from the enzyme proteins. The sequences of the other amylases were all predicted from the nucleotide sequences of the mRNA or cDNA by molecular cloning techniques.

Mammalian Amylase

Hog Pancreatic α-Amylase Isozyme I : This is one of the two very similar hog pancreatic α-amylase isozymes and has been designated isozyme I. Its amino acid sequence was determined by sequence analysis of the fragments generated by cyanogen bromide cleavage of the native enzyme protein. Gln-1 is pyrrolidone carboxylic acid (1). Cys-103 and Cys-119 are free sulfhydryl groups. A disulfide bond links residues between Cys-70 and Cys-115. Four other disulfide bonds exist but their bridge positions have not been determined yet (2).

Mouse Pancreas, Salivary Gland and Liver α-Amylase Precursors : The sequences of these α-amylases were translated from the nucleotide sequence of their mRNAs. Salivary and liver α-amylases were coded by the amylase genes, respectively. However their mRNAs have different 5'-untranslated sequences. Both amylase precursors are identical and consist of 511 amino acid residues (3). Two different mouse α-amylase mRNAs are known to exist : one in pancreas and the other in salivary gland. The comparable portions of these mRNAs are homologous by 89%. The mRNA in pancreas codes α-amylase precursor consisting of 508 amino acid residues (4). Residues 1-15 are the signal sequence. Gln-16 is pyrrolidone carboxylic acid in the mature protein of mouse pancreas, salivary and liver α-amylases. Mouse pancreatic α-amylase differs in 12% of its amino acid sequence from salivary and liver enzymes. This fact explains the previously observed difference in the net charge and antigenic properties of these three tissue specific amylases.

Rat Pancreas α-Amylase Precursor : The amino acid sequence was translated from the cDNA sequence. The precursor protein consists of 503 amino acid residues. Residues 1-10 are the putative signal sequence. The amino end of the mature enzyme is acetylated. The sequence homologies of rat pancreatic α-amylase with corresponding α-amylases of mouse, pig, and human are 93%, 83% and 83%, respectively.

Human Salivary Gland and Pancreas α-Amylase Precursor : Human pancreas α-amylase gene *Amy2* and salivary α-amylase gene *Amy1* were cloned and the nucleotide sequences of their cDNAs determined. The nucleotide sequences of the α-amylase cDNAs are homologous by as much as 92% in the coding regions. The precursor proteins of both α-amylases consist of 512 amino acid residues. Residues 1-12 are the putative signal sequence (6). The homology of the predicted amino acid sequences is 94%. The homology between human and mouse pancreatic α-amylases is 83% and that between these mammalian salivary gland α-amylases, 82%.

Bacterial and Mold Amylases

Taka-amylase A (*Aspergillus oryzae*) : The complete amino acid sequence of Taka-amylase A has been made clear by sequence analysis of peptides generated by cyanogen bromide cleavage of reduced-carboxymethylated amylase and tryptic digestion of the methylated enzyme protein. The enzyme consists of 478 amino acid residues. Asn-197 links to the N-acetylglucosamine of the carbohydrate. Cys-227 is the sole free sulfhydryl group. Four disulfide bonds are present ; Cys-30—Cys-38 ; Cys-150—Cys-164 ; Cys-240—Cys-283 ; and Cys-439—Cys-474 (7). X-Ray crystallographic analysis of Taka-amylase A suggests that Glu-230 and Asp-297 are possible catalytic residues, serving as general acid and base catalysts, respectively (8).
Bacillus subtilis α-Amylase Precursor : Three amylase genes from different strains of *Bacillus subtilis* have been cloned and characterized. The α-amylase gene *AmyE* from *Bacillus subtilis* strain 1A289 has been cloned and the nucleotide sequence of cDNA, determined. The precursor protein of α-amylase deduced from the cDNA sequence consists of 660 amino acid residues. Residues 1-32 are the putative signal sequence (9). The α-amylase genes *amyR2* and *amyE* from α-amylase hyperproducing *Bacillus subtilis* strain NA64 have been cloned and their nucleotide sequences determined. The M-type α-amylase encoded by *amyEm* was characterized. The precursor protein consists of 592 amino acid residues and residues 1-41 are the putative signal sequence (10). Also, the amylase genes *amyR2* and *amyEn* coding for N-type α-amylase from *Bacillus subtilis* strain N7 have been cloned and their complete nucleotide sequences, determined. The amino acid sequence of N-type α-amylase precursor translated from the cDNA sequence consists of 477 amino acid residues. Residues 1-41 are the putative signal sequence (11). The molecualr weights of α-amylase-N7 and -NA64 estimated by SDS-PAGE are 47,000 and 64,000, respectively. The electrophoretic mobility of α-amylase-N7 toward anode in 7.5% polyacrylamide gel (pH 8.3) is greater than that of α-amylase-NA64. Both amylases are identical in the immunological properties. However, the major hydrolysis products formed by α-amylase-NA64 from soluble starch are glucose and maltose. In contrast, those formed by α-amylase-N7 are maltose and maltotriose.

Bacillus amyloliquefaciens α-Amylase Precursor : α-Amylase gene from *Bacillus amyloliquefaciens* has been cloned and its complete nucleotide sequence, determined. The amino acid sequence of the precursor protein deduced from the cDNA sequence consists of 514 amino acid residues. Residues 1-31 are the signal sequence. The mature enzyme is shorter by 31 amino acid residues, with a molecular weight of 54,778. This α-amylase lacks any cysteine residue (12). The partial sequence of the amino-terminal 191 residues (residues 32-222) has been determined by sequence analysis of the fragments obtained by cyanogen bromide cleavage of the protein (13). However, this sequence differs from that deduced from the cDNA, because Ile-54, Leu-64, Asp-79, and Ser-84 are obtained by the sequence analysis.

Bacillus licheniformis α-Amylase Precursor : The amylase gene coding the heat-stable α-amylase of *Bacillus licheniformis* strain 584 has been cloned and its complete nucleotide sequence, determined. The amino acid sequence of the α-amylase precursor deduced from the cDNA sequence consists of 512 amino acid residues, having a signal sequence of 29 amino acid residues. The mature enzyme comprises 483 amino acid residues, and the molecular weight is 55,200. The amino acid sequence of *Bacillus licheniformis* α-amylase shows 65.4% and 80.3% homologies with those of the heat-stable *Bacillus stearothermophilus* α-amylase and the less heat-stable *Bacillus amyloliquefaciens* α-amylase, respectively (14).

Bacillus stearothermophilus α-Amylase Precursor : The amino acid sequence of the precursor of this α-amylase has been deduced from the cDNA sequence. The precursor protein consists of 548 amimo acid residues having a signal sequence of 34 amino acid residues. The amino-terminal sequence of the mature enzyme (*Bacillus stearothermophilus* strain DY5) was determined by automated Edman degradation using a spinning cup sequencer (JEOL LAS-47K). The mature enzyme contains two cysteine residues, which play an important role in maintaining the stable protein conformation (15).

Streptomyces hygroscopicus α-Amylase Precursor : The α-amylase gene (*amy*) from *Streptomyces hygroscopicus* strain AA69-4 has been cloned and its complete nucleotide sequence, determined. The *amy* nucleotide sequence indicated that it coded a protein of 52 kilodaltons (478 amino acid residues). The size of the predicted gene product (MW 47,980) agrees well with the molecular weight of α-amylase estimated by SDS-PAGE (MW 47,500). The amino acid sequence deduced from the nucleotide sequence agreed with the partial amino acid sequence and also with the amino acid composition of the isolated enzyme. The 30-residues leader sequence is similar to the signal sequences of other bacterial α-amylases (16). A comparison of the amino acid sequence of Taka-amylase A with mammalian α-amylases and those of three bacterial liquefying α-amylases are shown in Fig. 1 and Fig. 2, respectively.

Plant Amylase

Barley α-Amylase Precursor : The formation of α-amylase in barley seeds is known to be hormonally regulated. Germinating barley embryos produce gibberellic acid which stimulates the aleurone cells for the endosperm to produce α-amylase. Barley α-amylase comprises two isozymes whose pI values are different so they are

Fig. 1. Comparison of the Amino Acid Sequences of Taka-amylase A and Mammalian α-Amylases.

The amino acid sequences of α-amylases from *Aspergillus oryzae* (TAA), hog pancreas, rat pancreas, mouse pancreas and salivary gland are aligned on the basis of that of Taka-amylase A. The identical residues among five sequences are boxed with solid lines.

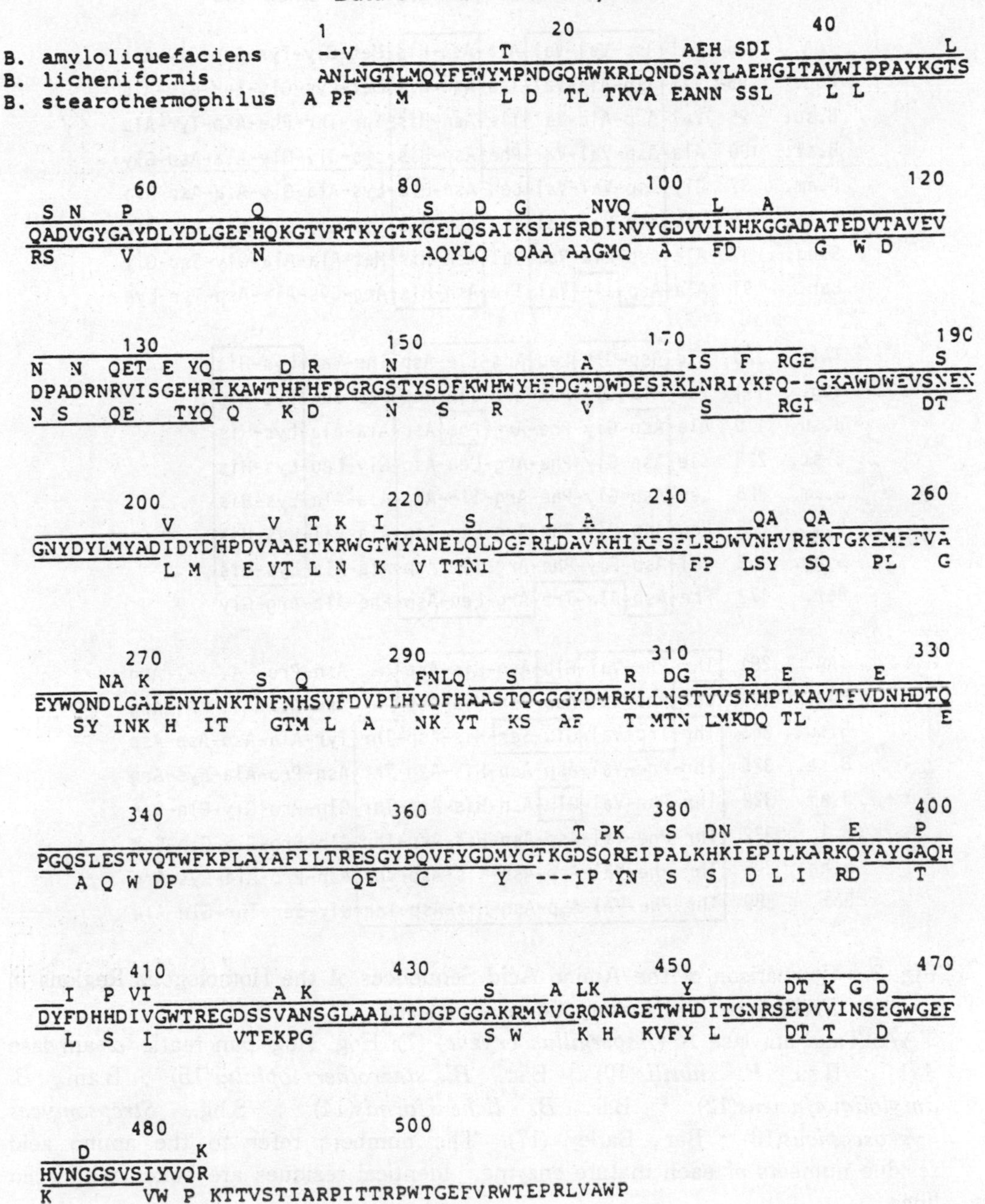

Fig. 2. Comparison of the Amino Acid Sequences of Three Bacterial Liquefying α-Amylases of the Mature Form.

The numbers refer to those of the amino acid residues of *B. licheniformis* enzyme. Lines indicate the segments showing a homology of more than 80% with each other (14).

TAA.	116	Val-Asp-Val-Val-Ala-Asn-His-Met-Gly-Tyr-Asp-Gly-Ala
Hog.	95	Val-Asp-Ala-Val-Ile-Asn-His-Met-Cys-Gly-Ser-Gly-Ala
B.su.	96	Val-Asp-Ala-Val-Ile-Asn-His-Thr-Thr-Phe-Asp-Tyr-Ala
B.st.	100	Ala-Asp-Val-Val-Phe-Asp-His-Lys-Gly-Gly-Ala-Asp-Gly
B.am.	97	Gly-Asp-Val-Val-Leu-Asn-His-Lys-Ala-Gly-Ala-Asp-Ala
B.li.	99	Gly-Asp-Val-Val-Ile-Asn-His-Lys-Gly-Gly-Ala-Asp-Ala
S.hg.	87	Ala-Asp-Ala-Val-Val-Asn-His-Met-Ala-Ala-Gly-Ser-Gly
Bar.	91	Ala-Asp-Ile-Val-Ile-Asn-His-Arg-Cys-Ala-Asp-Tyr-Lys

TAA.	200	Ile-Asp-Gly-Leu-Arg-Ile-Asp-Thr-Val-Lys-His
Hog.	191	Val-Ala-Gly-Phe-Arg-Ile-Asp-Ala-Ser-Lys-His
B.su.	170	Ala-Asp-Gly-Phe-Arg-Phe-Asp-Ala-Ala-Lys-His
B.st.	228	Ile-Asp-Gly-Phe-Arg-Leu-Asp-Gly-Leu-Lys-His
B.am.	225	Leu-Asp-Gly-Phe-Arg-Ile-Asp-Ala-Ala-Lys-His
B.li.	225	Leu-Asp-Gly-Phe-Arg-Leu-Asp-Ala-Val-Lys-His
S.hg.	168	Val-Asp-Gly-Phe-Arg-Ile-Asp-Ala-Ala-Lys-His
Bar.	172	Phe-Asp-Ala-Trp-Arg-Leu-Asp-Phe-Ala-Arg-Gly

TAA.	291	Thr-Phe-Val-Glu-Asn-His-Asp- - Asn-Pro - - Arg
Hog.	294	Val-Phe-Val-Asp-Asn-His-Asp- - Asn-Gln - - Arg
B.su.	263	Thr-Trp-Val-Glu-Ser-His-Asp-Thr-Tyr-Ala-Asn-Asp-Asp
B.st.	325	Thr-Phe-Val-Asp-Asn-His-Asp-Thr-Asn-Pro-Ala-Lys-Arg
B.am.	322	Thr-Phe-Val-Glu-Asn-His-Asp-Thr-Gln-Pro-Gly-Gln-Ser
B.li.	322	Thr-Phe-Val-Asp-Asn-His-Asp-Thr-Gln-Pro-Gly-Gln-Ser
S.hg.	325	Thr-Phe-Val-Asp-Asn-His-Asp-Thr-Asn-Pro-Ala-Lys-Arg
Bar.	289	Thr-Phe-Val-Asp-Asn-His-Asp-Thr-Gly-Ser-Thr-Gln-Ala

Fig. 3. Comparison of the Amino Acid Sequences of the Homologous Regions in Prokaryotic and Eukaryotic α-Amylases.

TAA, Taka-amylase A (*Aspergillus oryzae*) (7); Hog, Hog pancreatic α-amylase I (1) ; B.su., *B. subtilis*(10) ; B.st., *B. stearothermophilus*(15) ; B.am., *B. amyloliquefaciens*(12) ; B.li., *B. licheniformis*(14) ; S.hg., *Streptomyces hygroscopicus*(16) ; Bar., Barley (17). The numbers refer to the amino acid residue numbers of each mature enzyme. Identical residues are boxed with solid lines.

separable by isoelectric focusing. This fact indicates the presence of at least two populations of α-amylase mRNA molecule being due to separate structural genes which are differently influenced by gibberellic acid in aleurone cells. Recently, two types of mRNAs of barley α-amylase have been isolated. Type A mRNA is present in a relatively large amount in unstimulated aleurone cells and increases about 20-fold by stimulation with gibberellic acid. In contrast, type B mRNA is present at a low level in unstimulated aleurone cells and increases at least 100-fold by stimulation with the hormone. Barley α-amylase cDNA clones for both of type A and type B isozyme have been isolated and their nucleotide sequences determined. The amino acid sequence of isozyme A precursor predicted from the cDNA sequence consists of 438 amino acid residues. Residues 1-23 are the putative signal sequence (17). The isozyme B precursor predicted from the cDNA sequence consists of 427 amino acid residues and residues 1-23 are likely the signal sequence. The comparison of type A and type B isozymes shows 80% sequence homology for the mature coding sequences but only 44% homology for the signal peptide regions (residues 1-23) (18).

Sequence Homology of α-Amylases

The sequence homology among α-amylases described above is not significantly high. However, as for mammalian α-amylases, high homologies up to 80% are seen. Other similar sequences are found among three bacterial liquefying α-amylases, as shown in Fig. 2. *B. stearothermophilus* α-amylase shows 64% and 67% homology with those of *B. amyloliquefaciens* and *B. licheniformis*, respectively. On the contrary, the sequence homology between Taka-amylase A and hog pancreatic α-amylase isozyme I is only 24% in their over all sequences. However, three regions of highly homologous sequence exist between the two amylases as mentioned previously (7). The amino acid sequences of conserved regions in Taka-amylase A are also found in other α-amylases from various sources. These conserved residues are highly possibly the enzymically functional site of α-amylase. Fig. 3 shows a comparison of the amino acid sequence of the homologous regions of prokaryotic and eukaryotic α-amylases.

(Hiroko Toda)

References

1. Kluh, I. (1981) *FEBS Lett. 136*, 231-234
2. Pasero, L., Mazzei, Y., Abadie, B., Moinier, D., Fougereau, M. & Marchis-Mouren, G. (1983) *Biochem. Biophys. Res. Commun. 110*, 726-732
3. Hagenbuchle, O., Tosi, M., Schibler, U., Bovey, R., Wellaue, P. K. & Young, R. A. (1981) *Nature 289*, 643-646
4. Hagenbuchle, O., Bovey, R. & Young, R. A. (1980) *Cell 21*, 179-187
5. MacDonald, R. J., Crerar, M. M., Swain, W. F., Pictet, R. L., Thomas, G. & Rutter, W. J. (1980) *Nature 287*, 117-122
6. Nakamura, Y., Ogawa, M., Nishide, T., Emi, M., Kosaki, G., Himeno, S. & Matsubara, K. (1984) *Gene 28*, 263-270
7. Toda, H., Kondo, K. & Narita, K. (1982) *Proc. Japan Acad. 58B*, 208-212
8. Matsuura, Y., Kusunoki, M., Harada, W. & Kakudo, M. (1984) *J. Biochem. 95*, 697-702
9. Yang, M., Galizzi, A. & Henner, D. (1983) *Nucl. Acids. Res. 11*, 237-249
10. Yamazaki, H., Ohmura, K., Nakayama, A., Takeichi, Y., Otozai, K., Yamasaki, M.,

Tamura, G. & Yamane, K. (1983) *J. Bacteriol. 156*, 327-337
11. Yamane, K., Hirata, Y., Furusato, T., Yamazaki, H. & Nakayama, A. (1984) *J. Biochem. 96*, 1849-1858
12. Takkinen, K., Pettersson, R. F., Kalkkinen, N., Palva, I., Soderlund, H. & Kaariainen, L. (1983) *J. Biol. Chem. 258*, 1007-1013
13. Chung, H. S. & Friedberg, F. (1980) *Biochem. J. 185*, 387-395.
14. Yuuki, T., Nomura, T., Tezuka, H., Tsuboi, A., Yamagata, H., Tsukagoshi, N. & Udaka, S. (1985) *J. Biochem. 98*, 1147-1156
15. Ihara, H., Sasaki, T., Tsuboi, A., Yamagata, H., Tsukagoshi, N. & Udaka, S. (1985) *J. Biochem. 98*, 95-103.
16. Hoshiko, S., Makabe, O., Nojiri, C., Katsumata, K., Satoh, E. & Nagaoka, K. (1987) *J. Bacteriol. 169*, 1029-1036.
17. Rogers, J. C. & Milliman, C. (1983) *J. Biol. Chem. 258*, 8169-8174
18. Rogers, J. C. (1985) *J. Biol. Chem. 260*, 3731-3738

Ⅱ.1.d.2. Three-Dimensional Structure of α-Amylases

The three-dimensional structures of Taka-amylase A (abbreviated as TAA) and porcine pancreatic α-amylase (abbreviated as PPA) determined by X-ray crystallography are described.

Taka-amylase A : The structure of TAA has been determined at 3.0Å resolution (1,2,3). The molecule of TAA folds into two globular domains, the main and C-terminal ones which are connected by a single peptide chain as shown in Fig. 1 (domain is a globularly folded subregion of a polypeptide chain). The main domain is composed of residues 1-376 and the C-terminal domain, 385-478 with a peptide (residues 377-384) connecting the two domains.
The main domain has an eight-fold α/β barrel like triose phosphate isomerase(4).
This α/β barrel is made up of two co-axial cylindrical layers. The inner layer is a parallel eight-stranded β-barrel. The outer layer consists of eight parallel α-helices whose peptide chain directions are opposite to those of the β-barrel. The main domain also contains one extra α-helix and long loops connecting the β-strands to the next α-helices along the peptide chain in the α/β barrel. The α-helices and β-strands alternate along the peptide chain in the α/β barrel.
The C-terminal domain is composed of an eight-stranded β/β structure or β-sandwich and has the same folding topology as domains of γ-Ⅱ crystallin (5). The C-terminal domain consists of two four-stranded β-sheets.

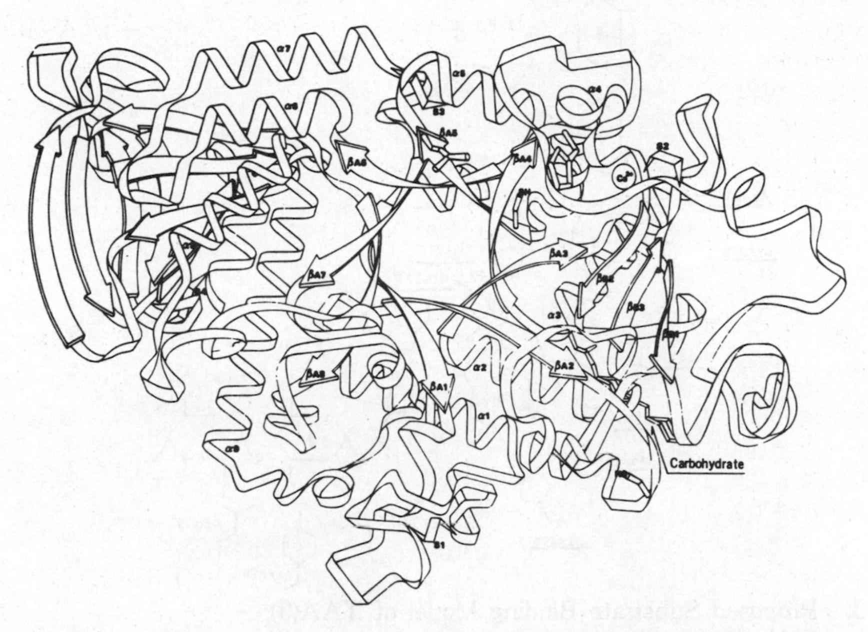

Fig. 1. Schematic Drawing of the Peptide Chain of TAA Viewed Down the α/β Barrel Axis from the Carboxyl End of the Parallel β-sheet (2).
Flat arrows indicate β-strands. S1 through S4 are the four disulfide bonds.

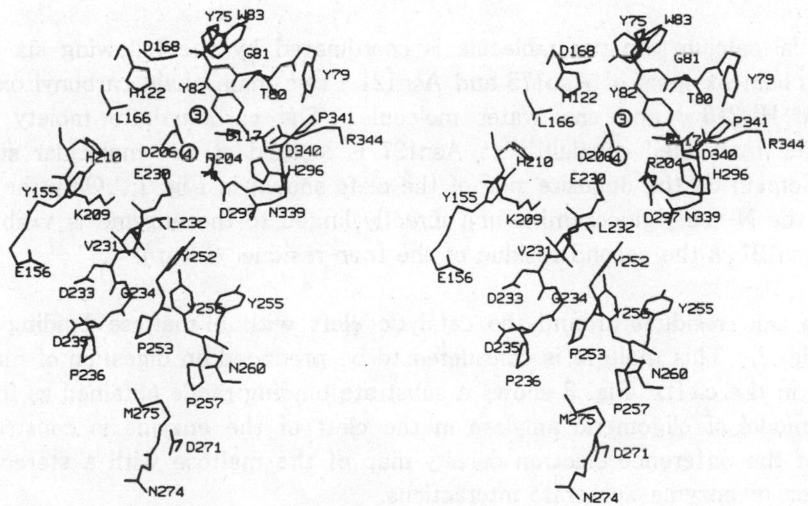

Fig. 2. Stereo-Diagram of the Amino Acid Residues in and around the Catalytic Cleft in the Main Domain of TAA.
The circles are the maltose binding sites of the TAA-substrate complex crystal (3). The numbers in the circles correspond to those of the substrate binding sites in Fig. 3.

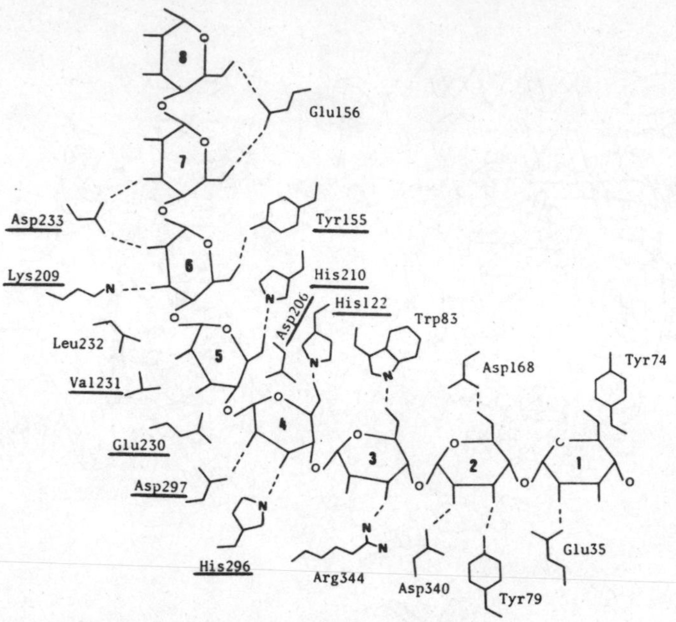

Fig. 3. Proposed Substrate Binding Model of TAA(3).
The numbers on the glucose residues indicate the subsite numbers. The dashed lines represent possible hydrogen bonds between the substrate and the enzyme. The underlined letters indicate the amino acid residues that have been shown to be conserved in TAA and PPA as described in section II.1.d.1.

One essential calcium ion per molecule is coordinated by the following six ligands : three side chain oxygens of Asp175 and Asn121 ; two main chain carbonyl oxygens of Gln162 and His210 ; and one water molecule. The carbohydrate moiety of eight hexose units (6) bonded covalently to Asn197 is located at the molecular surface of the main domain on the opposite side of the cleft shown in Fig. 1. Only the electron density of the N-acetylglucosamine unit directly linked to the enzyme is visible in the crystal. Asn197 is the second residue of the four-residue β-turn.

The amino acid residues around the catalytic cleft with a maltose binding site are given in Fig. 2. This maltose is considered to be produced on digestion of maltotriose and bound in the cleft. Fig. 3 shows a substrate binding mode obtained as follows : a molecular model of oligomeric amylose in the cleft of the enzyme is constructed on the basis of the difference electron density map of the maltose with a stereochemical consideration of enzyme-substrate interactions.

The catalytic site is located at the C-terminal side of the parallel β-strands of the α/β barrel and also roughly on the axis of the α/β barrel in the cleft of the main domain. The catalytic site or cofactor binding site is located at this position for all the α/β barrel enzymes whose three-dimensional structures are known (4). Asp297 and Glu230 are proposed to be catalytic residues from a pH-activity relation and a stereo chemical consideration based on the structure of the TAA-maltotriose complex crystals (3). This residue pair corresponds to Asp300 and Glu233 in the case of PPA.

Porcine pancreatic α-amylase : The molecular structure of PPA is essentially the same as that of TAA (7,8). The molecule is composed of three domains, e.g., domains A (residues 1-100 and 169-407), B (residues 101-168), and C (residues 408-496). The domain A is an α/β barrel structure. The domains A and B of PPA correspond to the main domain of TAA. The domain C corresponds to the C-terminal domain of TAA. The location of the essential Ca^{2+} and the residues liganded to it are similar to those of TAA. The catalytic residues are proposed to be Asp197 and Asp300. This residue pair corresponds to Asp206 and Asp297 of TAA. The activator Cl^- ion is located on the axis of the β-barrel of the domain A and near its carboxyl side.

(Masami Kusunoki and Yoshiki Matsuura)

References

1. Matsuura, Y., Kusunoki, M., Date, W., Harada, S., Bando, S., Tanaka, N. & Kakudo, M. (1979) *J. Biochem. 86,* 1773-1783.
2. Kusunoki, M., Matsuura, Y., Tanaka, N. & Kakudo, M. (1983) *J. Jpn. Soc. Starch Sci.* (in Japanese) *30,* 141-148
3. Matsuura, Y., Kusunoki, M., Harada, W. & Kakudo, M. (1984) *J. Biochem. 95,* 697-702.
4. Muirhead, H. (1983) *Trends Biochem. Sic. 8,* 326-330
5. Wistow, G., Turnell, B., Summers, L., Slingsby, C., Moss, D., Miller, L., Lindley, P. & Blundell, T. (1983) *J. Mol. Biol. 170,* 175-202
6. Yamaguchi, H., Ikenaka, T. & Matsushima, Y. (1971) *J. Biochem. 70,* 587-594
7. Payan, F., Haser, R., Pierrot, M., Frey, M., Astier, J. P., Abadie, B., Duée, E. & Buisson, G. (1980) *Acta Cryst. B36,* 416-421
8. Buisson, G., Duée. E., Haser, R. & Payan, F. (1987) *EMBO J. 6,* 3909-3916

II.1.e. Structure of α-Amylase Genes of *Bacillus* Genera

Bacilli are well known for their ability to produce a large amount of extracellular α-amylase. These α-amylases are quite important enzymes in food and starch industries. The α-amylase genes of *B. subtilis, B. amyloliquefacines, B. licheniformis,* and *B. stearothermophilus* have been cloned and their structures analyzed. *B. subtilis* α-amylase belongs to saccharifying type α-amylases and the other three α-amylases, to liquefying type enzymes. The nucleotide and predicted amino acid sequences of these liquefying type α-amylases are similar to each other, but they entirely differ from those of the saccharifying type enzyme. The amino acid sequences of only the three short peptide fragments, which are considered to constitute the active center of the enzymes, show a high homology (1). Structure and relatedness of the α-amylase genes are described.

Genetic Relatedness of *B. subtilis, B. amyloliquefaciens, B. licheniformis* and *B. stearothermophilus* : The genetic relatedness of the *Bacilli* has been investigated by measuring the GC contents of their DNAs, ratios of DNA-DNA hybridization, and abilities to transform *B. subtilis*. The results are summarized in Table I.

Table 1. Genetic Relatedness of *Bacilli*

Strains	GC content (%)	Transformation of *B. subtilis** (Trp^+ or Arg^+)	DNA-DNA hybridization (%)
B. subtilis Marburg wild	43.0	100	100
B. amyloliquefaciens	44.9	0	14.7-15.4
B. licheniformis	46.0	0	—
B. stearothermophilus	43.5	0	—

—, Not determind. **B. subtilis* Marburg (trp^- arg^-) was used as the DNA recipient.

B. subtilis, which produces saccharifying α-amylase, has little genetic relation to the other *Bacilli,* which produce liquefying type α-amylase (2,3). In contrast, the chromosomal DNAs of *B. natto (subtilis)* IAM 1212 and *B. subtilis* var. *amylosaccharitiicus* show 80 and 85% homologies to the DNA of *B. subtilis* Marburg 6160 when determined by the DNA-DNA hybridization analysis. The DNAs of these two strains have ability to transform the nutritional requirements of *B. subtilis* to prototrophic (4).

Genetic Mapping of α-Amylase Genes on the *B. subtilis* Chromosomal DNA : The hyper-producing genetic character of extracellular α-amylase and the α-amylase structural genes of *B. natto (subtilis)* IAM 1212 and of *B. subtilis* var. *amylosaccharitiicus* could also be transferred into *B. subtilis* Marburg 6160 by transformation. The extracellular α-amylases from these three strains were different from each other in their molecular weight, hydrolysis products from soluble starch, electrophoretic mobility in a 7.5% polyacrylamide gel and thermostability. Thus they are distinguishable from each other. On the basis of the different characteristics of the three α-amylases, the α-amylase structural genes of *B. natto* IAM 1212, *B. subtilis* var. *amylosaccharitiicus* and *B. subtilis* 6160 have been designated as *amyEn, amyEs* and *amyEm* respectively. The three α-amylases showed a cross-reaction with rabbit antisera against α-amylase from *B. subtilis* var. *amylosaccharitiicus* (5).

When the α-amylase hyper-producing character of *B. natto (subtilis)* IAM 1212 (*amyR2 amyEn*), was transferred into *B. subtilis* Marburg 6160 (*AmyR1 amyEm*), the ratio of *amyEn/amyEm* in *amyR2* transformants was 69/8. This result suggested that the genetic factor (*amyR2*) causing a hyper-producing character could be segregated from the α-amylase structural gene but that it was closely linked to the structural gene (*amyEn*) in the chromosomal DNA (6). The two genes are mapped near the *aro I* locus which is located at 24 position on the chromosomal DNA. The distances from *aro I* to *amyEm*, *amyEn*, and *amyEs* are indistinguishable from each other. They are allelic genes. By using tunicamycin-resistant mutants (*tmrA* and *tmrB*) of *B. subtilis*, the genetic map around *amyE* was determined as follows (7) :

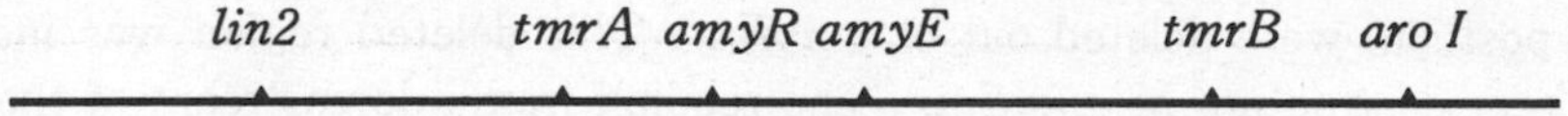

lin2 *tmrA* *amyR* *amyE* *tmrB* *aro I*

tmrA is a tunicamycin-resistant mutation that simultaneously causes a hyper-production of α-amylase by a gene amplification of an 16 kb fragment containing *amyR* *amyE* and *tmrB* but not *aro I* (8). The chromosomal DNA of *B. amyloliquefaciens* has only 14.7 to 15.4% homology to that of *B. subtilis* Marburg. When *B. subtilis* 207-21, which had lost a modification system of DNA and acquired a high frequency transformation ability, was used as the DNA recipient, some nutrient requirements in the *B. subtilis* strain were transformed to prototrophic by the DNA of *B. amyloliquefaciens*. By using the DNA of *B. amyloliquefaciens* (earlier *Thermophile* V2, now corrected to *B. amyloliquefaciens*) (9,10), the α-amylase gene of this DNA was transferred into the *B. subtilis* chromosomal DNA. The gene was mapped near *pyrA* locus, which is located at 129 on the chromosome. This location was completely different from that of *amyE* locus (11).

Cloning of α-Amylase Genes of *Bacilli* : The *amyEm* gene of *B. subtilis* NA64, an *amyR2* transformant of *B. subtilis* Marburg 6160, was cloned into *B. subtilis* temperate phage ρ11, and recloned in *B. subtilis* plasmid pUB110, which was originated from *Staphylococcus aureus*. The constructed plasmid was designated as pTUB4, in which a 2.3 kb DNA fragment was inserted. Approximately 2600 bp of the DNA nucleotide sequence of the *amyR2 amyEm* region in pTUB4 was determined. An open reading frame starting from an initiation codon of AUG was composed of 1776 bp (592 amino acids). Among the 1776 bp, 1674 bp (558 amino acids) were found in the cloned DNA fragment and 102 bp (34 amino acids) were in the DNA of the vector pUB110 (12). On the other hand, the *amyEm* of *B. subtilis* Marburg 168 (*amyR1 amyEm*) was cloned in the *E. coli* vector system (13). In this case, an intact *amyEm* was cloned which was composed of 1980 bp (660 amino acids). The 1674 bp sequence in pTUB4 was included in this frame. The comparison of the NH_2-terminal amino acid sequence of the extracellular *B. subtilis* α-amylase with the sequence deduced from the nucleotide sequence suggests that the first 41 amino acids constitute a signal peptide involved in secretion of exported proteins. Upstream from the promoter region exists an AT-rich inverted repeat structure (-21.2 kcal/mol). This structure may correspond to the hyper-producing character. The nucleotide sequence of this region was quite different from the corresponding region of the *amyR1* gene of *B. subtilis* Marburg 168. Transcription activity to *amyEm* promoter was enhanced approximately 6 fold by the addition of starch in the medium and less repressed by glucose in the presence of the inverted repeat structure (14).

The *amyEn* gene of *B. natto (subtilis)* IAM 1212 was also cloned in plasmid pUB110. Almost all the DNA nucleotide sequence of *amyEn* was identical with that of *amyEm*. Analysis of the nucleotide sequence of *amyEn* revealed that the α-amylase gene was composed of 1431 bp (477 amino acids). 32 bp of *amyEm* (1406 to 1437 nucleotide position) were deleted out of *amyEn*. This deleted region was included in the direct repeat structure of *amyEm*. The reading frame downstream of the deletion region of *amyEn* shifted and a new termination codon appeared at 26 bp downstream from that of the *amyEm*. Thus the difference in M-type and N-type α-amylases (*amyEm* and *amyEn*) seems to originate with the different amino acid sequences due to this 32 bp deletion (15) in the COOH-terminal region.

The α-amylase gene of *B. amyloliquefaciens* was at first cloned in *B. subtilis* temperate phage φ3T. Then the gene was directly cloned into plasmid pUB110 after the chromosomal DNA was partially digested by restriction enzyme *Mbo* I, and 2 to 5 kb DNA fragments were isolated by the sucrose gradient ultracentrifugation. The cloned gene was well expressed in *B. subtilis* cells and so a large amount of extracellular α-amylase was produced. An open reading frame, starting from an AUG initiation codon and comprising a total 1542 bp (514 amino acids), was found in the inserted DNA. The NH_2-terminal portion of the gene encodes a 31 amino acid-long signal peptide (16).

The α-amylase gene of *B. licheniformis* was cloned in *E. coli* vector system. The chromosomal DNA of *B. licheniformis* was partially digested by *Mbo* I and ligated with plasmid pBR322 after it had been cleaved by *BamHI*. Then the constructed plasmid was transferred into *E. coli* HB101. The gene was composed of 1536 bp (512 amino acids) starting from an AUG initiation codon. Out of the deduced amino acid sequence, a 29 amino acid sequence of the NH_2-terminal region was a signal peptide for secretion (17).

The α-amylase gene of *B. stearothermophilus* was at first found in a plasmid. The gene was able to hybridize with the genes of the other strains of *B. stearothermophilus*, in which their chromosomal DNA carried these genes (18). The genes of strains CU21, DY-5 and A631 were cloned in *B. subtilis* plasmid pAT5 and plasmid pBR322. More than 98% of the DNA nucleotide sequences and the predicted amino acid sequences of the three α-amylase genes were identical with each other. The genes were composed of a total 1647 bp (549 amino acids), starting from a GUG codon as methionine. A sequence of 34 amino acids in the NH_2-terminal region was thought to be a signal peptide. The predicted signal peptides were almost identical in the three α-amylases (1).

Comparison of the α-Amylase Genes of *Bacilli* : Some properties of the extracellular α-amylases from *B. amyloliquefaciens, B. licheniformis* and *B. stearothermophilus* are similar. They belong to the liquefying type α-amylases and their hydrolysis products from soluble starch are mainly maltose and maltotriose. The amino acid sequence of *B. stearothermophilus* α-amylase has 65 and 66% homologies to those of *B. licheniformis* and *B. amyloliquefaciens* α-amylases, respectively. Furthermore, more than 80% of the amino acid sequence of *B. licheniformis* α-amylase is homologous to that of the *B. amyloliquefaciens*. Therefore, it is possible to conclude that the three α-amylase genes may be originated from a common ancestor gene. However, the predicted amino acid sequences of their signal peptides were completely different from each other. In contrast, the enzymic properties and amino acid sequences of the saccharifying type α-amylases, produced by *B. subtilis* and related strains, were entirely different from those of the liquefying type α-amylases except for three short amino acid sequences (1).

Recently, the amino acid sequences of more than 10 kinds of α-amylases from bacterial to human have been determined or predicted from the DNA nucleotide sequences. Three common amino acid sequences (A, B and C regions), which may constitute the active centers of the enzymes, have been found in various α-amylases. The common amino acid sequences of α-amylases of *Bacilli* were compared to those of human salivary α-amylase (Table II). The three amino acid sequences are also found in the amino acid sequences of cyclodextrin glucanotransferases (CGTase) (19,20).

The locations of the three common regions in α-amylases and CGTase (CGTase from an alkalophilic *Bacillus* sp. #1011) are summarized in Fig. 1. The A, B and C regions are located approximately 100, 200 and 300 amino acid positions from each NH_2-terminal. The amino acid sequence of CGTase has 16-17% homology to that of *B. subtilis* α-amylase and more than 30% homology to that of *Aspergillus oryzae* α-amylase (20).

Table 2. Amino Acid Sequence Homology among the Three Common Regions (A, B and C) of α-Amylases from *Bacilli* and Human Saliva

Origin of α-amylases	A-region	B-region	C-region
Human salivary	--DAVINH--	--GFRIDASKH--	--VFVDNHD--
B. amyloliquefaciens	--DVVLNH--	--GFRIDAAKH--	--TFVENHD--
B. licheniformis	--DVVINH--	--GFRLDAVKH--	--TFVDNHD--
B. stearothermophilus	--DVVFDH--	--GFRLDAVKH--	--TFVDNHD--
B. subtilis	--DAVINH--	--GFRFDAAKH--	--TWVESHD--

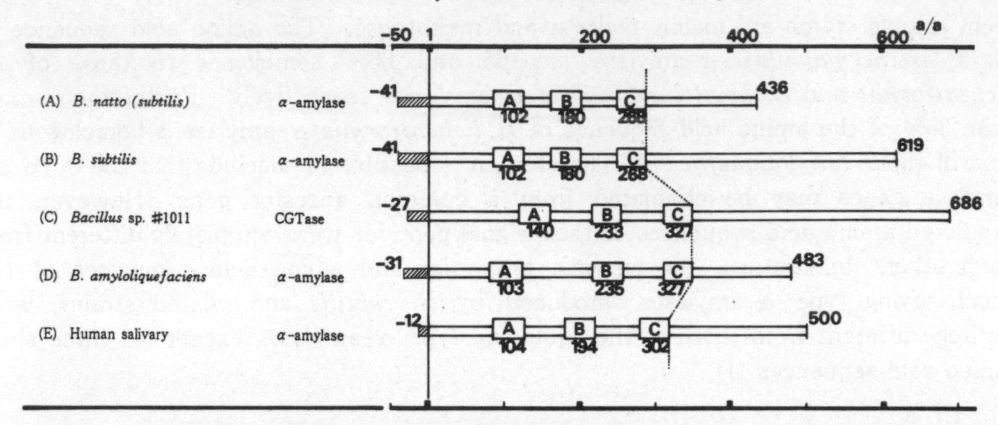

Fig. 1. Schematic Drawing of Location of the Three Common Regions (A,B and C) in α-Amylases and CGTase.
Amino acid position 1 corresponds to the NH_2-terminus of each mature enzyme. Numbers at the COOH-terminus of each enzyme refer to the amino acid length of each enzyme. Numbers under the boxes (A, B and C) indicate the amino acid positions of His residue in the three common regions. ▨, signal peptides.

(Kunio Yamane)

References

1. Yamane, K. & Sohma, A. (1986) *Hakko to Kougyo 44,* 2-11
2. Takahashi, H., Saito, H. & Ikeda, Y. (1966) *J. Gen. Appl. Microbiol. 12,* 113-118
3. Marmur, J., Seaman, E. & Levine, J. (1963) *J. Bacteriol. 85*, 461-467
4. Yoneda, Y., Yamane, K., Yamaguchi, K., Nagata, Y. & Maruo, B. (1963) *J. Bacteriol. 120,* 1144-1150
5. Matsuzaki, H., Yamane, K., Yamaguchi, K., Nagata Y. & Maruo, B. (1974) *Biochim. Biophys. Acta 365,* 235-247
6. Yamaguchi, K., Matsuzaki, H. & Maruo, B. (1969) *J. Gen. Appl. Microbiol. 15,* 97-107
7. Nomura, S., Yamane, K., Sasaki, T., Yamasaki, M., Tamura, G. & Maruo, B. (1978) *J. Bacteriol. 136,* 818-821
8. Hashiguchi, K., Tanimoto, A., Nomura, S., Yamane, K., Yoda, K., Harada, S., Mori, M., Furusato, T., Takatsuki, A., Yamasaki, M. & Tamura, G. (1985) *Agric. Biol. Chem. 49,* 545-550
9. Shinomiya, S., Yamane, K. & Ohshima, T. (1980) *Biochem. Biophys. Res. Commun. 96,* 175-179
10. Shinomiya, S., Yamane, K., Kawamura, F., Yoshikawa, H., Kazami, J. & Saito, H. (1984) *Agric. Biol. Chem. 48,* 1339-1341
11. Yamane, K. & Shinomiya, S. (1982) *Microbiology* 8-11
12. Yamazaki, H., Ohmura, K., Nakayama, A., Takeichi, Y., Otozai, K., Yamasaki, M., Tamura, G. & Yamane, K. (1983) *J. Bacteriol. 156,* 327-337

13. Yang, M., Galizzi, A. & Henner, D. (1983) *Nucleic Acids Res. 11*, 237-249
14. Takano, J., Kinoshita, T. & Yamane, K. (1987) *Biochem. Biophys. Res. Commun. 146*, 73-79
15. Yamane, K., Hirata, Y., Furusato, T., Yamazaki, H. & Nakayama, A. (1984) *J. Biochem. 96*, 1849-1858
16. Takkinen, K., Petterson, R. F., Kalkkinen, N., Palva, I., Söderlund, H. & Kääriäinen, L. (1983) *J. Biol. Chem. 258*, 1007-1013
17. Yuuki, T., Nomura, T., Tezuka, H., Tsuboi, A., Yamagata, H., Tsukagoshi, N. & Udaka, S. (1985) *J. Biochem. 98*, 1147-1156
18. Mielanz, J. R. (1983) *Proc. Natl. Acad. Aci. USA 80*, 5975-5979
19. Kimura, K., Kataoka, S., Ishii, Y., Takano, T. & Yamane, K. (1987) *J. Bacteriol. 169*, 4399-4402
20. Sakai, S., Kubota, M., Yamamoto, K., Nakada, T., Torigoe, K., Ando, O. & Sugimoto, T. (1987) *J. Jpn. Soc. Starch Sci.* (in Japanese) *34*, 140-147

II.2. β-Amylases (EC 3.2.1.2. α-1,4-Glucan Maltohydrolase)

Kuhn (1924) and Ohlsson (1930) named a β-maltose-producing enzyme in malt as β-amylase. The prefix, β-, does not refer to the substrates but to the stereochemical specificity of the products ; namely, anomer type of the product. This naming for the enzyme still prevails and, therefore, α-maltose-producing enzymes are ruled out of the category of β-amylases (1). Distribution of β-amylase has long been believed to be limited to higher plants such as barley, wheat, soybeans, sweet potato, some vegetables, etc. singly or co-existently with α-amylase (1). In 1974, however, a β-amylase was first reported in a bacterium. Since then, the existence of β-amylase has been reported in *Bacillus* (2-5), *Pseudomonas* (3), *Streptomyces* (3), etc. These enzymes attracted much attention to fermentative production of bacterial β-amylases instead of plant β-amylases.

Reaction :

$$\text{Soluble starch} + n\, H_2O \longrightarrow n\ \beta\text{-Maltose} + \text{Dextrin.}$$

β-Amylases consist of a single protein except for the sweet potato β-amylase having the subunit structure (6). All the β-amylases need no coenzyme, no metal ion, nor prosthetic group for the enzyme action. The action pattern of β-amylase was first observed as what we call "single chain type", but later was corrected into "multiple attack" system including both single- and multi-chain types (1). This was also supported by evaluation of subsite affinity in the active center (7). β-Amylase hydrolyzes amylose from its non-reducing end and releases β-maltose; accordingly, the extent of hydrolysis reaches 100%. The hydrolysis of amylopectin stops at certain places near the branched points of the α-1,6-glucosidic linkages leaving β-limit dextrins. One of the unelucidated problems on β-amylase action is how the "Walden inversion" happens during release of β-maltose from the α-1,4-glucosidic bonded substrates.

Specificity : The active center of β-amylase had been believed to contain a SH group (s) since the activity is inhibited with SH reagents like *p*-chloromercury benzoate, N-ethylmaleimide, etc. (1). However, recent studies revealed the real role of the SH group of β-amylase and they suggest no direct involvement of SH groups in the enzyme reaction (6,8-10). In addition to such β-amylase inhibitors as inorganic and organic SH reagents, *Streptomyces* was reported to excrete a glycoprotein, glucosyl deoxynojirimycin, a strong inhibitor of β-amylase (11). Chemically synthesized β-maltosylamine is also a reversible inhibitor of β-amylase and protects the modification of SH groups with N-ethylmaleimide (12).

Assay Methods : The assay methods of β-amylase are mainly colorimetric methods with 3,5-dinitrosalicylic acid reagent by Noelting and Bernfeld (13) and by Bernfeld (14), and with Somogyi's copper reagent by Nelson (15). A method for determination of β-amylase activity in the presence of α-amylase was reported by Krueger and Mordmann (16). The principle of this method is that maltotetraose is used as substrate which is hardly hydrolyzed by α-amylase. The activity determined is usually expressed in amounts of β-maltose (μg or mg) produced by an enzyme preparation for a given time at a given temperature.

Industrially, β-amylases are used in the fields of food and fermentation industries such as production of malt syrup, whisky production, bread making, etc. Production of maltose by β-amylase is now very important in pharmaceutical industry. β-Amylase preparation with α-amylase is used as medicine for digestion (17).

(Ryu Shinke)

References

1. Thoma, J. A., Spradlin, J. E. & Dygert, S. (1971) *Plant and Animal Amylases* in *The Enzymes,* Vol. 5 (Boyer, P. D., ed.) Academic Press, pp. 115-189
2. Higashihara, M. & Okada, S. (1974) *Agric. Biol. Chem. 38,* 1023-1029
3. Shinke, R., Nishira, H. & Mugibayashi, N. (1974) *Agric. Biol. Chem. 38,* 665-666
4. Takasaki, Y. (1976) *Agric. Biol. Chem. 40,* 1515-1522, 1523-1530
5. Ohyama, K. & Murao, S. (1979) *Hakkokogaku, 57,* 1-5
6. Thoma, J. A., Koshland, D. E. Jr., Shinke, R. & Ruscica, J. (1965) *Biochemistry, 4,* 714-722
7. Suganuma, T., Ohnishi, M., Hiromi, K. & Morita, Y. (1980) *Agric. Biol. Chem. 44,* 1111-1117
8. Nanmori, T., Shinke, R., Aoki, K. & Nishira, H. (1983) *Agric. Biol. Chem. 47,* 941-947
9. Mikami, B. & Morita, Y. (1983) *J. Biochem. 93,* 776-786
10. Nitta, Y., Kunikata, T. & Watanabe, T. (1983) *J. Biochem. 93,* 1195-1201
11. Murao, S. (1981) *Nippon Nogei-Kagaku-Kaishi* (in Japanese) *55,* 503-513

12. Shinke, R. (1986) *β-Amylase* in *Amylases* (Nakamura, M., ed.) (in Japanese), Gakkai-Shuppan Center, Tokyo, pp. 69-97
13. Noelting, G. & Bernfeld, P. (1948) *Helv. Chim. Acta 31*, 286-290
14. Bernfeld, P. (1955) *Amylases, alpha and beta* in *Methods in Enzymology* (Colowick, S. P. & Kaplan, N. O., eds.) Academic Press, pp. 149-158
15. Nelson, N. (1944) *J. Biol. Chem. 153*, 375-380
16. Krueger, E. & Nordmann, A. (1983) *Mschr. Brauwiss. 1*, 30-36
17. Aizawa, T., Ono, M., Tezuka, T. & Yanagida, T. (1980) *β-Amylase* in *Handbook of Enzyme Utilization* (Ozaki, M., ed.) (in Japanese) Chijinshokan, Tokyo, pp. 55-59

II.2.a. Plant β-Amylases

II.2.a.1. Malt β-Amylases

The amylolytic enzyme was first observed in malt extracts by Payen and Persoz (1833) (1) and in barley extracts by Kjeldahl (1879) (2). This was the first enzyme separated scientifically from nature and named "diastase" (1). Diastase was later found to be composed of two components, α- and β-amylase named by Kuhn (1924) and Ohlsson (1930) (3). The former is characterized by decrease in viscosity or liquefaction and the latter, by formation of fermentable sugars or saccharification. Reychler (1889) found a difference in the activity between water and dilute acid extracts of barley (3). Ford and Guthrie (1908) (4) reported two forms of amylases; namely, free amylase (active form) and zymogen amylase (inactive form). The β-amylase contents in barley and other cereals were compared by Kneen (1944). Table 1 shows a comparison of the amylolytic activities of dormant and germinated cereals.

Table 1. Amylolytic Activities of Dormant and Germinated Cereal Seeds (5)

	β-Amylase				α-Amylase			
	Native Grain		Germinated Grain		Native Grain		Germinated Grain	
Cereal	Free Enzyme	Total[a] Enzyme	Free Enzyme	Total Enzyme	Free Enzyme	Total Enzyme	Free Enzyme	Total Enzyme
Barley	10.7	29.8	26.5	34.4	0.045	0.058	90.5	94.0
Wheat	7.5	25.1	16.5	23.7	0.050	0.063	197.3	214.7
Rye	9.1	17.8	15.4	17.6	0.089	0.111	93.2	119.8
Oats	0.7	2.4	–	–	0.262	0.297	53.1	60.3
Maize	–	–	–	–	0.101	0.249	31.1	35.6
Sorghum	–	–	–	–	0.031	0.127	73.4	75.5
Rice	–	–	–	0.2	0.075	–	1.4	2.3

[a] The difference between the free and total enzyme is due to the latent enzyme.

The activation mechanism of the inactive form of barley amylase attracted many investigators' attention and many different hypotheses for the enzyme activation were postulated by terms like zymogen β-amylase, latent β-amylase, bound β-amylase, etc. (3). The hypotheses can be classified into two categories; proteolytic activation and activation by disulfide bond cleavage. However, all the hypotheses require further experimental results before they can be accepted.

Specificity : Shinke, *et al.* (1971) found four types of zymogen β-amylases in barley; salt-soluble and salt-insoluble zymogen β-amylases, papain-soluble and papain-insoluble zymogen β-amylases (6). The two types of salt-soluble zymogen β-amylases were first isolated and purified. Fig. 1 shows a fractionation pattern of salt-soluble zymogen β-amylases in Sephadex G-75 gel filtration. Fraction A is a heteropolymer type of zymogen β-amylase with the molecular weight of about 280,000 which is activated by both proteolysis and disulfide bond cleavage. Fraction B is a homopolymer type of zymogen β-amylase with the molecular weight of about 160,000 which is activated only by disulfide bond cleavage to form an active type of β-amylase with the molecular weight of 56,000 (Fraction C). From these results obtained a complete activation of zymogen β-amylases in barley seems to require both proteolytic and disulfide bond cleaving activation. Fig. 2 shows the schematic outline of a possible mechanism *in vitro* of activation by proteolysis with papain and disulfide bond cleavage with 2-mercaptoethanol (2-ME) (7).

Studies of the activation mechanism of zymogen β-amylases *in vivo* also showed that both activations by malt proteinase and by disulfide bond cleaving enzyme, protein disulfide reductase in malt operated during germination of barley (8). As for the biosynthesis of zymogen β-amylases during ripening of barley, zymogen β-amylases were found to appear even in the early stages of ripening and increase rapidly in the later stages with correlative decrease of the active β-amylase (9).

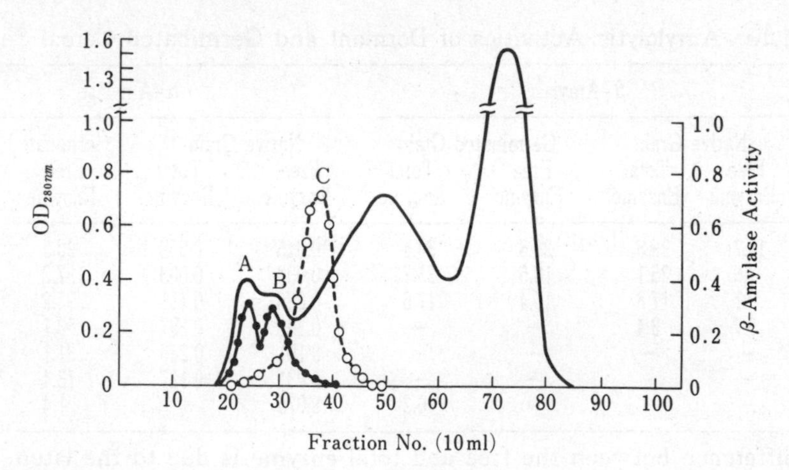

Fig. 1. Fractionation of Salt-soluble Zymogen β-Amylases on a Sephadex G-75 Column.
A: heteropolymer, B: homopolymer, C: active β-amylase

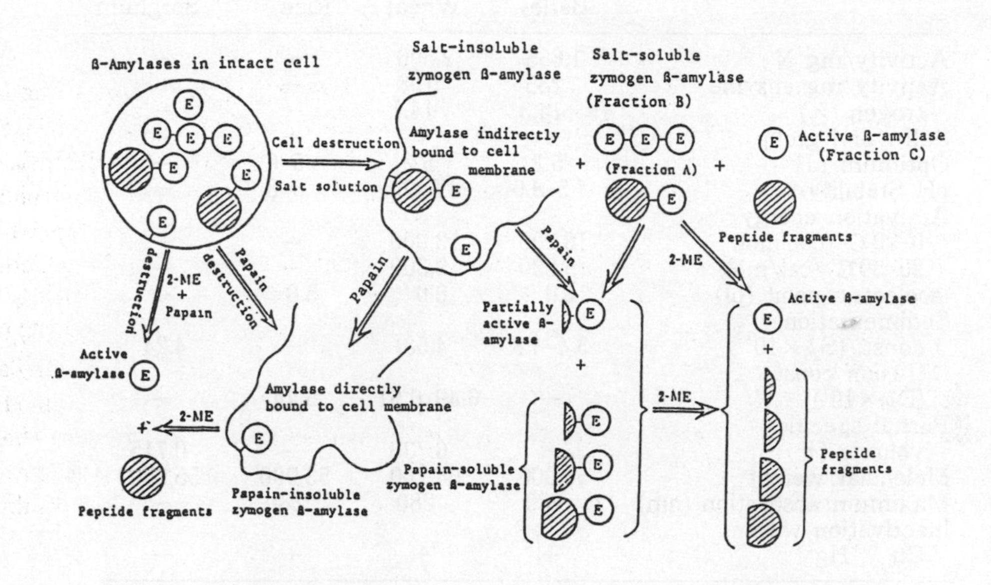

Fig. 2. A Possible Activation Mechanism of Zymogen β-Amylase in Barley Seed.

Though zymogen β-amylases were found to be activated during germination, *de novo* synthesis of active β-amylase did not occur in a sharp contrast to the case of α-amylase (10). Accordingly, β-amylase in barley and malt is composed of active β-amylase (MW 56,000), basic unit of barley β-amylase and inactive β-amylase of various molecular weights. The reaction between disulfide and sulfide bonds is concerned with polymerization and depolymerization of the enzyme protein and with the control of β-amylase activity in barley and malt (7). The biochemistry of malting has been reviewed in detail with changes of amylolytic enzymes (11-15). The primary structure of barley endosperm β-amylase has been deduced from the nucleotide sequence of a cloned full-length cDNA (16).

Enzymic and Proteochemical Properties : Table 2 shows the physicochemical and enzymic properties of the purified β-amyases so far reported (7). The molecular weight of rice β-amylase (53,000) is the smallest and those of sorghum (55,950), barley (56,000) and wheat (64,000) increase in the order. The optimum pH values for the reaction are 5.0 to 6.0 and those for the stability are 4 to 9, respectively.
All the β-amylases are inactivated with cupric and mercuric ions.

Table 2. Physicochemical and Enzymic Properties of β-Amylases

	Barley	Wheat	Rice	Sorghum
Activity/mg N	1,665	2,500	—	—
Activity/mg enzyme	235	198	—	—
Nitrogen (%)	14.3	14.1	—	—
Sulfhydryl group	+	+	+	—
Optimum pH	5.2	5.2	5–6	—
pH Stability	4.5–8.0	4.5–9.2	6–8	—
Activation energy				
0–20℃ (cal/mol)	16,200	13,000	—	—
20–50℃ (cal/mol)	5,530	9,300	—	—
Isoelectric point (pI)	6.0	6.0	5.0	—
Sedimentation const. ($S_{20} \times 10^{13}$)	3.4–4.8	4.58	—	4.24
Diffusion const. ($D_{20} \times 10^{7}$)	—	6.49, 6.51	6.44	—
Partial specific volume ($\bar{v}$)	—	0.733	—	0.715
Molecular weight	56,000	64,200	53,000	55,950
Maximum absorption (nm)	280	280	280	—
Inactivation with Cu^{2+}, Hg^{2+}	+	+	+	—

(Ryu Shinke)

References

1. Payen, A. & Persoz, J. (1833) *Ann. Chim. Phys. 53,* 73
2. Kjeldahl, J. (1879) *Compt. rend. Trav. Lab. Carlsberg. 1,* 129–134
3. Harris, G. (1962) *The Enzyme Content and Enzymic Transformation of Malt* in *Barley and Malt* (Cook, A. H., ed.), Academic Press, pp. 583–624
4. Ford, J. S. & Guthrie, J. M. (1980) *J. Inst. Brew. 14,* 61–87
5. Kneen, E. (1944) *Cereal Chem. 21,* 304
6. Shinke, R. & Mugibayashi, N. (1971) *Agric. Biol. Chem. 35,* 1381–1390.
7. Shinke, R. (1986) *β-Amylase* in *Amylases* (Nakamura, M. ed.) (in Japanese), Gakkai-Shuppan Center, Tokyo, pp. 69–97
8. Shinke, R. & Mugibayashi, N. (1969) *Nippon Nogei-Kagaku Kaishi (*in Japanese) *43,* 556–563
9. Shinke, R. & Mugibayashi, N. (1971) *Agric. Biol. Chem. 35,* 1391–1397
10. Shinke, R. & Mugibayashi, N. (1972) *Agric. Biol. Chem. 36,* 378–382
11. LaBerge, D. E. & Meredith, W. O. (1971) *J. Inst. Brew. 77,* 436–440
12. Numi, M., Vilhunen, R. & Enari, T. M. (1965) *Proc. of Eur. Brew. Conv. Congress Stockholm,* pp. 52–69
13. Hejgaard, J. (1978) *J. Inst. Brew. 84,* 43–46.
14. Hough, J. S., Briggs, D. E. & Stevens, R. (1971) *The Biochemistry of Malting* in *Malting and Brewing Science,* Chapman and Hall Ltd., pp. 54–107
15. Briggs, D. E., Hough, J. S., Stevens, R. & Young, T. W. (1981) *The Biochemistry of Malting Grain* in *Malting and Brewing Science,* Vol. 1, pp. 57–107

16. Kreis, M., Williamson, M., Buxton, B., Pywell, J., Heigaad, J. & Svendsen, I. (1987) *Eur. J. Biochem.* *169*, 517-525

II.2.a.2. Soybean β-Amylase

Soybean β-amylase is one of the most investigated β-amylases. Since the enzyme was first crystallized by Fukumoto and Tsujisaka (1), proteochemical (2,3), kinetic (4) and technological investigations have been carried out. Soybean β-amylase is industrially utilized in the production of highly purified maltose (see, part VI.3). The enzyme is relatively stable and not contaminated with α-amylase. Soybean seeds contain several isozymes of β-amylase which have different isoelectric points (5). Most of soybean seeds contain isozyme 2 (pI =5.25) as a major component of β-amylases. The behavior of the isozymes has been well investigated from the genetic and physiological points of view (6,7). As soybean seeds which contain very little starch do not require β-amylase for their growth, the biological function of β-amylase is still obscure (7). The details of the structure of this enzyme and the mechanism of its inactivation by SH-reagents are described in section II.2.c.

Enzymic and Proteochemical Properties : Table 1 shows the properties of soybeen β-amylase isozyme 2 (3,5). The enzyme is composed of a single polypeptide containing no carbohydrate. The enzyme reaction is inhibited by glucose (mixed type, K_i= 320 mM), maltose (competitive, 5.8 mM) and cyclohexaamylose (competitive, 0.39 mM) at pH 5.4. Glucose also acts as a competitive inhibitor at pH 8.0 (K_i = 34 mM) (8).

Table 1. Enzymic and Proteochemical Properties of Soybean β-Amylase Isozyme 2

Molecular weight	57,000
N (%)	16.1
E (1%, 280nm)	17.0
Isoelectric point	5.25
Helix content (%)	26
Cysteine residues	6
Optimum pH	5.4(5–6)
*K*m (amylopectin, %)	0.225
Activity (units/mg protein)	860
pH Stability (25°C, 3 h)	3.5–10
Thermal stability (pH 5.5, 30 min)	55°C

As shown in Table 2 the isozymes (isozyme 1,2,4 and 6) have the same molecular weight and enzymic properties but have slightly different isoelectric points and amino acid compositions (5).

Isolation and Purification : Soybean seeds are divided into two types, low pI and high pI types, by their β-amylase isozyme pattern (5). The low pI type contains isozymes

1 and 2 while the high pI type contains isozymes 3 and 4 as major components of β-amylase. Presumably, the isozymes are coded by two small multigene families under the codominant control of alleles which regulate the two pI types (gene symbols Sp_1^b and Sp_1^a for low and high pI types, respectively) (9). The commercial defatted soybean flour or whey (supernatant of isoelectric precipitation of defatted soybean flour) contains at least seven isozymes of β-amylase. Usually, isozyme 2 is the most predominant isozyme (about 50% of total β-amylase) because most soybean seeds are low pI type.

Table 2. Comparison of Properties of Soybean β-Amylase Isozymes

Isozymes	p I	MW	N-Terminal amino acid	C-Terminal amino acid	K_m (mg/ml amylopectin)	Immuno-precipitation (for antibody 2)
1	5.15	57,000				+
2	5.25	57,000	Blocked	-Gly	2.25	+
4	5.55	57,000	Blocked	-Val-Ala/-Gly	2.25	+
6	5.93	57,000	Blocked	-Gly	1.65	+

For purification and storage, 20 mM 2-mercaptoethanol and 1 mM EDTA should be added to the buffer solution in order to protect the SH groups of the enzyme from oxidation or mercaptide formation (5,10). The defatted soybean flour is suspended in 10 mM acetate buffer, pH 5.4 (10 liter/1 kg) for 2 h and the suspension is centrifuged. The pH of the supernatant is adjusted to 5.0 to remove the globulin fraction. Heat treatment (60°C, 20 min) is also effective in this step. After the precipitate is removed by filtration, the enzyme is fractionated by 25-60 % saturation of $(NH_4)_2SO_4$. The precipitated protein is dialyzed against 10 mM acetate buffer pH 5.0 and adsorbed on a CM-Sephadex C-50 column. The enzyme is eluted by pH gradient formed by 50 mM acetate buffer, pH 5.0 and 6.1. Four active fractions consisting of each isozyme 1, 2, 3, and 4 are eluted. Each fraction is further purified by ion-exchange chromatography on CM- and DEAE-Sephadex columns and by gel-filtration on Sephadex G-100. About 1 g purified isozyme 2 can be obtained from 7 kg defatted soybean flour. The purified enzyme is stored in a form of $(NH_4)_2SO_4$ precipitate or in 40% glycerol solution at pH 5.4. Freezing (lyophilization) inactivates the purified enzyme with its coagulation.

Assay Method : β-Amylase activity is readily measured by the Bernfeld method (11). For substrate, 1 g of amylopectin, Schoch's B fraction (12) is dissolved in 100 ml of 0.1 M acetate buffer, pH 5.4. The reaction is performed at 37°C for 3 min and terminated by the addition of 3,5-dinitrosalicylate solution. After boiling for 5 min and dilution with water, the absorbance due to the reduction product of 3,5-dinitrosalicylate is measured at 540 nm. One amylase unit is defined as the amount of enzyme which liberates 1 μmol maltose per min in the reaction mixture at 37°C.

(Bunzo Mikami and Yuhei Morita)

References

1. Fukumoto, J. & Tsujisaka, Y. (1954) *Kagaku to Kogyo*, Osaka (in Japanese) *28*, 282-287
2. Gertler, A. & Birk, Y. (1965) *Biochem. J. 95*, 621-627
3. Morita, Y., Yagi, F., Aibara, S. & Yamashita, H. (1976) *J. Biochem. 79*, 591-603
4. Suganuma, T., Ohnishi, M., Hiromi, K. & Morita, Y. (1980) *Agric. Biol. Chem. 44*, 1111-1117
5. Mikami, B., Aibara, S. & Morita, Y. (1982) *Agric. Biol. Chem. 46*, 943-953
6. Hildebrand, D. F. & Hymowitz, T. (1980) *Crop Sci. 20*, 165-168
7. Hildebrand, D. F. & Hymowitz, T. (1981) *Physiol. Plant 53*, 429-434
8. Nomura, K., Mikami, B. & Morita, Y. (1986) *J. Biochem. 100*, 1175-1183
9. Nakamura, I. & Futsuhara, Y. (1985) *Japan J. Breed. 35*, 153-159
10. Morita, Y., Aibara, S., Yamashita, H., Yagi, F., Suganuma, T. & Hiromi, K. (1975) *J. Biochem. 77*, 343-351
11. Bernfeld, P. (1955) *Methods in Enzymol.* (Colowick, S. P. and Kaplan, N. O., eds.) Vol. 1, pp. 149-158, Academic Press, New York
12. Schoch, T. J. (1957) *Methods in Enzymol.* (Colowick, S. P. and Kaplan, N. O., eds.) Vol. 3, pp. 5-17, Academic Press, New York

II.2.a.3 Sweet Potato β-Amylase

In 1920, Gore showed the occurrence of β-amylase in sweet potato (*Ipomoea batatas*) (1). It usually contains a large amount of the enzyme, as high as about 13% of the total soluble protein, although a few varieties of sweet potato contain very low β-amylase activity (2). Sweet potato β-amylase is the first amylase obtained in a crystalline form (3). Its physicochemical properties, substrate specificity, and action pattern and mechanism have been investigated in detail (4-7). The enzyme from this origin is very useful for the structural analyses of starch (amylose and amylopectin), glycogen and their degraded products, and is used for the determination of the degree of gelatinization and retrogradation of starch pastes (8).

Reaction : It has been suggested that besides the hydrolytic action, the enzyme catalyzes the condensation of β-maltose to maltotetraose (9) and the transfer of maltosyl residues from β-maltosyl fluoride to another molecule of the substrate by α-1,4-glucosidic bond (10).

Specificity : The enzyme releases maltose of β-anomeric form by successively attacking the non-reducing terminal maltosyl residues of α-1,4-glucan (11), and completely degrades it to maltose. The smallest maltooligosaccharide which is readily hydrolyzed is maltotetraose. Maltotriose is degraded at a very slow rate to maltose (G2) and D-glucose. G2 and cyclodextrins are not hydrolyzed. Methyl-α-maltotetraoside, methyl-α-maltotrioside (4), aryl-β-maltotriosides (12) and α-maltosyl fluoride (10) are hydrolyzed to G2 and methyl-α-maltoside, G2 and methyl-α-D-glucoside, G2 and aryl-β-D-glucosides, and G2 and hydrogen fluoride, respectively. The hydroxymethyl group on C-5 of the methyl-α-D-pyranoside residue of methyl-α-maltotrioside and methyl-α-maltotetraoside is essential for the enzymic action, while the hydroxyl group on C-2 is unessential (13). The hydroxyl group on C-4 of the non-reducing residue of α-1,4-glucan appears to be unnecessary (14).

The α-1,6-glucosidic linkages of amylopectin and glycogen are not hydrolyzed and act as a barrier against hydrolysis. Even on exhaustive hydrolysis of starch by the enzyme, dextrins of high molecular weights remain, that is, β-limit dextrins. Amylopectins and glycogens are degraded to the limited extents of 55-60% and 35-40%, respectively. Amylose preparations prepared from various starches by the conventional method are readily degraded, but usually incompletely (75-91%) due to the presence of certain limited numbers of branch linkages in the amyloses (15). The structures around the outermost branch linkages of β-limit dextrin are as follows (16) ;

```
O-O              O-O             O-O-O         O-O-O
 ↓                 ↓                 ↓               ↓
O-O-O---Ø,    O-O-O-O---Ø,      O-O-O---Ø,   O-O-O-O---Ø
```

The β-amylolysis ceases in the vicinities of 3- and 6-phosphoglucosyl residues found in potato amylopectin, e.g., one or no glucosyl residues remain attached to 6-phosphoglycosyl residues as indicated below (17) :

```
  P6             P6               O               O
  |              |                ↓               ↓
O-O-O----Ø*,    O-O----Ø**,     O-O-O----Ø*,    O-O-O----Ø**
```

(* and **; odd and even numbers of glucosyl residues at the non-reducing side, respectively.)

These actions are similar to those on branched oligosaccharides with a glucosyl stub (18). 3-Phosphoglucosyl residues in potato amylopectin are more obstructive against the enzyme action than 6-phosphoglucosyl residues (17). The enzyme does not attack partially-methylated (19) or 6-deoxy (20) amyloses. The enzyme cleaves the C1-O-C'4 bond in α-1,4-glucan between C1 and the oxygen atom (21). The "induced-fit" hypothesis (22) accounts for the substrate and inhibitor specificity. The mode of hydrolysis of amylose by the enzyme is not necessarily the pure single-chain attack, but a mixed type of single- and multi-chain attack (multiple attack) (5).

Table 1. Properties of Sweet Potato β-Amylase

N Content (%)	15.1 (3), 16.5 (23), 16.2 (24)
E (1%, 280nm)	17.1 (3), 17.7 (23)
Molecular weight[1)]	152,000[a] (25), 197,000[b] (26), 206,000[c] (27), 215,000[b] (28)
Subunit molecular weight	47,500[b] (26), 50,000[b,d,e] (24,28), 56,000[f], 57,000[e] (29)
Sedimentation coefficient	S_{20W} = 8.9s (25), 9.3s (30)
Diffusion coefficient	$D_{20} = 5.77 \times 10^{-7} cm^2 sec^{-1}$(25)
pI	4.77 (4)
Amino acid composition (31) (mol/50,000g)	Asp_{48}, Thr_{12}, Ser_{14}, Glu_{34}, Pro_{23}, Gly_{31}, Ala_{30}, Val_{25}, Met_{14}, Ile_{18}, Leu_{29}, Tyr_{17}, Phe_{17}, Lys_{25}, His_7, Arg_{15}, Try_{15}, Cys_5. (374 residues)
Carbohydrate	+ (24,32)
N-Terminal amino acid sequence (33)	Ala-Pro-Ile-Pro-Gly-Val-Met-Pro-Ile-Gly-Asn-Try-Val-Val-Leu-Try-Val-Met-Leu-Pro-
C-Terminal amino acid sequence (24)	-Pro-Gly and /or -X-Pro
Possible functional groups and amino acids in the active center	
Catalytic effect	Carboxyl group
Substrate binding	His, Tyr, Try, amino group (34)
Possible number of subsites in the active center (22) : 4	
Optimum pH	5-6 (36)
Optimum temperature	50-55℃ (4)
Inhibitor (6)	Glucose(NC[2)]), maltose (NC) α-Methyl glucoside (NC) β-Maltosylamine (50% inhibition at 0.3 mM, uncompetitive) (36) α-Cyclodextrin (C[3)], K_i = 0.17mM) β-Cyclodextrin (C) Thiol reagents (4-chloromercuribenzoate, iodoacetoamide) Heavy metal ions (Cu^{2+}, Hg^{2+}, Ag^{2+})

1) Determined by the sedimentation velocity method (a), sedimentation equilibrium method (b), X-ray analysis (c), equilibrium dialysis with [^{14}C]-β-cyclodextrin (d), N-terminal amino acid analysis (e), and SDS-gel electrophoresis (f). 2) and 3); Non-competitive and competitive inhibitions, respectively.

Enzymic and Proteochemical Properties : In Table 1 are summarized the proteochemical and enzymic properties of sweet potato β-amylase. The values of the Michaelis constant, Km and molecular activity, k_O, are 13.8 mM and 0.96 s^{-1} for maltotriose, and 1.4 mM and 950 s^{-1} for maltotetraose, respectively (12). Fig. 1 shows the dependence of Km and Vmax on the outer chain lengths of amylopectin (37).

Purification (23) : Sweet potato (1.7 kg) is mashed in a home juicer after the peel and cambium are removed. The removal of the cambium prevents possible contamination of a yellow substance in the final enzyme preparation. Unless otherwise specified, the operations are conducted in an ice bath, and centrifugation is performed at 0℃ and 10,000×g for 5 min. The juice obtained is centrifuged for 7 min to remove starch and some insoluble materials. To inactivate α-amylase the supernatant (crude extract, 220 ml) is slowly acidified to pH 3.6 with 1M HCl. After standing for 10 min, the coagulated materials are centrifuged off. Immediately, the pH of the resulting supernatant is adjusted to 4.8 with 3% aq. NH_4OH. This operation should be rapidly conducted to avoid inactivation of the enzyme.

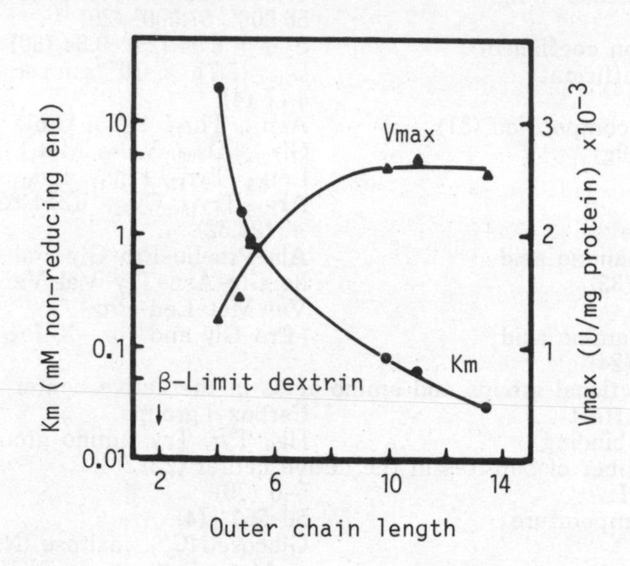

Fig. 1. Dependence of *K*m and *V*max of Sweet Potato β-Amylase on Outer Chain Lengths (number of glucosyl residues) of Amylopectin.

To the solution (210 ml), cold acetone (0°C) is gradually added to give 47% (v/v). The precipitate formed is immediately collected by centrifugation, and suspended in water (about 20 ml). The insoluble materials are centrifuged off. To the supernatant (20 ml) is added the saturated $(NH_4)_2SO_4$ solution (4°C) until 47.5% saturation with stirring. After standing for 10 min, the precipitate is collected by centrifugation. The supernatant should be completely removed because $(NH_4)_2SO_4$ interferes with the crystallization of β-amylase in the following step. The precipitate is suspended in 10 ml of distilled water. Insoluble materials are removed by centrifugation. To the solution (12 ml), cold acetone is added slowly to make it 45% (v/v) followed by immediate centrifugation. The precipitate is dissolved in water (7 ml). If insoluble materials remain, they are removed by centrifugation. Cold acetone is added slowly to the solution to give 40% (v/v). The resulting precipitate is immediately collected by centrifugation and dissolved in water (2.5 ml) at room temperature. The solution should contain as much as 40 mg protein per ml, and is cooled in an ice bath. During standing for several min β-amylase crystallizes out in a plate shape. The crystals are collected by centrifugation (3000×g) and washed by centrifugation in ice-cold water (2.5 ml). The washed crystals are resuspended in cold water (2.5 ml). Crystalline enzyme (60 mg) with a specific activity of 2500 U/mg protein (37°C) is obtained in a yield of 19% from the crude extract. The crystalline enzyme is free from α-amylase and maltase activities, and can be stored at 4°C for 3 months without appreciable loss of activity. This preparation is easily recrystallized from $(NH_4)_2SO_4$ solution by the method of Balls *et al.* (3). A purification procedure on ion-exchange chromatographies was recently reported (32).

Assay Method (23) : To 0.1 ml of 2% soluble starch in 50 mM acetate buffer, pH 4.8, is added 0.1 ml of the enzyme solution appropriately diluted with 50 mM acetate buffer, pH 4.8, containing 0.04% Triton X-100 or bovine serum albumin (36). Soluble starch with a low reducing power (less than 0.03 μmol of reducing sugar equivalent to maltose per 2 mg of soluble starch) should be used. After a 10 min incubation at 37°C, the reaction is terminated by adding 2 ml of 2-fold diluted Somogyi's reagent (38). This mixture is heated for 10 min in a boiling-water bath. After cooling, 1 ml of Nelson's reagent is added (39). The contents are filled up to 10 ml with water. The developed color is read with a photometer using a filter of 540 nm or with a spectrophotometer at 500 nm using maltose (0.556 μmol) as the standard. The activity is proportional to the amount of enzyme until the released maltose level reaches 0.556 μmol. One unit of the activity is defined as the enzyme amount to produce one μmol of maltose per min under the conditions described above.

(Yasuhito Takeda)

References

1. Gore, H. C. (1920) *J. Biol. Chem. 44*, 19-20
2. Baba, T., Nakama, H., Tamaru, Y. & Kono, T. (1987) *Nippon Shokuhin Kogyo Gakkaishi* (in Japanese), *34*, 249-253
3. Balls, A. K., Walden, M. K. & Thompson, R. R. (1948) *J. Biol. Chem. 173*, 9-19
4. French, D. (1960) *The Enzymes*, 2nd ed., Vo. 4, pp. 345-368 Academic Press, New York
5. Greenwood, C. T. & Milne, E. A. (1968) *Advances in Carbohydrate Chemistry 23*, 281-366
6. Thoma, J. A., Spradlin, J. E. & Dygert S. (1971) *The Enzymes*, 3rd ed., Vol, 5, pp. 115-189, Academic Press, New York
7. Takeda, Y. & Hizukuri, S. (1976) *Shokubutukoso Tampakushitu Kenkyuho* (in Japanese) pp. 438-443, Kyoritu Shuppan, Tokyo
8. Kainuma, K., Matsunaga, A., Itagawa, M. & Kobayashi, S. (1981) *J. Jap. Soc. Starch Sci.* (in Japanese) *28*, 235-240
9. Abdullah, M. & French D. (1966) *Nature 210*, p. 200
10. Hehre, E. J., Brewer, C. F. & Genghof, D. S. (1979) *J. Biol. Chem. 254*, 5942-5950
11. French, D., Levine, M. L., Norberg, E., Nordin P., Pazur J. H. & Wild, G. M. (1954) *J. Am. Chem. Soc. 76*, 2387-2390
12. Suetsugu, N., Takeo, K., Sanai, Y. & Kuge, T. (1978) *J. Biochem.* (Tokyo) *83*, 474-478

13. Weill, C. E. & Rebhahn R. (1966) *Carbohydr. Res. 3,* 242–245
14. Wing, R. E. & BeMiller, J. N. (1969) *Carbohydr. Res. 10,* 371–377
15. Takeda, Y., Hizukuri, S., Takeda, C. & Suzuki, A. (1987) *Carbohydr. Res. 165,* 139–145
16. Summer, R. & French, D. (1956) *J. Biol. Chem. 222,* 469–477
17. Takeda, Y. & Hizukuri, S. (1981) *Carbohydr. Res. 89,* 174–178
18. Kainuma, K. & French, D. (1970) *FEBS Lett. 6,* 182–186
19. Husemann, E. & Lindeman, E. (1954) *Stärke 6,* 141–148
20. Bines, B. J. & Whelan, W. J. (1960) *Chem. Ind.* (London) 997–998
21. Mayer, F. C. & Larner, J. (1959) *J. Am. Chem. Soc. 81,* 188–193
22. Thoma, J. A. & Koshland, D. E. Jr. (1960) *J. Am. Chem. Soc. 82,* 3329–3333
23. Takeda, Y. & Hizukuri, S. (1969) *Biochim. Biophy. Acta 185,* 469–471
24. Uehara, K., Mizoguchi, T. & Mannen, S. (1970) *J. Biochem.* (Tokyo) *68,* 359–367
25. Englard, S. & Singer, T. P. (1950) *J. Biol. Chem. 187,* 213–219
26. Spradlin, J. & Thoma, J. A. (1970) *J. Biol. Chem. 245,* 117–127
27. Colman, P. M. & Matthews, B. W. (1971) *J. Mol. Biol. 60,* 163–168
28. Thoma, J. A., Koshland, D. E. Jr., Rusica, J. & Baldwin, R. (1963) *Biochem. Biophys. Res. Commun. 12,* 184–188
29. Takeda, Y., Hizukuri, S. & Murakami, T. (1971) *Agric. Biol. Chem. 35,* 778–780
30. Uehara, K., Mizoguchi, T. & Mannen, S. (1970) *Seikagaku* (in Japanese) *42,* p. 500
31. Thoma, J. A., Koshland, D. E. Jr. & Shinke, R. (1965) *Biochemistry 4,* 714–722
32. Roy, F. & Hegde, M. V. (1985) *J. Chromatography 324,* 489–494
33. Shim W. M. (1984) *Ph. D. Thesis,* Chung-Ang University (Korea)
34. Hollo, J., Laszlo, E., Hoschke, A., Hawary, F. E. & Banky, B. (1982) *Stärke 34,* 304–308
35. Takeda, Y. & Hizukuri, S. (1972) *Biochim. Biophys. Acta 268,* 175–183
36. Walker, D. E. & Axelrod, B. (1979) *Arch. Biochem. Biophys. 195,* 392–395
37. Hizukuri, S., Nakahara, K., Tabata, S. & Takeda, Y. (1975) *J. Jap. Soc. Starch Sci.* (in Japanese) *22,* p. 55
38. Somogyi, M. (1952) *J. Biol. Chem. 195,* 19–23
39. Nelson, N. (1944) *J. Biol. Chem. 153,* 375–380

II.2.b. Bacterial β-Amylases (*B. cereus, B. polymyxa,* etc.)

β-Amylases occur in various plants and their enzymic properties have been made clear in detail. However, very little information has been known about the properties of bacterial β-amylase. Robyt *et al.* reported that an amylase from *B. polymyxa* liberated mainly β-maltose from starch, though its catalytic mechanism seemed to be different from those of plant β-amylases and α-amylases (1). Recently, extracellular amylases from *B. megaterium* and *B. cereus* have been identified as β-amylases from the result of measurements of the change in optical rotation of maltose released (2,3). The amylase isolated from the culture filtrates of *B. polymyxa* was also found to be a β-amylase (4,5).

Reaction : Starch + n H_2O ⟶ G2 + β-Limit dextrin

Similarly to plant β-amylases, bacterial β-amylases successively liberate β-maltose from the nonreducing ends of starch, glycogen and maltooligosaccharides.

Specificity : In contrast to plant β-amylases, bacterial β-amylases are adsorbed onto raw starch and degrade it. The degradation extent of wheat and corn starch by *B. cereus* β-amylase at 40℃ is 5% and 2%, respectively. The hydrolysis extents are much lower than those of α-amylase or glucoamylase I (6). However, the hydrolysis of raw starch of bacterial β-amylase is enhanced by pre-treatment of the starch in warm water (60℃, 2 min). In this case, the degradation extent reached about 60% at 60℃ under the standard reaction conditions (7).

Table 1. Enzymic and Proteochemical Properties of β-Amylases

	B. cereus	*B. polymyxa*	*B. megaterium*
Molecular weight	5.8×10^4	7.0, 5.8, 4.2×10^4	5.8×10^4
E (1%, 280nm)	19.7		
Optimum pH	7.0	7.5	6.5
Optimum temp. (℃)	40	45	40–55
*K*m (soluble starch, %)	0.4	?	?
Specific activity (u/μg)	13.0	?	?
Isoelectric point	8.3	8.35, 8.59	9.1
S_{20w}	4.8	?	?
Cysteine contents (11,12)	3	3	?
Free SH	1	?	?
–S–S–	1	?	?
Inactivation by PCMB	+	+	+
Reaction with DTNB (presence of 0.1% SDS)	+	?	?
Amino acid sequence of N-terminal areas (11,12)	Ala-Val-Asn-Gly-Lys-Gly-Met-Asn-Pro-	Ala-Val-Ala-Asp-Asp-Phe-Gly-Ala-Ser-Val-Met-Gly-Pro-	?
Inactivation by 2',3'-epoxypropyl α-D-Glucopyranoside (α-EPG) (13)	+	?	?

Enzymic and Proteochemical Properties : The molecular weights of bacterial β-amylases are estimated to be 5.8 × 10^4 (*B. cereus*), 5.8 × 10^4 (*B. megaterium*) and 7.0 × 10^4 (*B. polymyxa*), respectively by SDS-polyacrylamide gel electrophoresis (8). But their molecular weights determined on a Sephadex column may be slightly different due to their interaction with the Sephadex gels (9). The isoelectric points of the bacterial β-amylases are not on the acidic side like those of plant β-amylases but on the alkaline side such as pH 8.3 (*B. cereus*). The molecules of bacterial β-amylases contain some amounts of sugars (10). These are characteristic features of bacterial β-amylases in comparison with those of plant β-amylases.

Antiserum against *B. cereus* BQ10-Sl SpoII β-amylase was prepared (8). The β-amylases from other *B. cereus* mutants reacted equally with the anti-β-amylase serum. Therefore, all β-amylases from *B. cereus* were immunologically identical.

Table 2. Amino Acid Composition of Bacterial β-Amylases

	B. cereus	*B. polymyxa*
Asp	58	67
Thr	32	30
Ser	30	46
Glu	50	34
Pro	25	23
Gly	40	58
Ala	36	43
Val	30	19
Met	14	7
Ile	27	25
Leu	39	31
Tyr	28	33
Phe	21	21
Lys	40	37
His	8	7
Trp	12	13
Arg	12	8
Half-Cys	3	3
Total	505	505

Table 3. Rate Parameters for β-Amylase Hydrolysis of Maltooligosaccharides at pH 7.0 and 25℃

	Km ($\times 10^{-3}$M)	*V*max ($\times 10^{-6}$M/sec)	e_0(M)	k_0(sec^{-1})	k_0/Km (M^{-1}sec^{-1})
G3	4.61	1.75	9.76×10^{-6}	0.18	3.90×10^{1}
G4	2.90	2.84	3.38×10^{-8}	84.0	2.90×10^{4}
G5	1.31	2.65	1.39×10^{-9}	1906	1.45×10^{6}
G6	1.04	3.50	1.41×10^{-9}	2482	2.39×10^{6}
G7	0.64	2.88	1.13×10^{-9}	2549	3.98×10^{6}

The antiserum showed a precipitin line with the β-amylase from β-amylase from *B. megaterium* by the Ouchterlony technique. However, the spur was formed on the Ouchterlony plate between the line of immunoprecipitin of the β-amylase from *B. cereus* and that from *B. megaterium*. On the other hand, no immuno-reaction occurred with the β-amylase from *B. polymyxa* and those from higher plants.

The amino acid compositions of β-amylases from *B. cereus* (12) and *B. polymyxa* (11) are shown in Table 2. The number of total amino acid residues for these β-amylases is assumed as 505. It was communicated that *B. polymyxa*-β-amylase was produced in multiforms with higher molecular weight (11). Rate parameters for *B.*

cereus β-amylase hydrolysis of maltooligosaccharides are shown in Table 3 (14). Intracellular immunoprecipitates of *B. cereus* homogenates were prepared. Immunoreactive protein was detected on nitrocellulose sheet by western-blotting method. The molecular weight of the immunoreactive protein in intracellular fractions was about 12×10^4. It was also found that this protein band appeared even in the case of SDS-PAGE after the treatment of the intracellular fraction at 100°C for 10 min in a solution containing 1% SDS, 2-ME and 8 M urea. These date suggest that this protein may not be a dimer of the β-amylase. But no protein band was detected when the nitrocellulose sheet was incubated with the unsensed rabbit serum with the purified β-amylase in place of the anti-β-amylase serum. From these results, it seems that the protein has the same immunological reactive site as β-amylase. It is possible that the protein is a precursor-like protein of β-amylase within the cell (8). Recently, Udaka and co-workers have isolated extracellular β-amylases with larger molecular weight (about 16×10^4). They have also suggested that the larger β-amylases were then proteolyzed to form mature β-amylase and that all exo-β-amylases were products of a single β-amylase gene (11). These data support the existence of a preform of bacterial β-amylase with high molecular weight in the cell.

Production and Purification : *B. cereus* BQ10-S1 was isolated after UV irradiation of a wild-type strain BQ10 (15). Strain BQ10-S1 secretes about 10-fold more β-amylase (about 300 units/ml) than its parent does. The enzyme production of strain BQ10-S1 is repressed by the addition of maltose to the medium. The β-amylase production of strain BQ10-S1 was observed to cease at the sporeforming stage. Isolation of an asporogenous mutant (*B. cereus* BQ10-S1 SpoⅡ) was performed (8,16). This rifampsin resistant mutant was capable of producing four-fold more β-amylase than strain BQ10-S1, reaching approximately 1,200 units/ml (8). However, as in the case of strain BQ10-S1, β-amylase production of strain BQ10-S1 SpoⅡ was observed to decrease in the presence of glucose or maltose in the medium and was not increased by addition of soluble starch to the culture medium. A further attempt was made to isolate rifampsin resistant mutants of strain BQ10-S1 capable of secreting much larger amounts of β-amylase even in the presence of glucose, maltose, or soluble starch. Such a strain *B. cereus* BQ10-S1 SpoⅢ was successfully isolated. The amylase production of strain BQ10-S1 SpoⅢ markedly increased to the highest level (about 6,700 units/ ml) when the concentration of soluble starch was 2%. The amounts of secreted β-amylase were determined by an immunological method, and it was found that strain BQ10-S1 SpoⅢ secreted about 382 μg/ml of β-amylase protein (Table 4) (17).

β-Amylase produced by *B. cereus* was purified by salting out with ammonium sulfate, and column chromatography on Sephadex G-100 and CM-Sephadex C-50. The β-amylase adsorbed to the CM-Sephadex C-50 equilibrated with 50mM acetate buffer (pH 5.0) was separated by NaCl gradient elution (9). β-Amylase from *B. megaterium* was purified in a similar way, except for the use of SE-Sephadex instead of the CM-Sephadex (2). In the case of *B. polymyxa* β-amylase, ion-exchange chromatography and gel filtration, as well as adsorption on corn starch was used in purification procedures. Two isozymes of *B. polymyxa* β-amylases were separated by isoelectric focusing (5).

Assay Method : β-Amylase activity is usually estimated by measuring the amounts of reducing ends of maltose formed. In general, the assay is carried out by using 3,5-dinitrosalicylic acid which reacts with a reducing end of the maltose to form a colored-complex (Bernfeld method). The protocol for the assay of the β-amylase in our laboratory is as follows ; the enzyme solution (0.25 ml) is added to 0.25 ml of substrate solution (1% soluble starch). After incubation for 3 min at 40°C, the reaction is stopped by adding 0.5ml of 3,5 DNS reagent. The reaction mixture is heated at 100°C for 10 min and cooled. Then, 5 ml of water is added. The optical density is measured at 530nm. The enzyme activity producing one mg maltose in the reaction mixture at 40°C for 60 min was expressed as one unit.

Table 4. Growth and Production of Exo-β-Amylase of Strain BQ10-S1 SpoIII and Its Parental Strains

Strain	Soluble starch	β-Amylase characteristics			
		Growth (A_{660})	Amount (μg/ml)	Activity (u/ml)	Sp. act. (u/μg)
BQ10-S1	+	31	23	250	10.9
	−	30	25	300	12.0
BQ10-S1 SpoII	+	20	115	1,500	13.0
	−	22	95	1,200	12.6
BQ10-S1 SpoIII	+	31	385	6,700	17.5
	−	23	152	1,800	11.4

Each strain was cultured in basal medium (3% Polypepton, 1% meat extract, pH 7.2) with (+) or without (−) 2% soluble starch. Amount ; amount of β-amylase. Sp. act. ; specific activity of β-amylase.

(Takashi Nanmori)

References

1. Robyt, J. & French, D. (1964) *Arch. Biochim. Biophys. 104*, 338-345
2. Higashihara, M. & Okada, S. (1974) *Agric. Biol. Chem. 38*, 1023-1029
3. Shinke, R., Nishira, H. & Mugibayashi, N. (1974) *Agric. Biol. Chem. 38*, 665-667
4. Marschall, J. J. (1974) *FEBS Lett. 46*, 1-4
5. Murao, S., Ohyama, K. & Arai, M. (1979) *Agric. Biol. Chem. 43*, 719-726
6. Nanmori, T. (1983) *Doctoral Dissertation* 17-18
7. Shinsuke, M., Higashihara, M. & Okada, S. (1986) *J. Jpn. Soc. Starch Sci.* (in Japanese) *33*, 238-243
8. Nanmori, T., Shinke, R., Nakano, S., Kitaoka, S. & Nishira, H. (1985) *Appl. Microbiol. Biotechnol. 21*, 383-389
9. Nanmori, T., Shinke, R., Aoki, K., & Nishira, H. (1983) *Agric. Biol. Chem. 47*,

941-947
10. Hosoe, T., Nanmori, T., Aoki, K. & Shinke, R. (1985) *Abstracts of Papers, Annual Meeting of Agric. Chem. Soc. of Japan* p. 366
11. Kawazu, T., Nakanishi, Y., Uozumi, N., Sasaki, T., Yamagata, H., Tsukagoshi, N. & Udaka, S. (1987) *J. Bacteriol. 169,* 1564-1570
12. Nanmori, T., Yoneda, I., Shinke, R., Mikami, B., Nomura, K. & Morita, Y. *Abstracts of Joint Meeting of Agric. Chem. Soc. of Kansai and Nishinihon Branches* p. 70 (1987)
13. Isoda, K., Takeo, K. & Nitta, Y. *Seikagaku 58,* p. 1057 (1986)
14. Hosoe, T., Yoneda, Y., Shinke, R., Aoki, K., Nanmori, T., Ohnishi, S., Hiromi, K., *Abstracts of Papers, Annual Meeting of Agric. Chem. Soc. of Japan* p. 656 (1986)
15. Shinke, R., Kunimi, Y., Aoki, K. & Nishira, H. (1977) *J. Ferm. Technol.,* 55, 103-109
16. Nanmori, T., Shinke, R., Aoki, K., & Nishira, H. (1983) *Agric. Biol. Chem.,* 47, 609-611
17. Nanmori, T., Numata, Y. & Shinke, R. (1987) *Appl. Environ. Microbiol.,* 55, 768-771

II.2.c Amino Acid Sequence and Three Dimensional Structure of β-Amylase

Recently, the amino acid sequences of β-amylases from soybean (1) and *Bacillus polymyxa* (2) have been deduced from the base sequence of their genes. The amino acid sequencing of the purified sweet potato β-amylase is also under way (3). Though the plant type β-amylase differs from the bacterial enzyme in the enzymic properties, both of them are inactivated by SH-reagents such as iodoacetamide, *p*-chloromercuribenzoic acid and N-ethylmaleimide. Chemical modification study of soybean β-amylase elucidated the mechanism of its inactivation (4,5). The modified SH groups were positioned on the amino acid sequence (6,7). In contrast to the development of determination of amino acid sequences, no detailed knowledge about the three dimensional structure of β-amylase is available except for the low resolution X-ray crystallographic study of soybean enzyme (8). The molecule of soybean β-amylase is composed of a large and a small domain separated by a cleft.

Amino acid sequence : Soybean β-amylase cDNA was isolated from soybean cotyledonary cDNA clones (1). The sequence contains an open reading frame of 1,485 nucleotides encoding a polypeptide composed of 496 amino acid residues. The N-terminal sequence analysis of soybean β-amylase isozyme 2 (9) revealed that Ala next to the initiation Met is the N-terminal amino acid which was acetylated in the mature enzyme by post or co-translational process. The C-terminal amino acid, Gly (10), and the partial amino acid sequence around 6 cysteine residues (7) coincided with the deduced amino acid sequence. The calculated amino acid composition and molecular weight (56 kDa) were also coincident with those obtained from the protein analyses (11). The gene encoding β-amylase of *Bacillus polymyxa* was cloned and sequenced by Kawazu *et al.* (2). The sequence determined (3.1 kilobase) contained an open reading frame of 2,808 nucleotides without any translational stop codon. The deduced amino acid sequence was composed of 936 amino acid residues containing a signal sequence of 33 or 35 residues at the N-terminal end. It was suggested that the precursor β-amylase larger than 100 kDa is hydrolyzed to 70, 56 and 42 kDa proteins by a proteolytic process. As these three β-amylases process the same N-terminal sequence, their C-termini are proposed to be the processing sites. Comparison of the deduced amino acid sequences of soybean and *Bacillus polymyxa* β-amylases revealed a marked similarity in their N-terminal regions (residues 1–495 of soybean enzyme and residues 1–458 of the bacterial enzyme) as shown in Fig. 1 (1). The amino acid identity on the alignable sequences reaches 32%. As soybean and *Bacillus polymyxa* β-amylases are evolutionally far related, they have different isoelectric points, pH optima and the hydrolytic abilities of raw starch granules as described in sections Ⅱ.2.a and b. Thus, the highly homologous regions in the two sequences boxed in Fig. 1 must be important for the activity of β-amylase.

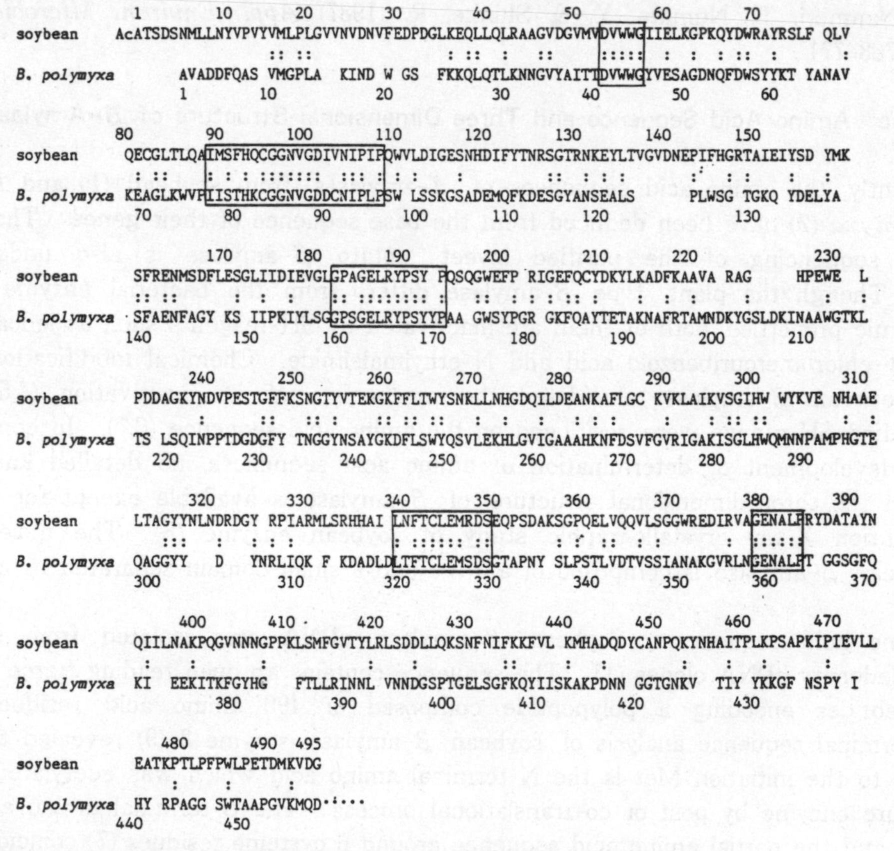

Fig. 1. Comparison of Amino Acid Sequences of Soybean and *Bacillus polymyxa* β-Amylases.

Chemical modification of sulfhydryl groups : Chemical modification of sulfhydryl (SH) groups in soybean β-amylase revealed that the second reactive SH group (SH2) is responsible for the inactivation by SH-reagents (4).

SH2 has been identified as Cys_{95} and the most reactive residue (SH1) as Cys_{448} (1,7). As to the role of SH groups, Spradlin and Thoma (12) first concluded that SH groups in sweet potato β-amylase do not participate in the catalysis of the enzyme.

Later, Mikami *et al.* (4) demonstrated that cyanilation of SH2 in DTNB ((5,5'-dithiobis-)2-nitrobenzoic acid) - modified soybean β-amylase resulted in restoration of enzymic activity, and that the inactivation depends upon the size of substituents at SH2, indicating that SH2 is not directly concerned with the catalysis or substrate binding.

The interaction of soybean β-amylase and maltose or cyclohexaamylose is not affected by the modification of SH2 with iodoacetamide and DTNB (13). In contrast, the binding of glucose is strongly affected by the modification of SH2 (5,14). Table 1 represents the effect of modification of SH2 on the activity and the dissociation constant (K_d) of the enzyme-glucose complex. The decrease of *V*max and the simultaneous increase of K_d values depend upon the size of substituents at SH2, but the *K*m values are almost constant. It is proposed from the subsite structure of soybean β-amylase (15) that both the position of SH2 and the glucose binding site are around subsite 1, where the nonreducing ends of the substrate bind productively.

Table 1. Effect of Modification of SH2 in Soybean β-Amylase on the Enzyme Activity and Binding of Glucose

β-Amylase derivatives	*V*max[a] (%)	*K*m[a] (mg/ml)	*K*d for glucose[b] complex (mM)
Native	100	2.3	45
Cyanide-modified	65	2.3	85
MDPS[c]-modified	9	2.6	540
Iodoacetamide-modified	2	2.7	>400

[a] Measured at pH 5.4 using amylopectin,
[b] measured at pH 8 by difference spectrum method,
[c] methyl 2,4-dinitrophenyl disulfide.

These results of chemical modification suggest that SH2 (Cys_{95}) is not essential for the activity of soybean β-amylase but that the proper conformation around the SH group is vital for the productive binding of substrate. As shown in Fig. 1, the sequence around Cys_{95} of soybean β-amylase has strong similarity to the corresponding region of *Bacillus polymyxa* enzyme. The cysteine residue corresponds to Cys_{83} of the bacterial enzyme. Figure 1 also shows another concerned region around Cys_{343} of soybean enzyme and Cys_{323} of the bacterial enzyme. Our recent results of chemical modification of SH group in soybean enzyme have also suggested the participation of Cys_{343} in the inactivation caused by N-ethylmaleimide and *p*-chloromercuribenzoic acid. The exact role of the cysteine residues is not known but the sequences around Cys_{95} and Cys_{343} must be one of the candidates for the active center of this enzyme. As to the catalytic residues of β-amylase, carboxyl, amino or imidazolium groups are proposed to be concerned in the catalytic action of this enzyme (16). Recently, Isoda and Nitta (17) suggest the participation of carboxyl group in the catalysis of β-amylase by using 2', 3'-epoxypropyl α-D-glucopyranoside as a proposed affinity labeling reagent.

X-ray crystallography : Preliminary X-ray crystallographic data of β-amylase was first obtained by Colman and Matthews (18) from tetragonal crystals of sweet potato enzyme. Later, Morita *et al.* used the crystals of β-amylase from Japanese radish (19) and soybean (11,20). These crystallographic data are summarized in Table 2. Soybean β-amylase was crystallized from ammonium sulfate solution in two forms, trigonal at pH 5.4 and hexagonal at pH 4. The latter crystal was first prepared by Fukumoto and Tsujisaka (21) in 1954.

Table 2. Crystallographic Data of β-Amylases

	Soybean		Japanese radish	Sweet potato
Conditions				
$(NH_4)_2SO_4$ saturation	45%	40%	40%	60%
pH	5.4	4.0	6.0	4.0
temperature	4℃	4℃	4℃	cold
Crystal structure	trigonal	hexagonal	hexagonal	tetragonal
Space group	$P3_121$ or $P3_221$	$P6_122$	$P6_1$	$P4_122$ or $P4_322$
Unit cell				
a (Å)	86.1	106	67	210.7
c (Å)	144.4	226	208	157.0
V ($Å^3 \times 10^{-6}$)	0.927	2.20	0.813	6.79
Molecular weight	57,000		58,000	206,000
Mass/asymmetric unit	116,000	138,000	102,000	676,000
Molecule/asymmetric unit	1	1	1	1
Protein content (%)	49	41	57	29

Aibara *et al.* (8) elucidated the molecular structure of soybean β-amylase at 4.5 Å resolution by the isomorphous replacement method using the trigonal crystals. Figure 2 shows the shape of the enzyme molecule and the mercury binding sites. The molecule appears to be composed of a large (55 × 50 × 45 Å) and a small (35 × 25 × 20 Å) domain. The mercury (*p*-chloromercuribenzene sulfonate) binds at two sites, one on the surface of the larger domain and the other, on the smaller domain near a cleft formed between the two domains. These two mercury binding sites are considered to be SH1 and SH2 on the basis of the specificity and the occupancy of the mercurial reagent. In order to follow the amino acid sequence on the molecular structure of β-amylase, a high resolution X-ray study is now in progress using the hexagonal crystals of soybean β-amylase.

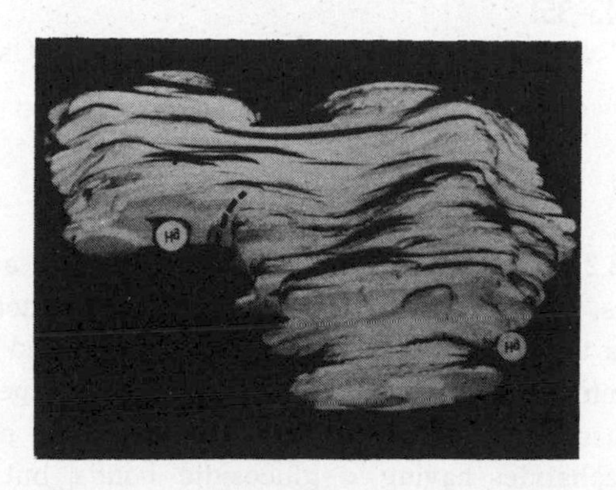

Fig. 2. A Balsa Wood Model of Soybean β-Amylase Structure at 5 Å Resolution.
The model is shown as viewed along the a-axis. The dotted line indicates the cleft. The two mercurial sites are indicated by small spheres.

(Bunzo Mikami, and Yuhei Morita)

References

1. Fukazawa, C., Kikuchi, T., Totsuka, A., Mikami, B., Nomura, K. & Morita, Y. (1988) in preparation
2. Kawazu, T., Nakanishi, Y., Uozumi, N., Sasaki, T., Yamagata, H., Tsukagoshi, N. & Udaka, S. (1987) *J. Bacteriol. 169*, 1564-1570
3. Toda, H. personal communication
4. Mikami, B., Aibara, S. & Morita, Y. (1980) *J. Biochem. 88*, 103-111
5. Nomura, K., Mikami, B., Nagao, Y. & Morita, Y. (1987) *J. Biochem. 102*, 333-340
6. Mikami, B. & Morita, Y. (1983) *J. Biochem. 93*, 777-786
7. Nomura, K., Mikami, B. & Morita, Y. (1987) *J. Biochem. 102*, 341-349
8. Aibara, S., Yamashita, H. & Morita, Y. (1984) *Agric. Biol. Chem. 48*, 1575-1579
9. Mikami, B., Nomura, K. & Morita, Y. (1986) *J. Biochem. 100*, 513-516
10. Mikami, B., Aibara, S. & Morita, Y. (1982) *Agric. Biol. Chem. 46*, 943-953
11. Morita, Y., Yagi, F., Aibara, S. & Yamashita, H. (1976) *J. Biochem. 79*, 591-603
12. Spradlin, J. E. & Thoma, J. A. (1970) *J. Biol. Chem. 245*, 117-127
13. Mikami, B., Nomura, K. & Morita, Y. (1980) *J. Biochem. 94*, 107-113
14. Nomura, K., Mikami, B. & Morita, Y. (1986) *J. Biochem. 100*, 1175-1183
15. Suganuma, T., Ohnishi, M., Hiromi, K. & Morita, Y. (1980) *Agric. Biol. Chem.*

44, 1111-1117
16. Nitta, Y., Kunitaka, T. & Watanabe, T. (1979) *J. Biochem.* *85*, 41-45
17. Isoda, Y. & Nitta, Y. (1986) *J. Biochem.* *99*, 1631-1637
18. Colman, P. M. & Matthews, B. W. (1971) *J. Mol. Biol.* *60*, 162-168.
19. Aibara, S., Yamashita, H. & Morita, Y. (1978) *Agric. Biol. Chem.* *42*, 179-180
20. Morita, Y., Aibara, S., Yamashita, H., Yagi, F., Suganuma, T. & Hiromi, K. (1975) *J. Biochem.* *77*, 343-351
21. Fukumoto, J. & Tsujisaka, Y. (1954) *Kagaku to Kogyo*, Osaka (in Japanese) *28*, 282-287

II.3. α-Glucosidases

α-Glucosidases (EC 3.2.1.20, α-D-glucoside glucohydrolase) are a group of typical exo-type carbohydrases, which release α-glucose from the nonreducing end side of the substrate. Various types of α-glucosidases are distributed widely in microorganisms, plants and animal tissues (1). Their substrate specificities are greatly diverse. Many α-glucosidases are capable of hydrolyzing not only synthetic α-glucosides and oligosaccharides having α-glucosidic bonds, but also α-glucans such as soluble starch and glycogen. Maltose, soluble starch and glycogen are degraded at a single active site in plant and mammalian acid α-glucosidases (2,3). However, only mammalian intestinal α-glucosidase consists of two α-glucosidase components. They are synthesized as a polypeptide and referred to as sucrase and isomaltase having a respective active site. Recently the primary structure of the sucrase-isomaltase complex was demonstrated (4). Oligo-1,6-glucosidase (EC 3.2.1.10) and sucrose α-glucosidase (EC 3.2.1.48) assigned in Enzyme Nomenclature (1984) also come into the category of α-glucosidases.

α-Glucosidase has often been called transglucosidase, because it catalyzes transglucosylation as well as hydrolysis. The reactions are shown as the following schemes :

$$\text{G-O-R} + \text{H-OH} \xrightleftharpoons{\text{Hydrolysis}} \text{G-OH} + \text{H-O-R}$$

$$\text{G-O-R} + \text{H-OA} \xrightleftharpoons{\text{Transglucosylation}} \text{G-OA} + \text{H-O-R}$$

where G, R and H-OA represent the glucosyl residue, aglycone and acceptor, respectively. Both hydrolysis and transglucosylation are exchange reactions between glucosyl residues and protons of water or acceptor. In these reactions, the glucose released is α-anomer (5) and the configuration of anomeric carbon of the substrate is retained in the transglucosylation product. Therefore, α-glucosidase is distinguished from glucoamylase which produces β-glucose.

α-Glucosidases are conventionally classified into the following three types, based on their substrate specificities (6). The first group comprises typical α-glucosidases which hydrolyze heterogeneous substrates such as phenyl α-glucoside and sucrose more rapidly than maltose. The second group comprises so-called maltases, which show a high activity toward homogeneous substrates such as maltooligosaccharides with feeble or no activity toward synthetic α-glucosides and sucrose. The third group comprises α-glucosidases capable of attacking also α-glucans, but the substrate specificity only differs from that of the second group in that the former type of enzyme exhibits activity toward α-glucan.

(Seiya Chiba)

References

1. Gottschalk, A. *The Enzymes* edited by 1st ED., Academic Press, 1950, Vol. 1, p 551-582
2. Chiba, S., Hibi, N., Kanaya, K. & Shimomura, T. (1977) *Agric. Biol. Chem. 41,* 1245-1248
3. Matsui, H., Sasaki, M., Takemasa, E., Kaneta, T. & Chiba, S. (1984) *J. Biochem. 96,* 993-1004
4. Hunziker, W., Spless, M., Semenza, G. & Lodish, H. F. (1986) *Cell 46,* 227-234
5. Chiba, S., Kimura, A. & Matsui, H. (1983) *Agric. Biol. Chem. 47,* 1741-1746
6. Chiba, S. & Shimomura, T. (1978) *J. Jap. Soc. Starch Sci.* (in Japanese) *25,* 105-112

II.3.a Mammalian α-Glucosidases (Mammalian Blood Serum and Human Urine)

α-Glucosidases are widely distributed in mammalian tissues and body fluids and classified into two main types from their pH optima: one is acid α-glucosidase with an optimal pH of 4 to 5 and the other, neutral α-glucosidase with an optimal pH of 6 to 7. Acid α-glucosidase is localized in lysosome of mammalian tissue. Since Hers (1) reported in 1963 that the patients with a certain type of a glycogen storage disease (Pompe's disease) lack this enzyme in the tissue and that the enzyme plays an important physiological role in metabolism of glycogen, a large number of researchers have studied the enzymic properties of the enzyme purified from various kinds of mammalian tissues. On the other hand, the physiological role of neutral α-glucosidase remains unclarified, though the enzyme exists in many mammalian tissues. This neutral α-glucosidase is also found in various mammalian blood sera but there is an extremely wide variety among enzyme amounts in sera of several mammalian species (2), as shown in Table 1. In particular, swine blood serum contains a large amount of neutral α-glucosidase. Swine α-glucosidase was isolated in a highly purified state (3) and its properties investigated in detail (4,5). Human urine contains at least three different types of α-glucosidase classified from their optimal pH and molecular weight (6). One of them, the one with the highest molecular weight, is shown to originate exclusively from kidney and to localize in the proximal convoluted portion and Henle's loop (7).

Table 1. α-Glucosidase Activities in Serum and Plasma of Several Mammals

Mammals	α-Glucosidase activity (mu/ml)	
	Serum	Plasma
Human	1.3	0.8
Horse	2,556	—
Cattle	3,689	2,800
Sheep	586	254
Swine	24,591	19,400
Dog	6,589	—
Cat	11	18
Guinea pig	2	0.9
Hamster	2,877	2,107
Rat	1,815	1,243
Rabbit	8	—

(Yamamoto, T. (2)).

Specificity : The specificity of neutral α-glucosidase of swine blood serum was detailed by Chiba *et al.* (4). The enzyme showed a broad specificity on various glucans having not only α-1,4-glucosidic linkages but also other types of glucosidic linkages. The enzyme hydrolyzes the α-glucosidic linkages of maltose and phenyl-α-maltoside at a rapid rate but has very little or no effect on phenyl-α-glucoside. This enzyme also hydrolyzes the α-1,2-glucosidic linkages of kojibiose and the α-1,3-glucosidic linkages of nigerose at a slower rate than it does the α-1,4-glucosidic linkages. The hydrolysis rate of the α-1,6-glucosidic linkages in isomaltose and panose is very slow (Fig. 1). The enzyme degrades several α-glucans as well as maltooligosaccharides to produce α-glucose. The extent of hydrolysis to α-glucans is about 80% for soluble starch, amylopectin and amylose, and about 50% for glycogen and β-limit dextrins (4).

Human urine α-glucosidase contains three different types of enzyme (F-1, F-2 and F-3). Each type hydrolyzes the α-1,4-glucosidic linkages of maltose and maltotriitol to produce α-glucose, but has very little effect on maltitol which is the hydrolysate of maltotriitol by the enzyme (8). The action of urine α-glucosidases on maltooligosaccharides has not been made clear in detail. All the enzymes are more active in hydrolyzing maltose than in hydrolyzing starch, though the ratios at an initial reaction velocity of maltose hydrolysis activity to that of starch hydrolysis activity is 9.0, 4.5 and 4.5 for F-1, F-2 and F-3, respectively. Although they also degrade β-limit dextrins, F-3 shows the highest activity and its extent of hydrolysis of β-limit dextrin is about 70%, suggesting that F-3 α-glucosidase hydrolyzes the α-1,6-glucosidic linkages of β-limit dextrin (6). Mammalian acid α-glucosidase is a higher than neutral α-glucosidase in the activity of hydrolyzing α-glucan. Table 2 presents a substrate specificity of an acid α-glucosidase from rabbit muscle (9).

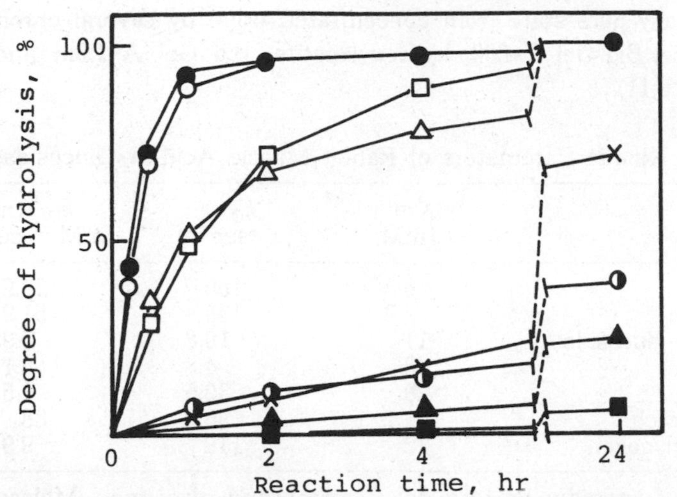

Fig. 1. Hydrolysis of Several Oligosaccharides with Swine Serum Neutral α-Glucosidase.

Substrate : ●, maltose; ○, phenyl-α-maltoside; △, nigerose; □, kojibiose; ◑, panose; ×, turanose ; ▲, isomaltose; ■, phenyl-α-glucoside. (Chiba *et al.* (4))

Purification : The neutral α-glucosidase of swine serum is purified by the method of Hibi *et al.* (3). The precipitate obtained from swine serum by ammonium sulfate fractionation (30 to 80% satn.) is dialyzed against Briton-Robinson buffer (used after dilution four times with deionized water, pH 6.0). The dialysate is centrifuged to remove insoluble materials and the supernatant is applied to a column of DEAE-Cellulose buffered with the same buffer. After the column is washed with Briton-Robinson buffer doubly diluted with water, pH 4.0, the α-glucosidase adsorbed is eluted with a linear gradient of NaCl in Briton-Robinson buffer of pH 6.0. The active fractions are combined and dialyzed against M/20 acetate buffer of pH 6.0. The dialysate is chromatographed on DEAE-Sephadex A-50 buffered with an acetate buffer of pH 5.6. The enzyme adsorbed on the column is eluted with a linear gradient of NaCl in the same buffer. The active fractions are combined and concentrated. The concentrate is chromatographed on Bio-Gel P-300 and subsequently on Sephadex G-200 in M/100 acetate buffer of pH 5.6. Neutral α-glucosidase is obtained in a highly purified state with a yield of about 60% of the activity.

One of the three types of human urine α-glucosidases (F-3), which have the lowest molecular weight originating to lysosome, is efficiently purified from a concentrate of human urine in a highly purified state by the purification procedure using two affinity chromatographies on Con A-Sepharose (adsorption ; M/20 Na-phosphate buffer containing M/2 NaCl, pH 6.5 ; elution, 1 M methyl-α-D-glucoside) and subsequently on Sephadex G-200 with which the enzyme has a weak affinity (M/50 Na-acetate buffer containing M/40 NaCl and M/1000 EDTA, pH 5.2) (10). α-Glucosidase (F-1) with the highest molecular weight which originates to kidney is purified in a electrophoretically pure state from concentrated urine by several chromatographies on DEAE-Cellulose, Bio-Gel P-200, hydroxylapatite, Bio Gel A-1.5m and finally heated glutinous rice (6,11).

Table 2. Kinetic Parameters of Rabbit Muscle Acid α-Glucosidase

Substrate	*K*m (mM)	k_O (sec^{-1})	k_O/*K*m (mM^{-1} sec^{-1})
Maltose	6.3	180	28.6
Maltotriose	2.6	135	51.9
Phenyl α-glucoside	11	10.8	0.98
Isomaltose	48	4.8	0.1
Panose	20	30.5	1.5
Soluble starch	1.1	96.9	88
Shellfish glycogen	12	119	9.9

*K*m, mM of nonreducing terminal in starch and glycogen. Molecular activity k_O, expressed for the cleavage of α-glucosidic linkage of nonreducing terminal. Reaction, 37℃.

Table 3. Some Enzymic and Proteochemical Properties of α-Glucosidases of Swine Blood Serum and Human Urine

	Swine serum	Human urine		
	neutral	F-1	F-2	F-3
Molecular weight (×10^5)	2.7	3.3	1.7	0.7
Isoelectric point	4.0	4.2	4.4	4.6
E (1%,280 nm)	8.1	—	—	—
$S_{20,w}$	10.7s	—	—	—
pH Stability	6-9	5-8	5-8	5-8
Thermal stability (℃)	55	53	45	53
Optimum pH				
(maltose)	7.0	5.6	6.2	5.0
(starch)	7.0	5.6	6.2	5.0
*K*m				
(maltose, mM)	2.1	0.84	0.65	0.65
(maltotriose, mM)	0.28	—	—	—
(maltotriitol, mM)	—	0.25	0.25	0.42
(starch, mg/ml)	9.8	1.1	0.56	0.63
(glycogen, mg/ml)	55.6	—	—	—
Limit hydrolysis (%)				
(starch)	100	64	70	83
(glycogen)	60	50	64	71
(β-limit dextrin)	70	30	31	68

Assay Method : Several substrates such as maltose, starch, glycogen (determination of reducing power), *p*-nitrophenyl-α-D-glucoside (colorimetry), and 4-methyl umbelliferyl-α-D-glucoside (fluorometry) are available, but maltotriitol is considered to be the most convenient substrate for assay of α-glucosidases of mammalian blood serum and human urine (8), because this substrate is not affected by the presence of α-amylase. One ml of the enzyme solution is incubated with 1 ml of 0.5% maltotriitol in M/20 buffer (phosphate buffer of pH 7.0 for swine blood serum enzyme; acetate buffer of pH 5.6 for urine F-1 and acetate buffer of pH 4.5 for urine F-3). After 5 to 10 min incubation at 37℃, the glucose liberated is determined by the Somogyi-Nelson method. One unit of enzyme activity is defined as the amount of enzyme which produces 1 μmol of glucose per min under the above conditions.

(Noshi Minamiura)

References

1. Hers, H. G. (1963) *Biochem. J. 86,* 11-16
2. Yamamoto, T. (1978) *J. Jap. Soc. Starch Sci.* (in Japanese) *25,* 79-81
3. Hibi, N., Chiba, S. & Shimomura, T. (1976) *Agric. Biol. Chem. 40,* 1805-1812
4. Chiba, S., Hibi, N. & Shimomura, T. (1976) *Agric. Biol. Chem. 40,* 1813-1817
5. Matsui, H., Yamada, T., Someya, Y. & Chiba, S. (1983) *Agric. Biol. Chem. 47,* 1817-1822
6. Minamiura, N., Chiura, H., Tsujino, K. & Yamamoto, T. (1975) *J. Biochem. 77,* 1015-1022
7. Nishinaka, H., Minamiura, N., Furusawa, M. & Yamamoto, T. (1982) *J. Hitochem. Cytochem. 30,* 1186-1189
8. Tsujino, K., Kida, J., Kano, K., Minamiura, N. & Yamamoto, T. (1976) *Jap. J. Clin. Chem. 5,*98-102
9. Matsui, H., Sasaki, M., Takemasa, E., Kaneta, T. & Chiba, S. (1984) *J. Biochem. 96,* 993-1004
10. Oudeelferink, R. P. J., Brouwer-Kelder, E. M., Surya, I., Strijland, A., Kroos, M., Reuser, A. J. J. & Tager J. M. (1984) *Eur. J. Biochem. 139,* 489-495
11. Minamiura, N., Matoba, K., Nishinaka, H. & Yamamoto, T. (1982) *J. Biochem. 91,* 809-816

II.3.b. Plant α-Glucosidases

α-Glucosidases occur widely in plant tissues (1), seeds, fruits, leaves and roots. The seeds of rice, corn, buckwheat, millet, barley, and sugar beet contain a relatively large amount of α-glucosidase, but in beans this enzyme is observed only at the germination stage. In general, plant α-glucosidase hydrolyzes not only oligosaccharides but also α-glucans. In germinated seeds α- and β-amylases produce oligosaccharides and limit dextrins by degrading starch. α-Glucosidase may hydrolyze these products to glucose. α-Glucosidase can't degrade raw starch, though the enzyme shows a high activity toward soluble starch. Plant α-glucosidases have been obtained in the homogeneously purified state from the seeds of rice (2), corn (3), buckwheat (4) and sugar beet (5).

Specificity : Sugar beet α-glucosidase has a wide substrate specificity and the highest hydrolytic activity toward soluble starch among the α-glucosidases so far reported.

Table 1. Kinetic Parameters of α-Glucosidases from Seeds of Buckwheat (6) and Sugar Beet (7)

Substrate	Buckweat			Sugar beet		
	*K*m (mM)	k_0 (sec^{-1})	k_0/Km ($mM^{-1} \cdot sec^{-1}$)	*K*m (mM)	k_0 (sec^{-1})	k_0/Km ($mM^{-1} \cdot sec^{-1}$)
G2	6.3	123	19.5	20	149	7.25
IG2	74	2.5	0.03	11	33.7	3.1
NG	12	124	10.3	17	195	11.5
KJ	3.2	21	6.6	1.3	16	12.3
φG	40	4.9	0.12	1.7	5.4	3.2
φM	4.9	130	27.7	4.1	83.4	20.3
G3	4.0	136	34.0	3.7	135	36.5
SS	1.1	103	97.2	0.27	179	663

*K*m, mM of nonreducing terminal in SS; k_0, molecular activity, expressed as velocity for cleavage of α-glucosidic linkage of nonreducing terminal. G2, maltose; IG2, isomaltose; NG, nigerose; KJ, kojibiose; φG, phenyl α-glucoside; φM, phenyl α-maltoside; G3, maltotriose; SS, soluble starch. Reaction, 37℃.

The activity toward isomaltose of plant α-glucosidase is generally modest with the exception of sugar beet α-glucosidase (Table 1). However, the velocity for the cleavage of α-1,6-glucosidic linkage of panose is greater than that of isomaltose. Little activity is seen toward isomaltooligosaccharides of 3 or more polymerization degree.

Buckwheat α-glucosidase catalyzes a characteristic transglucosylation, that is, predominant formation of α-1,2-, α-1,3- or α-1,4-glucosidic linkage (8) in the transglucosylation. This enzyme synthesizes kojibiose, nigerose and maltose from soluble starch. Besides these disaccharides 2,4-di-α-glucosyl-glucose and 4-α-nigerosyl-glucose are produced from maltose. This enzyme, therefore, is useful for the preparation of kojibiose and nigerose. Fig. 1 shows the time course for formation of disaccharides from soluble starch (8).

Purification : Sugar beet α-glucosidase —— Powderized ungerminated seeds are suspended in 0.1 M sodium acetate solution. The precipitation obtained from the extract by salting-out with ammonium sulfate up to 0.9 saturation is dialyzed and then subjected to chromatography with CM-cellulose (0.01 M sodium acetate buffer, pH

5.4). The α-glucosidase is homogeneously purified by twice repeated gel chromatographies on Bio-Gel P-150 in a 30% yield (5).

Buckwheat α-glucosidase —— The extract with 0.1 M sodium acetate buffer (pH 5.0) from powderized ungerminated seeds is viscous and this viscous material is precipitated with 1% rivanol (6,9-diamino-ethoxyacridinium lactate). The rivanol remaining in the supernatant is removed by addition of a least amount of active carbon. The enzyme in the decolored solution is precipitated with ammonium sulfate. The precipitate obtained between 0.3 and 0.7 saturation is dialyzed and then subjected to chromatography with CM-Cellulose (0.02 M sodium phosphate buffer, pH 6.8). The enzyme is isolated in a highly purified state by gel chromatography on Bio-Gel P-150 with an activity recovery of 27% (4).

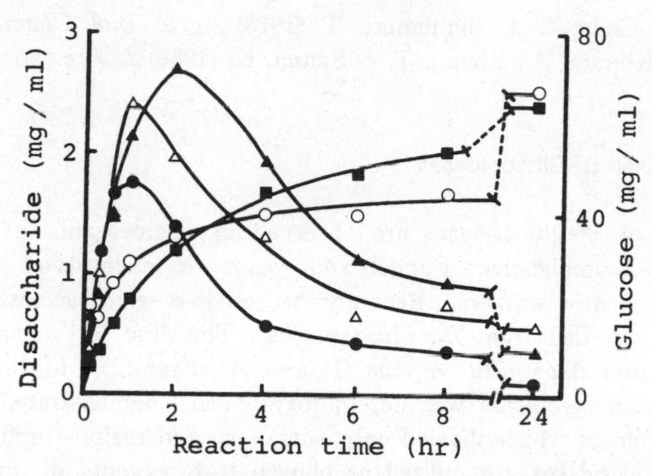

Fig. 1. Time Course of Formation of Transglucosylation-Oligosaccharides from Soluble Starch (10%) by Buckwheat α-Glucosidase.
Ordinate: mg of each product per ml of reaction mixture; ●, maltose; ○, isomaltose; ▲, kojibiose; △, nigerose; ■, glucose.

Table 2. Enzymic and Proteochemical Properties of α-Glucosidases from Seeds

Source of α-glucosidase	Molecular weight	Optimum pH
Buckwheat	88,000	5.0
Sugar beet	91,000	4.5
Rice (II)	100,000	4.0
Flint corn	65,000	3.6

(Seiya Chiba)

References

1. Hutson, D. H. & Manners, D. J. (1965) *Biochem. J. 94*, 783–789
2. Takahashi, N., Shimomura, T. & Chiba, S. (1971) *Agric. Biol. Chem. 35*, 2015–2024
3. Chiba, S. & Shimomura, T. (1975) *Agric. Biol. Chem. 39*, 1033–1040
4. Kanaya, K., Chiba, S., Shimomura, T. & Nishi, K. (1976) *Agric. Biol. Chem. 40*, 1929–1936
5. Chiba, S., Inomata, S., Matsui, H. & Shimomura, T. (1978) *Agric. Biol. Chem. 42*, 241–245
6. Chiba, S., Kanaya, K., Hiromi, K. & Shimomura, T. (1979) *Agric. Biol. Chem. 43*, 237–242
7. Matsui, H., Chiba, S. & Shimomura, T. (1978) *Agric. Biol. Chem. 42*, 1855–1860
8. Chiba, S., Kimura, A., Kobori, T. & Saitoh, K. (1985) *J. Jpn. Soc. Starch Sci. 32*, 213–216

II.3.c. Microbial α-Glucosidases

Various types of α-glucosidases are observed in microorganisms. Their substrate specificities are complicated. Fungal and yeast α-glucosidases have long been investigated by many workers. Recently bacterial α-glucosidases also have been homogeneously purified from *Bacillus* species. The first crystalline α-glucosidase was obtained from *Aspergillus oryzae* (1) and *A. niger* (2). In 1967 α-glucosidase of *Saccharomyces cerevisiae* was chromatographically demonstrated to be separated into two components which showed only isomaltase and maltase activities (3). These kinds of α-glucosidases are utilized as clinical test reagents of human α-amylase (4). The microbial α-glucosidase which shows a high hydrolytic activity toward α-1,6-glucosidic linkages may be available for hydrolysis of stubbed-glucose or isomaltose in oligosaccharides produced as a by-product in industrial glucose production. Recentrly the transglucosylation action of α-glucosidase has been applied for manufacture of oligosaccharides having α-1,6-glucosidic linkage.

Specificity : *A. oryzae* and *A. niger* α-glucosidases hydrolyze various maltooligosaccharides and isomaltooligosaccharides. These enzymes are known to show little or no activity toward phenyl α-glucoside, sucrose and soluble starch (1,2). Similar α-glucosidases are found in *S. logos* (6), *Bacillus subtilis* (7) and *B. cereus* (8). As shown in Table 1, however, another crystalline α-glucosidase from *A. niger* shows capability of splitting soluble starch (5). *Mucor javanicus* and *Penicillium purpurogenum* α-glucosidases show a relatively high activity toward soluble starch (9,10). α-Glucosidases of *Aspergillus* species specifically build α-1,6-glucosidic linkages by transglucosylation, and produce isomaltose and panose from maltose.

Table 1. Kinetic Parameters of α-Glucosidases from *A. niger* (4) and *S. logos* (5)

Substrate	*A. niger*			*S. logos*		
	Km (mM)	k_0 (sec^{-1})	k_0/Km ($mM^{-1} \cdot sec^{-1}$)	*Km* (mM)	k_0 (sec^{-1})	k_0/Km ($mM^{-1} \cdot sec^{-1}$)
G2	0.76	54.8	73.1	7.7	187	24.3
G3	0.69	69.3	100	8.2	219	26.7
G4	1.1	74.1	67.4	6.0	176	29.3
G5	1.9	56.4	29.7	6.0	93.5	15.6
NG	12	33.8	2.8	11.5	58.0	5.04
KJ	4.6	12.4	2.7	11.5	52.4	4.55
IG2	8.0	19.3	2.4	16.5	39.3	2.38
φG	0.36	0.7	1.9	8.7	11.2	1.29
SS	4.3	62.3	14.5	—	—	—

See Table 1 in Chapter II.3.b. (except for G4 (maltotetraose) and G5 (maltopentaose)).
Reaction, 37℃ for *A. niger* and 33℃ for *S. logos*.

α-Glucosidase from *Candida tropicalis* possesses a wide substrate specificity and a high transglucosylation activity (11). This enzyme can also degrade α-glucans. *S. italicus* α-glucosidase hydrolyzes synthetic α-glucosides such as phenyl and *p*-nitrophenyl α-glucosides. It also hydrolyzes maltose, sucrose and turanose, but does not hydrolyze isomaltose (12). Two kinds of α-glucosidases occur in *S. cerevisiae* ; one is maltase, and the other, isomaltase (3). Table 2 presents the substrate specificities of α-glucosidase-I (isomaltase) and -II (maltase) isolated from brewer's yeast. Phenyl α-glucoside is the most preferable substrate for them; isomaltase can't attack isomaltotriose or sucrose, and maltase shows little or no activity toward maltotetraose. *S. logos* α-glucosidase is similar to *Aspergillus* species enzymes in the substrate specificty, and shows no activity toward sucrose or soluble starch (5). Brewer's yeast α-glucosidase-II catalyzes a high glucosyl transfer action to D-fructose, L-sorbose, D-mannose as acceptor. D-Glucose is not so good an acceptor as the aforementioned sugars. Thus, the enzyme is useful for preparation of heterogeneous oligosaccharides consisting of those sugars.

B. thermoglucosidius α-glucosidase is an isomaltase, which hydrolyzes a series of isomaltooligosaccharides and also panose. As presented in Table 3, however, the most favorable substrate is a synthetic α-glucoside. The enzyme does not attack maltose (13). Two α-glucosidases of a thermophilic *Bacillus* sp. are highly specific for α-1,4-glucosidic linkage. Their substrate specificities are so narrow that the enzymes do not attack oligosaccharides other than maltooligosaccharides and phenyl α-maltoside (15). *B. stereothermophilus* α-glucosidase (16) has no activity toward isomaltose, but it readily hydrolyzes phenyl α-glucoside, maltooligosaccharides and α-glucans (Table 3).

Table 2. Kinetic Parameters of Brewer's Yeast α-Glucosidase I and II

Substrate	Isomaltase (I)			Maltase (II)		
	Km (mM)	k_0 (sec^{-1})	k_0/Km ($mM^{-1} \cdot sec^{-1}$)	*Km* (mM)	k_0 (sec^{-1})	k_0/Km ($mM^{-1} \cdot sec^{-1}$)
φG	2.9	310	107	0.8	212	265
SU	—	—	—	16.8	106	6.4
TU	—	—	—	11.7	142	12.1
G2	—	—	—	14.3	36	2.6
G3	—	—	—	6.2	34	5.5
IG2	28.6	27.9	0.98	—	—	—

See Table 1 in Chapter II.3.b. (except for SU (sucrose) and TU (turanose)). Reaction, 33℃.

Table 3. Kinetic Parameters of α-Glucosidases of *Bacillus* species (14,16)

Substrate	*Bacillus thermoglucosidius*			*Bacillus stereothermophilus*		
	Km (mM)	k_0 (sec^{-1})	k_0/Km ($mM^{-1} \cdot sec^{-1}$)	*Km* (mM)	k_0 (sec^{-1})	k_0/Km ($mM^{-1} \cdot sec^{-1}$)
PNPG	0.24	233	971	0.63	123	195
IG2	9.5	194	20.4	—	—	—
IG3	11	208	18.9	—	—	—
IG4	13	233	17.9	—	—	—
IG5	21	167	8.0	—	—	—
PN	3.3	74.2	22.5	—	—	—
G2	—	—	—	5.6	877	157
G3	—	—	—	1.1	1220	1110
G4	—	—	—	3.0	987	329
G5	—	—	—	3.0	791	264
AM	—	—	—	8.0	1150	144

See Table 1 in Chapter II.3.b. (except for PNPG (p-nitrophenyl α-glucoside), PN (panose), AM (amylose) and IG3~IG5 (isomaltotriose, -tetraose and -pentaose)). Reaction, 60℃.

Purification : α-Glucosidase of *A. niger* is easily purified to a crystal form from a commercial preparation, such as Transglucosidase Amano (Amano Pharmaceutical Co., Ltd., Japan). The crude enzyme is dissolved in 0.02 M phosphate buffer, pH 6.7, and dialyzed against the same buffer. The supernatant is subjected to chromatography with DEAE-Sepharose by which a large portion of impurities is removed. Then the enzyme is chromatographed by Toyopearl HW-55. The α-glucosidase is homogeneous and the activity recovery, about 40%. Saturated ammonium sulfate solution is added to the concentrated enzyme solution (10%) until a slight turbidity appears. The α-glucosidase is crystallized out in a needles form on standing for one week at 4℃ (5).

Table 4. Enzymic and Proteochemical Properties of α-Glucosidases of Various Microorganisms

Source of α-glucosidase	Molecular weight	Optimum pH
A. niger (5)	60,000	4.2
M. javanicus (9)	124,600	4.6
P. purpurogenum (10)	120,000	4.0
S. italicus (12)	85,000	6.8
S. cerevisiae (3) Maltase	68,500	7.0
Isomaltase	64,700	7.0
S. carlsbergensis (17)	63,000	6.7
Brewer's Yeast (13) Maltase	69,000	6.8
Isomaltase	52,000	6.5
S. logos (6)	270,000	4.6
B. subtilis (7)	33,000	6.0
B. thermoglucosidius (14)	60,000	6.0
B. cereus (8)	57,000	7.0
Bacillus sp. (15) Maltase I	53,000	6.8
Maltase II	43,000	6.8
B. stereothermophilus (16)	47,000	6.3

The numbers in parentheses indicate those of references.

Assay Method : Glucose liberated by α-glucosidase is usually determined by the Somogyi-Nelson method or by the Tris-glucose oxidase-peroxidase method (18,19). Tris (hydroxymethyl) aminomethane is a strong inhibitor of α-glucosidase.
Phenol and *p*-nitrophenol liberated from the aryl α-glucosides are simply determined by colorimetric methods : phenol, colored with 2,6-dibromoquinone-chloroimide (20), and *p*-nitrophenol, colored under alkaline conditions (21).

(Seiya Chiba)

References

1. Sugawara, S., Nakamura, Y. & Shimomura, T. (1961) *Agric. Biol. Chem. 25,* 358-361
2. Tsujisaka, Y. & Fukumoto, J. (1963) *Nippon Nogei-Kagaku Kaishi* (in Japanese) *31,* 747-752
3. Kahn, N. A. & Eaton, N. R. (1967) *Biochim. Biophys. Acta 146,* 173-180
4. Nakagiri, Y., Kanda, T., Otaki, M., Inomoto, K., Asai, T., Okada, S., & Kitahata, S (in Japanese) (1982) *J. Jpn. Soc. Starch Sci. 29,* 161-166
5. Kita, A., Matsui, H., Chiba, S. & Sakai, T. (1984) *Abstracts of Papers, Annual Meeting of Agric. Chem. Soc. of Japan,* p. 545
6. Chiba, S., Saeki, T. & Shimomura, T. (1973) *Agric. Biol. Chem. 37,* 1823-1836
7. Wang, L. -H. & Hartman, P. A. (1976) *Appl. Environ. Microbiol. 31,* 108-118
8. Yoshigi, N., Chikano, T. & Kamimura, M. (1985) *J. Jpn. Soc. Starch Sci. 32,* 273-279
9. Yamasaki, Y., Miyake, T. & Suzuki, Y. (1973) *Agric. Biol. Chem. 37,* 251-259

10. Yamasaki, Y. Suzuki, Y., & Ozawa, J. (1976) *Agric. Biol. Chem. 40,* 669-676
11. Sawai, T. & Hehre, E. J. (1962) *J. Biol. Chem. 237,* 2047-2052
12. Halvorson, H. & Ellias, L. (1958) *Biochim. Biophys. Acta 30,* 28-39
13. Suzuki, Y., Ueda, Y., Nakamura, N., & Abe, S. (1979) *Biochim. Biophys. Acta 566,* 162-166
14. Suzuki, Y., Ueda, Y., Nakamura, N. & Abe, S. (1979) *Biochim. Biophys. Acta 566,* 62-66
15. Suzuki, Y., Ikemoto, T. & Abe, S. (1978) *J. Ferment. Technol. 56,* 8-14.
16. Suzuki, Y., Shinji, M. & Eto, N. (1984) *Biochim. Biophys. Acta 278,* 281-289
17. Needleman, R. B., Federoff, H. J., Eccleshall, T. R., Buchferer, B. & Mamur, J. (1978) *Biochemistry 17,* 4657-4661
18. Papadopoulos, N. M. & Hess, W. C. (1960) *Arch. Biochem. Biophys. 88,* 167-171
19. Dahlqvist, A. (1961) *Biochem. J. 80,* 547-551
20. Robertson, J. J. & Halvorson, H. O. (1957) *J. Bacteriol. 73,* 186-198
21. Tanimura, T., Kitamura, K., Fukuda, T., & Kikuchi, T., (1979) *J. Biochem. 85,* 123-130

II.4. Glucoamylase

Glucoamylase (EC 3.2.1.3, 1,4-α-D-glucan glucohydrolase) is an exo-splitting enzyme that consecutively removes the glucose units from the nonreducing end of starch and glycogen. The enzyme is produced by various microorganisms, that is, *Saccharomyces, Schwnniomyces* and *Endomycopsis, Aspergillus, Penicillium, Monascus, Chalara, Mucor* and *Rhizopus, Clostridium* etc.. But, the commercially available enzyme preparations at present are obtained from either *Aspergillus* sp. or *Rhizopus* sp.

Other names than glucoamylase that have been used in literatures for this type of enzymic activity include amyloglucosidase, glucamylase, maltase, saccharogenic amylase and γ-amylase. The end product of the reaction is glucose, which clearly differentiates this enzyme from α- and β-amylases. α-Glucosidase is similar to glucoamylase, but the enzyme hydrolyzes the α-1,4-glucosidic linkages more rapidly in low molecular weight oligosaccharides than in starch. Another significant difference is that the glucose residue released by the action of α-glucosidase is α-configuration while the glucose produced by glucoamylase is β-configuration.

Glucoamylase hydrolyzes the α-1,6-glucosidic linkages of stubbed glucose residues in maltooligosaccharides at a rate slower than that of hydrolyzing the α-1,4-glucosidic linkages. Phosphate linked with certain glucose residues in tuber starches, such as potato starch and sweet potato starch, blocks the action of glucoamylase. Therefore, the combination of α-amylase, glucoamylase and phosphatase is necessary to hydrolyze tuber starch to glucose completely.

Fungal glucoamylase consists at least of two components. They are conventionally referred to as glucoamylase I and glucoamylase II. Glucoamylase I is more active in debranching action and is stronger in raw starch digestion than glucoamylase II. Glucoamylase I is readily adsorbed on raw starch, especially on cereal starches, but glucoamylase II is hardly adsorbed on those starches.

Glucoamylase is able to condensate glucose residues to produce oligosaccharides by the reverse reaction of hydrolysis. The main reversion products are maltose and isomaltose, though on a prolonged incubation at high substrate concentrations other oligosaccharides are observed.

The production of glucoamylase by certain kinds of yeast have been reported by some research groups, but their activities are very low. The use of plasmid or bacteriophage vectors to transfer the glucoamylase genes of fungi into fermentative yeast is being attempted to generate ethanol-producing yeast species with a strong amylase activity.

The most important application of glucoamylase is the production of high glucose syrup (90-97% D-glucose), which is used for production of crystalline glucose and high fructose syrup. Ethanol fermentation of starch materials without cooking is successfully performed by applying glucose preparation on an industrial scale.

(Seinosuke Ueda)

II.4.a. Fungal Glucoamylase (*Asp. niger, Asp. awamori, Rhi. delemar, Rhi. niveus et al.*)

It had long been thought that fungal amylolytic enzyme consisted of α-amylase, β-amylase and maltase, because the fungal enzyme hydrolyzes starch into glucose with a significant yield. Since around 1950, however, the study results achieved mostly by Japanese researchers elucidated that the enzyme that had been considered to be comparable with β-amylase shows the activity of converting starch into glucose completely and that it is a new enzyme different from β-amylase of plant origin (1,2,3,4). The enzyme was tentatively designated as a saccharogenic amylase or amyloglucosidase to distinguish it from β-amylase. In 1958, our research group successively succeeded in crystallizing the saccharogenic amylases of *Rhizopus delemar* and *Aspergillus niger* (5). Investigations with these crystalline enzyme preparations revealed that fungal saccharogenic amylase hydrolyzes not only the α-1,4-glucosidic linkages, but also the α-1,6-glucosidic linkages in starch. It was also shown that the enzyme exhibits a maltase activity. Later the enzyme came to be called glucoamylase because it releases glucose directly from starch and glycogen. In 1959 a Japanese corporation was the first in the world to start the production of glucose with the use of glucoamylase (of *Rhi. niveus*) on a industrial scale (6). The worldwide rise of the industry of glucose and high-fructose syrup production placed glucoamylase among the most important industrial enzymes.

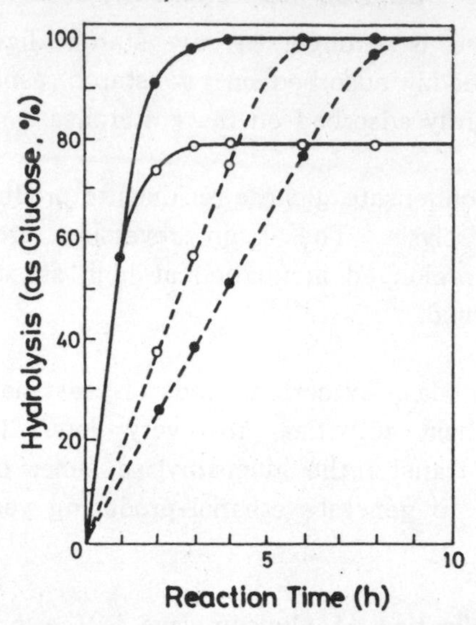

Fig. 1. Hydrolysis Profiles of Two Glucoamylase on Soluble Starch and Maltose. ●,*Rhi. delemar;* ○*Asp. niger;* —— Soluble Starch ; ------ Maltose.

Reaction : Glucoamylase cleaves the α-1,4-glucosidic linkages in α-glucans such as amylose, amylopectin and glycogen. The splitting reaction proceeds one by one from the non-reducing ends to release glucose of β-anomeric configuration. In case the substrate is amylopectin or glycogen, the amylase also cleaves the α-1,6-glucosidic linkages in the substrate. Glucoamylase that cleaves both α-1,4- and α-1,6-glucosidic linkage may be assumed to hydrolyze starch theoretically 100% into glucose.

As shown in Fig. 1, however, there are two types glucoamylase : one that hydrolyzes starch into glucose completely as is the case of *Rhi. delemar* glucoamylase, and another (*Asp. niger*) with which the hydrolysis is suspended at the extent of 80%. As shown in Table 1, *Rhi. delemar* glucoamylase hydrolyzes completely even branched glucans such as amylopectin, glycogen and β-limit dextrin. On the other hand, *Asp. niger* glucoamylase to a lesser extent hydrolyzes these branched substrates. This was especially the case with β-limit dextrin. Many researchers have proposed explanations for the different extents of starch hydrolysis between the two glucoamylases, but no clear explanations have so far been proposed (6,7,8).

Glucoamylase produces several oligosaccharides from glucose by the reverse reaction of hydrolysis. The higher the glucose concentration, the more intense the reverse reaction that takes place. Maltose is produced in the early stage of the reverse reaction but it does not increase in amount while saccharides that are difficultly hydrolyzable by glucoamylase, such as isomaltose and isomaltotriose, accumulate as the reaction proceeds (9). Many fungal glucoamylases hydrolyze raw starch into glucose. *Rhi. delemar* (10) and *Asp. awamori* (11) secrete several glucoamylase isozymes, but their activities of hydrolyzing raw starch vary greatly.

Hydrolyzable α-1,6 linkage

Panose **6^3-Glucosylmaltotriose** **α-Limit-dextrin** **Pullulan**

Unhydrolyzable α-1,6 linkage

Isomaltose **Isomaltotriose** **6^2-Isomaltosylmaltose** **Isopanose**

↓ α-1,6 linkage, — α-1,4 linkage

Fig.2. Specificity of Glucoamylase on Oligosaccharides Containing α-1,6-linkages.

Table 1. Specificities of Two Glucoamylases on Various Substrates

	Rhi. delemar		*Asp. niger*	
	Limit of hydrolysis(%)	Initial reaction rate	Limit of hydrolysis(%)	Initial reaction rate
Soluble starch	100	100	78	100
Amylopectin	100	105	75	126
Glycogen	100		66	
β-Limit dextrin	100	80	40	80
Amylose	100	27	100	30
Maltotriose	100	16	100	19
Maltose	100	8	100	15
Dextran	0	0	0	0
Isomaltotriose	0	0	0	0
Isomaltose	trace		trace	
α-Limit dextrin	100	55	100	69
6^2-Isomaltosylmaltose	0		0	
Panose	100	6	100	9

Initial reaction rate is expressed by designating that for soluble starch as 100%.

Specificity : In Table 1 are compared the extents and initial reaction rates of hydrolysis of various substrates with crystalline preparations of glucoamylases from *Rhi. delemar* and *Asp. niger*. As can be seen from Fig. 1 and Table 1, the two glucoamylases are different in the extent of hydrolysis of branched α-glucan, but are not different in the hydrolysis of linear maltooligosaccharides. Both glucoamylases hydrolyze α-1,4-linked oligo- and polysaccharides, such as maltose and amylose, completely. Further, as shown in Fig. 2, glucoamylase cleaves first the α-1,6-glucosidic linkages in such oligosaccharides as panose, 6^3-glucosylmaltotriose and then, attacks the α-1,4-glucosidic linkages in the intermedially produced oligosaccharides.

Glucoamylase, on the other hand, does not act on saccharides containing only α-1,6-glucosidic linkages or those which contain two or more repetitions of α-1,6-glucosidic linkages, such as isomaltose, isomaltotriose and 6^2-isomaltosylmaltose. These findings show that the α-1,6-glucosidic linkages sensitive to glucoamylase are limited to those which are present in glucose residues other than the residue that composes the reducing end of maltodextrin (minimum unit maltose), such as the α-1,6-glucosidic linkages present in panose. Glucoamylase thus also hydrolyzes pullulan which is a polysaccharide composed of panose units.

Production and Purification : At the early stages of glucoamylase research, the enzyme production was performed by cultivating the fungus by the solid culture: A medium consisting of one part wheat bran and 0.6 parts water is inoculated with the fungal strain and incubated at 27°C～30°C for about 72 h. The cultured material is then extracted with water. To the aqueous extract is added solid ammonium sulfate or a solvent such as ethanol and acetone to precipitate the enzyme. The final enzyme product is prepared into powder. The industrial production of glucoamylase of *Rhi. delemar* and *Rhi. niveus* still involves solid cultures. However, *Asp. niger* glucoamylase for industrial use is produced by submerged cultures. Purification at the early stages of glucoamylase research was very simple and primitive (5,6). The aqueous extract was carefully fractionated with ammonium sulfate and acetone, then subjected to ion exchange chromatography and the enzyme was finally crystallized from aqueous acetone. The enzymic and proteochemical properties are as follows : (Mol. wt, 7.0×10^4; opt. pH, 4.5～5.0; activity, 6.0 μmol glucose per min per mg protein at pH 4.9, 40°C ; *K*m, 0.121×10^{-3}M (maltoheptaose)) The great advances in enzyme purification technology of the recent years enabled separation of each of the multiple glucoamylases from fungal strains including those of *Rhi. delemar* and *Asp. awamori*. This study led to the elucidation that many isozymes of glucoamylase are present. The recent findings on the enzyme multiplicity of glucoamylase will be covered in the following section.

(Yoshio Tsujisaka)

Multiplicity : A single strain of fungus generally produces several forms of glucoamylase whose molecular sizes, amino acid and carbohydrate compositions as a glycoprotein vary from one to another. Also, they differ in the raw starch digestibility, hydrolysis curves on glycogen, and stability (12,13). This multiplicity of fungal glucoamylase is observed when the precursor of glucoamylase I is degraded stepwise with certain proteinases and glycosidases. However, a mutant strain which is negative in the production of proteinase and glycosidase, produces only a single form of glucoamylase (14,15) which is characterised by a high content of carbohydrate moieties and high activity in the raw starch digestion.

Aspergillus awamori var. *kawachi* glucoamylase usually exists in three forms : Raw starch adsorbable and raw starch digestive glucoamylase I (MW 90000), raw starch unadsorbable and raw starch nondigestive glucoamylases I (MW 83000) and II (MW 57000). Glucoamylase I is split with proteinases at the two sites of the peptide bonds "Val^{470}-Ala^{471}" and "Val^{515}-Ala^{516}", and converted to the raw starch nondigestive glucoamylase I with liberation of glycopeptide I. This glycopeptide is enzymatically inactive, but specifically adsorbed onto raw starch. The base sequence of the glycosyl peptide I is ATGGTTTATTTGSGGVTSTSKTTTTASKTSTTTSSTSCTT-PTAV. Some short mannoside chains exist through *O*-glycosidic linkages with hydroxy amino acids. The glycopeptide I region in glucoamylase I is located far from the active site, and constitutes a domain which shows a raw starch affinity to favor raw starch digestion.

(Sinsaku Hayashida)

Morita and his colleagues obtained several glucoamylase preparations (A-1, A-2, A-3, B-2) from *Asp. oryzae* cultured on steamed rice by the column chromatographic technique. The enzymes were distinguished from each other in the electrophoretic mobility and sedimentation constant. Also, they were different in the total sugar content. However, the molecular extinctions estimated at 280 nm were very close, especially so were those of A-1, A-2, and A-3 ranging from 12.46 to 12.94. Also, these three enzymes were shown to be almost similar in the amino acid composition. The B-2 preparation differed slightly from the others described above. Nevertheless, the enzymic properties of all the isozymes were almost the same. Morita and his colleagues suggest that the synthesis of these glucoamylases is controlled by the same gene (16~18).

Ottesen and his colleagues obtained two purified preparations of glucoamylase from a strain of *Asp. niger*. The two preparations (G I and G II) were different in the molecular weight. They also differed in the activities of hydrolysis of raw and soluble starch. However, despite a slight difference in the C-terminal amino acid sequence, no significant differences in the amino acid composition and N-terminal amino acid sequence of the native enzymes and even in that of the peptides produced by cyanogen bromide cleavage were observed between the two enzyme preparations (19). Most of the paper so far published on the multiplicity of fungal glucoamylase confirm the results by Ottesen and his colleagues.

Several isozymes of glucoamylase are also known to occur in *Rhiz.* fungi. Abe and Hizukuri observed that G III, one of the three glucoamylases from *Rhiz. delemar* was split by chymotrypsin into a new glucoamylase, G' (20). The molecular weights were 7.8×10^4 for G III and 7.2×10^4 for G', respectively. G' was less adsorbable on raw starch and less active in hydrolysis of glycogen than G III. The peptide(s) liberated by chymotrypsin was found to reduce the activity of G III. However, the addition of the peptide to G' gave no effect on the activity of G'. Similarly to the result reported by

Hayashida and his colleague, Hizukuri suggested that the peptide is the moiety or a fragment of the moiety of GⅢ molecule to responsible for adsorbability onto raw starch.

(Takehiko Yamamoto)

References

1. Kitahara, K. & Kurushima, M. (1949) *Hakko-Kogaku Zasshi* (in Japanese) *21*, *254-257*
2. Philips, L. L. & Caldwell, M. L. (1951) *J. Am. Chem. Soc. 73*, 3559-3562
3. Okazaki, H. (1955) *Nippon Nogei-Kagaku Kaishi* (in Japanese) *29, 273-283*
4. Ueda, S. (1957) *Bull. Agr. Chem. Soc. Japan 21*, 284-287
5. Tsujisaka, Y., Fukumoto, J. & Yamamoto, T. (1958) *Nature 181*, 770-771
6. Tsujisaka, Y. (1960) *Bull. Osaka Municipal Tech. Res. Inst.* (in Japanese) *28*, 59-66
7. Abe, J., Nagano, H. & Hizukuri, S. (1958) *J. Appl. Biochem. 7*, 235-247
8. Abe, J., Takeda, Y. & Hizukuri, S. (1982) *Biochim. Biophys. Acta 703*, 26-33
9. Hehre, E. J., Okada, G. & Genghof, D. S. (1969) *Arch. Biochem. Biophys 135*, 75-81
10. Ueda, S. & Kano, S (1975) Stärke *27*, 123-128
11. Yoshino, E. & Hayashida, S. (1978) *J. Ferment. Technol. 56*, 289-295
12. Ueda, S. (1957) *Bull. Agr. Chem. Soc. Japan 21*, 379-385
13. Hayashida, S. (1978) *J. Ferment. Technol. 56*, 289-295
14. Svensson, B., Larsen, K. & Gunnarsson, A. (1986) *Eur. J. Biochem. 154*, 497-502
15. Hayashida, S., Nakahara, K., Iwanada, S. & Sasaki, Y. (1988) *Agric. Biol. Chem. 52*, 273-275
16. Morita, Y., Shimizu, K., Ohga, M. & Korenaga, T. (1966) *Agric. Biol. Chem., 30*, 114-121
17. Ohga, M., Shimizu, K. & Morita, Y. (1966) *Agr. Biol. Chem. 30*, 967-972
18. Morita, Y., Ohga, M. & Shimizu, K. (1968) *Mem. Research Inst. Food Sci. Kyoto Univ., 29*, 18-23
19. Svensson, B., Pedersen. T. G., Svendsen, I., Sakai, T. & Ottesen, M. (1982) *Carlsberg Res. Communnm 47*, 55-69
20. Abe, J. & Hizukuri, S. (1988) *J. Jpn. Soc. Starch Sci., 35*, 43-47

Ⅱ.4.b. Bacterial and Yeast Glucoamylases

Glucoamylase occurs almost exclusively in fungi, and far less in bacteria and yeasts. *Clostridium acetobutylicum* produces α-amylase and glucoamylase. The latter enzyme was originally referred to as maltase, though it hydrolyzed starch to glucose completely (1). A thermoanaerobic bacterium, *C. thermohydrosulfuricum* produces an extremely thermoactive glucoamylase whose activity is best at 75℃, and at pH 5 to 6, and which is stable at 85℃ in a pH range from 4 to 6 (2).

As for enzymes from yeasts, *Saccharomycopsis fibuligera* (formerly *Endomyces* sp.) glucoamylase was first purified, and characterized (3). Diastatic strains of *Saccharomyces cerevisiae* (formerly *Saccharomyces diastaticus*) bear one or three polymeric *STA* genes, and release glucoamylase in the culture (4). The amylolytic enzymes of yeasts usually consist of both α-amylase and glucoamylase, as has been reported for *Saccharomycopsis fibuligera* (formerly *Endomycopsis fibuligera*), *Saccharomycopsis capsularis*, and more recently for *Schwanniomyces castellii*, *Schwanniomyces alluvius*, *Lipomyces kononenkoae*, *Lipomyces starkeyi*, *Candida antarctica* (formerly *Trichosporon oryzae*), *Trichosporon pullulans* and *Filobasidium capsuligenum* which have been partially characterized (5,6).

Some properties of yeast glucoamylase are very similar to those of fungal glucoamylase. They are glycoproteins and stable under almost identical conditions. The molecular weight ranges from 48000 to 300000 determined by SDS-gel electrophoresis (7). Amylolytic enzymes of some yeasts show so called debranching activity and hydrolyze α-1,4- and α-1,6-glycosidic linkages. Both activities are essential to hydrolyse starch, glycogen and various branched α-1,4- : α-1,6-oligosaccharides, completely. However the debranching activity of *S. diastaticus* glucoamylases is hardly detectable (8). The multiplicity of yeast glucoamylase has been reported for some yeasts. The raw starch digestion activity of yeast glucoamylase has been reported only for *S. fibuligera* and *C. antarctica*. Both the glucoamylases are known to be adsorbed on raw starch granules (9).

Recently, the amylolytic yeasts have attracted attention in production of single cell protein and alcohol fermentation directly from a starchy biomass. Also, there have been some attempts to introduce a foreign amylase gene into non-amylolytic but fermentative yeast cells such as *S. cerevisiae* by recombinant DNA technology.

A 1.5 kb-*Hind* III fragment carrying the entire intron-free glucoamylase structural gene of *A. awamori* was introduced into *S. cerevisiae* by use of an expression vector pACl which contained an *E. coli* origin of replication, the *bla* gene from pBR322, the yeast 2u origin of replication, and the yeast *LEU2* structural gene (10). The vector also contained the promoter and termination regions of the yeast enolase genes *ENO1*. A single *Hind* III site located at the junction between the promoter and termination was used to introduce the glucoamylase gene. A yeast transformed with the resultant plasmid was found to efficiently secrete glucoamylase into the medium. The natural leader sequence of the precursor of glucoamylase was processed correctly by the yeast and the enzyme secreted was glycosylated. Thus, they constructed a clone of *S. cerevisiae* capable of simultaneous saccharification and fermentation of starch.

Likewise, a yeast transformed with a plasmid carrying the glucoamylase gene from a *Rhizopus* cDNA library, secretes the glucoamylase into the medium (11). The glucoamylase gene from *S. diastaticus* genomic DNA was also cloned in *S. cerevisiae*. The halo-forming clone on the starch-containing medium, showed starch fermentation ability and produced glucoamylase into the medium (12).

(Shinsaku Hayashida)

References

1. Hockenhull, D. J. D. & Herbert, D. (1945) *Biochemical J. 39,* 102-106
2. Hyun, H. H. & Zeikus, J. G. (1985) *Appl. Environ. Microbiol. 49,* 1168-1173
3. Fukui, T. & Nikuni, Z. (1969) *Agric. Biol. Chem. 33,* 884-889
4. Kleinman, M. J., Wilkinson, A. E., Wright, I. P., Evans, I. H. & Beran, E. A. (1988) *Biochem. J. 249,* 163-170
5. De Mot, R. & Verachtert, H. (1985) *Appl. Environ. Microbiol. 50,* 1474-1482
6. De Mot, R. & Verachtert, H. (1987) *Eur. J. Biochem. 164,* 643-654
7. Wilson, J. J. & Ingledew, W. M. (1982) *Appl. Environ. Microbiol. 44,* 301-307
8. Modena, D., Vanoni, M., Englard, S. & Marmur, J. (1986) *Arch. Biochem. Biophys. 248,* 138-150
9. Ueda, S. & Saha, B. C. (1983) *Enzyme Microb. Technol. 5,* 196-198
10. Innis, M. A., Holland, M. H., McCabe, P. C., Cole, G. E., Wittman, V. P., Tal, R., Watt, K. W. K., Gelfand, D. H., Holland, J. P. & Meade, J. H. (1985) *Science 228,* 21-30
11. Ashikari, T., Nakamura, N., Tanaka, Y., Kiuchi, N., Shibano, Y., Tanaka, T., Amachi, T. & Yoshizumi, H. (1985) *Agric. Biol. Chem. 49,* 2521-2528
12. Yamashita, I. & Fukui, S. (1983) *Agric. Biol. Chem. 47,* 2689-2697

Ⅲ. Data on Individual Related Enzymes

Ⅲ.1. Debranching Enzymes

Debranching enzyme hydrolyzes the α-1,6-glucosidic linkages specifically in α-glucans such as starch, amylopectin, glycogen and pullulan which are fabricated with α-1,4- and α-1,6-glucosidic linkages. Enzymes of this kind are known to be present in higher plants and microorganisms. Isoamylase (glycogen 6-glucanohydrolase, EC 3.2.1.68), pullulanase (pullulan 6-glucanohydrolase, EC 3.2.1.41) and amylo-1, 6-glucosidase (amylo-1,6-glucosidase /1,4-α-glucan : 1,4-α-glucan 4-α-glycosyltransferase, EC 3.2.1.33) belong to this enzyme group. The debranching action was first found in yeast extracts by Nishimura (1), and the enzyme was named amylosynthase. Later, Maruo and Kobayashi re-examined the action of amylosynthase on rice starch, and proposed the name isoamylase for the enzyme. Pullulanase was first discovered in *Klebsiella pneumoniae* (traditional name, *Aerobacter aerogenes*). Amylo-1,6-glucosidase was found in rabbit muscle in 1951. These enzymes are known to catalyze the reactions of not only hydrolysis of α-1,6-glucosidic linkage, but also synthesis of the linkage by condensation and transfer action.

Isoamylase and pullulanase are useful for the fine structural analysis of starch, glycogen, pullulan and branched dextrins. The most important application of the enzymes is for industrial production of glucose or maltose in combination with glucoamylase or β-amylase, respectively. Recently, 6-O-α-maltosyl-, and 6-O-α-maltotriosyl-cyclodextrins are being produced in high yields by applying the condensation reaction or transfer reaction of the enzymes.

(Shigetaka Okada)

Ⅲ.1.a. Isoamylases (*Pseudomonas amyloderamosa,* etc)

"Debranching enzymes" which hydrolyze the α-1,6-glucosidic inter-chain linkages in branched α-D-glucans, are known to occur in higher plants and microorganisms. In 1930, Nishimura (1) reported that yeast extracts contained an enzyme, tentatively named amylosynthase which showed starch liquefying activity and accelerated the action of other amylases, but differed in thermolability from the α- and β-amylases. Further studies by Nishimura and Minagawa (2) showed that on glutinous rice starch, the enzyme caused an increase in the iodine color reaction. The enzyme was therefore regarded as a starch-synthesizing enzyme. Later, the enzyme action on rice starch was re-examined by Maruo and Kobayashi (3), and it was revealed that the reaction products were smaller in the molecular weight than the original starch. The products showed greater extents in β-amylolysis than the original starch, and were stained bluish-purple with iodine. Also, the reaction mixture showed a tendency to produce precipitates (retrogradation of starch) from the solution. These phenomena were considered to be due to an extensive debranching of starch and thus, the name of "isoamylase" was proposed for the enzyme. The action pattern of this enzyme is similar to that of R-enzyme (later called plant pullulanase) isolated from potato and broad bean. In 1961, a bacterial debranching enzyme was discovered in *Aerobacter aerogenes (Klebsiella pneumoniae* in the present classification) by Bender and Wallenfels, and named pullulanase.

In 1968, Harada *et al.* isolated a strain (*Pseudomonas* sp. SB-15) which gave a blue color reaction with iodine on a medium containing glutinous rice starch as the sole carbon source (4). This organism produced extracellular isoamylase and was identified as *Pseudomonas amyloderamosa*. Isoamylase is the only known debranching enzyme which debranches glycogen completely. The enzyme is also distinguished from pullulanase in that it shows no activity to hydrolyze pullulan (5). Another characteristic of isoamylase distinguishing it from pullulanase is its inability to remove two- and three-glucose units linked side chains in β-limit dextrin and α-limit dextrin of amylopectin or glycogen (6). Several debranching enzymes of this type were also found in bacteria, yeasts and plants: *Cytophaga* sp. (7), *Bacillus amyloliquefaciens* (8), *Escherichia coli* (9), *Flavobacterium* sp. (10), *Lipomyces kononenkoae* (11) and potato tubers (12). Only the intracellular enzyme of *Escherichia coli* differs from the other isoamylases with its inability to hydrolyze glycogen.

Isoamylase is useful for studying the inner structure of α-D-glucans such as glycogen (13). *Pseudomonas* isoamylase is also useful for industrial production of glucose or maltose from starch in combination with glucoamylase or β-amylase.

Reaction :

$$\alpha\text{-1,4 : 1,6-Branched glucan (amylopectin, glycogen)} + n\ H_2O \longrightarrow$$
$$(n+1)\ \alpha\text{-1,4-Glucan (amylose of short and long chain lengths)}$$

Isoamylase catalyzes the hydrolysis of the α-1,6-glucosidic bonds of starch, glycogen, and their degradation products. Consequently, the debranching of amylopectin results in an increase in the density of blue color developed by iodine reaction (complexing ability with iodine) and reducing power. Isoamylase completely splits the branches forked by α-1,6-glucosidic linkages in glycogen, but its activity of hydrolyzing pullulan is feeble. As shown in Table 1, the most sensitive substrates for isoamylase are amylopectin and glycogen. Maltosidic side chains in β-limit dextrins are hydrolyzed by the enzyme, but much more slowly than maltotriosidic side chains (14,16).

Isoamylase hydrolyzes both the inner and outer branching linkages of amylopectin. The enzyme also attacks the branching linkages of glycogen comparatively, while pullulanase is known to split only a few of them. The mode of action of isoamylase on amylopectin may be said to be endowise, while on glycogen it may be exowise (5).

Table 1. Relative Reaction Rate of *Pseudomonas* and *Cytophaga* Isoamylases on Various Branched Oligosaccharides and Polysaccharides (14,15)

Substrate	*Pseudomonas* isoamylase※	*Cytophaga* isoamylase※※
Amylopectin	100	—
Glycogen	124	100
Pullulan	<1	0.04
6^3- α -Maltosylmaltotriose	2.8	0.04
6^3- α -Maltosylmaltotetraose	6.9	0.27
6^3- α -Maltotriosylmaltotriose	9.7	0.17
6^3- α -Maltotriosylmaltotetraose	33	—

※ Rate relative to amylopectin = 100.
※※ Rate relative to glycogen = 100.

Specificity : The kinetic properties of isoamylases on various substrates are shown in Table 2 (16). As shown in Table 3, the difference in the rate of degradation by isoamylase observed between phosphorylase limit and phosphorylase- β -amylase limit dextrins is considered to be due to the difference in the size of side chains, namely, maltotetraose in the former limit dextrins, and maltose in the latter limit dextrins.

Table 2. Kinetic Properties of *Pseudomonas* Isoamylase on Various Substrates

Substrate	*Km*※	*V*max※※
Amylopectins		
Potato	0.14	225
Sweet potato	0.13	240
Wheat	0.14	250
Maize	0.13	280
Waxy maize	0.17	215
Glutinous rice starch	0.13	170
Glycogens		
Oyster	0.11	165
Rabbit liver	0.12	115
Phytoglycogen		
Sweet corn	0.10	240
Pullulan	2.0	1.1

※ Expressed as mg/ml.
※※ Expressed as μ mol aldehyde groups/min/mg enzyme protein.

Enzymic and Proteochemical Properties : The molecular weight of *Pseudomonas* isoamylase is about 90,000 and the pI, 4.4. The enzyme action is competitively inhibited by maltotriose, maltotetraose and other maltooligosaccharides, but is not affected by glucose and maltose (17). The enzyme action is also slightly inhibited uncompetitively by glucitol and maltitol in the case of amylopectin as substrate. *Aerobacter* pullulanase, however, is significantly inhibited by these sugar alcohols.

Table 3. Effect of *Pseudomonas* Isoamylase on β-Amylolysis of Amylopectin, Glycogen and Dextrins Derived therefrom (6)

Substrate	Conversion to maltose (%) by	
	β-Amylase alone	Isoamylase and then β-amylase
Waxy maize amylopectin	50	99
Oyster glycogen	38	102
Waxy maize amylopectin phosphorylase limit dextrin	21	95
Rabbit liver glycogen phosphorylase limit dextrin	28	94
Waxymaize amylopectin β-limit dextrin	0	80
Oyster glycogen β-limit dextrin	0	79
Waxy maize amylopectin phosphorylase β-limit dextrin	0	48
Rabbit liver glycogen phosphorylase β-limit dextrin	0	44

Cyclodextrins also inhibit competitively the isoamylase action on amylopectin substrate. But, the inhibition by cyclodextrins is not so strong as that observed for pullulanase (Yokobayashi, K., unpublished) (Table 4). The isoamylase is inhibited by heavy metal ions such as Hg^{2+}, Ag^{+} and Cu^{2+}. Several other properties of isoamylases isolated from various sources are shown in Table 5.

Table 4. Effects of Various Inhibitors on Debranching Action

Enzymes	*Km* for		*Ki* value					
			Competitive			Uncompetitive		
	amylopectin (mg/ml)	amylose (mg/ml)	α-CD (mM)	β-CD (mM)	γ-CD (mM)	glucitol (mM)	maltitol (mM)	maltotriitol (mM)
Pseudomonas isoamylase	0.46	5.0	6.8	2.8	1.9	210	310	12
Aerobacter pullulanase	7.42	2.1	0.57	0.013	0.051	11	0.25	0.0065

Production and Purification : *Pseudomonas amyloderamosa* is grown aerobically at 30℃. The enzyme is extracellular and inducibly produced by the presence of maltose, maltodextrin, starch or glycogen. No other amylases or α-glucosidases other than isoamylase are produced in the culture medium (4,19). A recommended medium consists of 2% maltose, 0.4% sodium glutamate, 0.3% diammonium hydrogenphosphate and several other inorganic salts of small quantities. The optimum pH for enzyme production is 5−6 ; the enzyme is stable at pH 2−6, but is labile at pH values over 7.0. The maximum production of the enzyme is obtained after 2 days under the above culture conditions (4). The bacterial cells are then removed by centrifugation at 10,000 x g for 30 min, and the clear supernatant is used for purification of the enzyme (6). Ammonium sulfate fractionation and DEAE- and CM-Cellulose column chromatographies are applied and isoamylase purified approximately 720-fold (specific activity, 59,000 U/mg protein) is obtained in a yield of 23%. The crystalline enzyme is obtained upon addition of ammonium sulfate solution dropwise to a concentrated solution of the enzyme in the cold (5).

Table 5. Comparison of Properties of Isoamylases from Various Sources

Source	Mol.wt. ($\times 10^3$)	Optimum pH	Optimum temp.(°C)	Inhibition by SH reagent※	Ref.
Pseudomonas amyloderamosa	90	3−4	52	+	6
Cytophaga	—	5.5	40		7
Escherichia coli	—	5.6−6.4	45−50	++	9
Flavobacterium	121	6.3	40	—	10
Saccharomyces cerevisiae	—	6	25	+	18
Lipomyces kononenkoae	65	5.6	30		11
Potato	—	5.5−6.0	50	++	12

※ The relative degree of inhibition is expressed by number of + signs.

Another purification procedure : Isoamylase produced in the culture broth of *Pseudomonas amyloderamosa* [strain SB-15 (ATCC 21262) or a mutant strain now being used industrially] is adsorbed on raw cereal starch (20) or cross-linked amylose gel (21). The kind of raw cereal starch, salt concentration, enzyme concentration, temperature, and time period for the processing have no influence on the enzyme adsorption. The enzyme adsorbed on raw starch is eluted by adding a solution of substrate of isoamylase such as amylopectin, glycogen or a solution containing the enzyme reaction products such as maltodextrin, maltotriose etc. The higher the temperature of the suspension of the enzyme adsorbing material, the more effective is the elution of isoamylase by maltodextrins. As an example, 15 g of raw corn starch /ℓ broth is added to a cultured broth containing isoamylase at pH 3.5 and the suspension, stirred for 30 min at 15℃. The starch having adsorbed the enzyme is then separated by filtration and dried (the adsorption efficiency is usually around 97%). The starch is suspended in a 2% solution of maltodextrins in a proportion of 100 g/ℓ and stirred for 30 min at 30℃. Ammonium sulfate is added to the eluate to 65% saturation and the resulting precipitate collected by centrifugation and dissolved in a least amount of water. The activity recovery of the enzyme solution thus obtained is 85%. This affinity adsorption method of the enzyme is simple and gives better results than the method described above.

Assay Method : The action of isoamylase results in an increase in the blue color density developed by adding iodine. Also, it causes an increase in the reducing power and in the extent of β-amylolysis of amylopectin or glycogen. The method based on the former reaction is relatively convenient and sensitive, though the phenomenon is not necessarily reflecting the number of α-1,6-glucosidic linkages hydrolyzed, whereas the latter method of determining the increase of reducing power is relatively accurate. In either method, some amylases may interfere the isoamylase assay on α-1,4-glucan substrate.

Iodine color reaction method : The reaction mixture, consisting of 0.5ml of 1.0% soluble glutinous rice starch or waxy maize starch, 0.1ml of 0.5M acetate buffer (pH 3.5) and 0.1ml of enzyme solution, is incubated at 40℃ for 1 h. A 0.5ml aliquot of the reaction mixture is taken out and mixed with 0.5ml of 0.01M iodine-0.1M potassium iodide solution and brought up to 12.5ml with water. This solution is allowed to stand at room temperature for 15 min and the absorbance at 610nm is measured against a reference solution containing heat-inactivated enzyme. One unit of the enzyme activity is defined as the amount of enzyme that increases one decimal unit absorption at 610nm in 1 h.

Reducing power method : The same enzyme reaction mixture as described above, is incubated for 15 min at 40℃. The reducing power produced in the mixture is determined, and one unit of enzyme activity is defined as the enzyme amount which produces reducing sugar equivalent to 1 μmol of glucose in 1 min. One unit enzyme activity defined by this method corresponds to 278 unit activity by the iodine color reaction method.

(Koji Yokobayashi)

References

1. Nishimura, S. (1930) *Nippon Nogei-Kagaku Kaishi* (in Japanese) *6*, 160-167.
2. Minagawa, T. (1932) *Nippon Nogei-Kagaku Kaishi* (in Japanese) *8*, 176-183.
3. Maruo, B. & Kobayashi, T. (1951) *Nature 167*, 606-607.
4. Harada, T., Yokobayashi, K. & Misaki, A. (1968) *Appl. Microbiol. 16*, 1439-1444.
5. Harada, T., Misaki, A., Akai, H., Yokobayashi, K. & Sugimoto. K. (1972) *Biochim. Biophys. Acta 268*, 497-505.
6. Yokobayashi, K., Misaki, A. & Harada, T. (1970) *Biochim. Biophys. Acta 212*, 458-469.
7. Gunja-Smith, Z., Marshall, J. J. & Whelan, W. J. (1970) *FEBS Lett. 12*, 96-100.
8. Urlaub, H. & Wober, G. (1975) *FEBS Lett. 57*, 1-4.
9. Jeanningros, R., Creuzet-Sigal, N., Frixon, C. & Cattaneo, J. (1976) *Biochim. Biophys. Acta 438*, 186-199.
10. Sato, H. H. & Park, Y. K. (1980) *Starch*/Stärke *32*, 132-136.
11. Spencer-Martins, I. (1982) *Appl. Environ. Microbiol. 44*, 1253-1257.
12. Ishizaki, Y., Taniguchi, H., Murayama, Y. & Nakamura, M. (1983) *Agric. Biol. Chem. 47*, 771-779.

13. Akai, H., Yokobayashi, K., Misaki, A. & Harada, T. (1971) *Biochim. Biophys. Acta* *237,* 422-429.
14. Kainuma, K., Kobayashi, S. & Harada, T. (1978) *Carbohydr. Res. 61,* 345-357.
15. Evans, R. M., Manners, D. J. & Stark, J. R. (1979) *Carbohydr. Res. 76,* 203-213.
16. Yokobayashi, K., Akai, H., Sugimoto, T., Hirao, M., Sugimoto, K. & Harada, T (1973) *Biochim. Biophys. Acta 293,* 197-202.
17. Kitagawa, H., Amemura, A. & Harada, T. (1975) *Agric. Biol. Chem. 39,* 989-994.
18. Gunja, Z. H., Manners, D. J. & Maung, K. (1961) *Biochem. J. 81,* 392-398.
19. Norrman, J. & Woeber, G. (1975) *Arch. Microbiol. 102,* 253-260.
20. Masuda, K. & Yokobayashi, K. (1973) *Japan Kokai Tokkyo Koho* 48-58188.
21. Kato, K., Konishi, Y., Amemura, A. & Harada, T. (1977) *Agric. Biol. Chem. 41,* 2077-2080.

Ⅲ.1.b Pullulanases of Several Microorganisms and Plants

Pullulanase is an enzyme which hydrolyzes the α-1,6-glucosidic linkages in pullulan. Pullulan is a polysaccharide extracellulary produced by *Aureobasidium pullulans* (traditional name, *Pullularia pullulans*). Pullulanase is classified among the debranching amylases which hydrolyze the α-1,6-glucosidic linkages of amylopectin and glycogen. Pullulanases are known to be produced by *Klebsiella pneumoniae* (traditional name, *Aerobactor aerogenes*) (1), *Escherichia intermedia* (2), *Streptococcus mitis* (3), *Bacillus acidopullulyticus* (4), *Streptomyces flavochromogenes* (5), *Oryza sativa* (6), *Hordeum valgare* (7), *Avena sativa* (6), *Vicia fava* (6), *etc.* Pullulanase from *K. pneumoniae* is a useful enzyme for studying the fine structure of starch and glycogen as well as the branched dextrins fabricated by α-1,4- and α-1,6-linkages. Also, pullulanases from *K. pneumoniae* and *B. acidopullulyticus* are practically applicable for saccharifying starch to produce glucose and maltose in an industrial scale.

Reaction :

Pullulan + n H_2O ⟶ n Maltotriose
Starch (glycogen) + n H_2O ⟶ n moles of various maltodextrins

Pullulanase hydrolyzes the α-1,6-glucosidic linkages in pullulan to produce maltotriose. A typical action of pullulanase on pullulan is shown in Fig. 1 (8). The viscosity of a pullulan solution rapidly decreases at a very early stage of the reaction, while the increase of reducing sugar is proportional to the time of incubation. At an early stage of the enzyme reaction, hexa- and nona-oligosaccharides are detected. The appearance of maltotriose is noticed at later reaction stage. Pullulanase is concluded to randomly cleave the α-1,6-glucosidic linkages of pullulan.

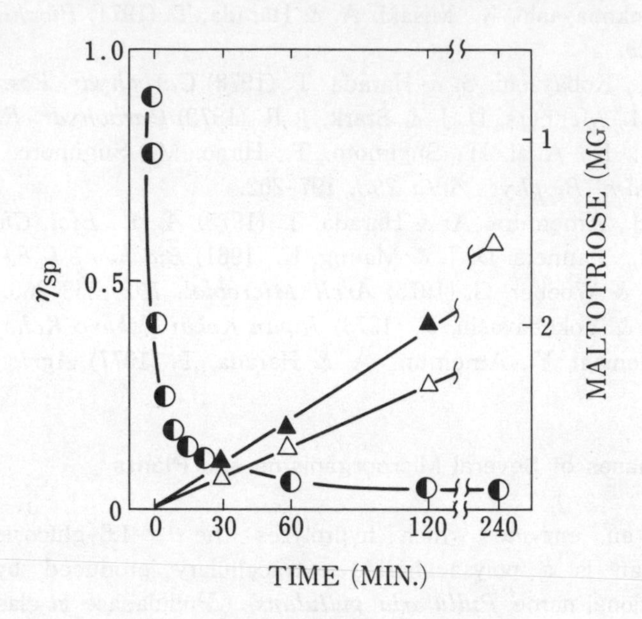

Fig. 1. Action of *K. pneumoniae* Pullulanase on Pullulan (8).
● and ○, Changes in the specific viscosity of pullulan on incubation with extracellular and intracellular pullulanase, respectively. ▲, Reducing sugar produced by extracellular pullulanase; △, reducing sugar produced by intracellular pullulanase.

Pullulanase also hydrolyzes the α-1,6-glucosidic linkages in starch and glycogen producing various maltooligosaccharides and increasing the blue color density with iodine. Consequently, pullulanase increases α-amylolysis of starch and glycogen (Table 1).

Table 1. Effect of *K. pneumoniae* Pullulanase on β-Amylolysis (9)

Substrate	β-Amylolysis (%)	
	Before debranching	After debranching
Amylopectin	50	95
Glycogen	38	46
Amylopectin β-LD	0	97
Glycogen β-LD	0	31

Specificity : *K. pneumoniae* pullulanase hydrolyzes the α-1,6-glucosidic linkages in branched oligosaccharides shown below (10) as well as those in pullulan, starch and glycogen.

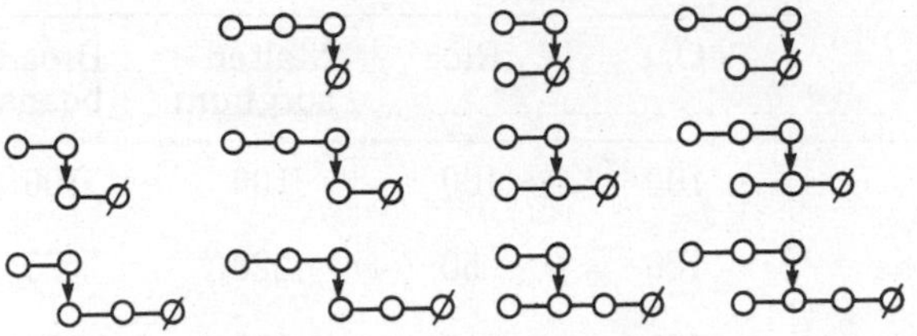

On the other hand, pullulanase does not cleave the α-1,6-glucosidic linkages in isopanose and various glucosyl stubbed oligosaccharides shown below.

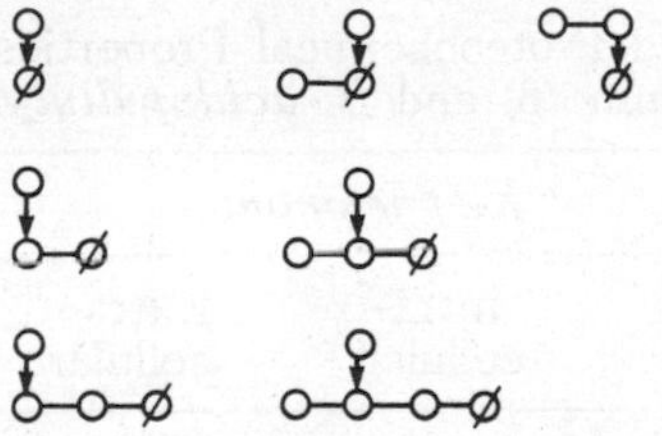

The substrate specificities of various pullulanases on α-1,4- and α-1,6-oligosaccharides are summarized in Tables 2 and 3. Plant pullulanase, the so-called R-enzyme, cleaves the maltosyl and maltotriosyl residues bonded by α-1,6-linkages more easily than microbial pullulanase.

Table 2. Relative Rates of Hydrolysis of α-1, 6-Glucosidic Linkages in Various Branched Oligosaccharides by *A. aerogenes* and *S. mitis* Pullulanases (3)

Substrate	*A. aerogenes*	*S. mitis*
Pullulan	100	100
6^2-α-Maltosylmaltotriose	23	3.8
6^3-α-Maltosylmaltotriose	55 (22)※	7.7
6^3-α-Maltotriosylmaltotriose	91 (162)	128
6^3-α-Maltosylmaltotetraose	171 (43)	23
6^3-α-Maltotriosylmaltotetraose	112 (146)	283
6^3-α-Maltotetraosylmaltotriose	57	208

※ Values in parenthese were quoted from reference 10.

Table 3. Relative Rates of Hydrolysis of α-1, 6-Glucosidic Linkages in Various Branched Oligosaccharides by Several Plant Pullulanases (R-enzymes) (6)

Substrate	Oat	Rice	Malted sorghum	Broad beans	Pea※
Pullulan	100	100	100	100	100
6^3-α-Maltosyl-maltotriose	120	50	220	75	90
6^3-α-Maltosyl-maltotetraose	290	210	300	195	230
6^3-α-Maltotriosyl-maltotriose	220	170	250	210	170
6^3-α-Maltotriosyl-maltotetraose	380	260	250	400	280

※ Yellowlees, D. (1980) *Carbohydr. Res.*, *83*, 109.

Enzymic and Proteochemical Properties of Pullulanases : The properties of pullulanase are summarizingly shown in Tables 4 and 5.

Table 4. Enzymic and Proteochemical Properties of Pullulanases from *K. pneumoniae* (8) and *B. acidopullulyticus* (4)

	K. pneumoniae		*B. acidopullulyticus*	
Pullulanase	Intra-cellular	Extra-cellular	Strain 11647	Strain 11777
Molecular weight	90,000	66,000	100,000	90,000
Activity/mg protein	7,000※1	7,200※1	100※2	50※2
Optimum pH	6.0	6.6	5.2	4.9
pH Stability	5.5-12	5.0-11.5	4-9※3	4-9※3
Thermal stability	40℃	50℃	55℃※3	55℃※3
Isoelectric point	3.88 4.46 7.6	3.72 4.35 7.7	5.0	4.9
Inactivation by CD	+	+	+※3	+※3
Action type	endo	endo	endo	endo

※1 Expressed as debranching amylase activity.
※2 Expressed as pullulan-hydrolyzing activity.
※3 Unpublished data (Sakano *et al.* (1987)).

Production of *K. pneumoniae* Pullulanase : *K. pneumoniae* produces extracellular (11) and intracellular (12) pullulanases. When the bacterium is cultured in a medium containing soluble starch as a carbon source and CH_3COONH_4 as a nitrogen source, pullulanase activity produced in the culture filtrate reaches about 500 units per ml of the culture medium in 120 h cultivation (30℃). On the other hand, when cultured in a medium containing $(NH_4)_2SO_4$ instead of CH_3COONH_4, the pullulanase produced is exclusively of the cell-bound type (120 units per ml of the culture medium in 18h cultivation). The medium suitable for extracellular enzyme production consists of 0.75% CH_3COONH_4, 1.2% soluble potato starch, 0.25% $NaNO_3$, 0.2% K_2SO_4, 0.1% K_2HPO_4, 0.05% each of $MgSO_4 \cdot 7H_2O$ and KCl, 0.04% peptone and 0.001% $FeSO_4 \cdot 7H_2O$. The medium for intracellular enzyme production is composed of 0.8% $(NH_4)_2SO_4$, 1.0% soluble potato starch, 0.1% K_2HPO_4, 0.05% each of $MgSO_4 \cdot 7H_2O$ and KCl, 0.05% peptone, 0.01% $Fe_2SO_4 \cdot 7H_2O$ and 0.7% $CaCO_3$.

Table 5. Enzymic and Proteochemical Properties of Plant Pullulanases (R-Enzymes)

Pullulanase		Oat	Rice	Malted sorghum	Broad beans	Pea※
Molecular weight		80,000	103,000	90,000	80,000	180,000
Optimum pH		5.0	5.0	5.0	6.6	5.8-6.2
Rate parameters						
amylopectin	*Km*	1.4	6	—	1.2	54
	V	26	35	14	10	—
amylopectin	*Km*	1.5	4	25	10	—
β-LD	*V*	160	150	360	65	—
glycogen	*Km*	—	7	30	17	—
β-LD	*V*	—	55	300	50	—

※ Yellowless, D. (1980) *Carbohydr. Res.*, *83*, 109.
Km, mg/ml ; *V*, relative activity to pullulan-hydrolyzing activity.

Purification and Crystallization of Extracellular Pullulanase (11) : The broth cultured for five days is centrifuged at 5,000 rpm for 15 min to remove the bacterial cells. The pH of the supernatant (about 9.2) is adjusted to 6.0-7.0 with 1N HCl, and solid ammonium sulfate is added to the solution to 20% saturation. After standing overnight at 7℃, the resulting precipitate is filtered off through Hyflosupercel under suction. To the filtrate is added solid ammonium sulfate to 80% saturation. The resulting precipitate is dissolved in a least amount of 0.02 M acetate buffer (pH 6.0). The solution is dialyzed against the same buffer, and the dialyzed solution is applied on a column of DEAE-Cellulose equilibrated with the same buffer, and the enzyme adsorbed is eluted with 0.02 M acetate buffer (pH 6.0) containing 0.3 M NaCl. The active fractions are combined, dialyzed against 0.02 M acetate buffer and then, applied on a column of Sephadex G-200. The active fractions collected are applied on a DEAE-Cellulose column equilibrated with 0.02 M acetate buffer (pH 6.0) and eluted with the same buffer containing 0.3 M NaCl. The active fractions are combined and dialyzed against 0.02 M acetate buffer (pH 6.0) for 2 days. The dialyzed solution is transferred into a seamless cellulose dialysis tubing and concentrated by polyethylene glycol in the cold until the protein concentration reaches around 2%. The enzyme solution is dialyzed against the same buffer and saturated ammonium sulfate solution is added dropwise into the dialyzed solution. As soon as the enzyme solution becomes opalescent, the addition of ammonium sulfate is stopped, and the solution is allowed to stand in the cold after it is shaken well. Afterwards, a drop of saturated ammonium sulfate solution is added into the enzyme solution. A silky thread-like turbidity appears in the solution after a week's standing. Then the solution is centrifuged and the supernatant removed. The crystalline enzyme precipitate is washed twice with 2 ml each of saturated ammonium sulfate solution, and dissolved in a least amount of 0.02 M acetate buffer (pH 6.0) and recrystallized by the same method as above. The second crystalline enzyme is dissolved in a least amount of the same buffer and applied on a Sephadex G-100 column as described above. The active fractions are collected, concentrated by polyethylene glycol and dialyzed. The third crystallization is performed by adding saturated ammonium sulfate solution to this dialyzed solution. The specific activity of the purified extracellular pullulanase is 7,200 units per mg protein.

Purification and Crystallization of Intracellular Pullulanase (13) : The cultured broth is subjected to centrifugation at 10,000 rpm for 30 min to collect the bacterial cells. After washing, the cells are suspended in 0.1 % SDS solution and continuously stirred at 30℃ for 30 h. To this SDS-extracted enzyme solution is added solid ammonium sulfate with stirring to achieve 5% saturation. After standing overnight at 7℃, the precipitate is filtered off through Hyflosupercel under suction, and ammonium sulfate is added to the filtrate to 60% saturation. The resulting precipitate is collected on Hyflosupercel and dissolved in a least volume of 0.02 M acetate buffer (pH 6.0). The solution is dialyzed against the same buffer and dialyzed solution is applied on a column of DEAE-Cellulose equilibrated with the same buffer. Elution is carried out stepwise with 0.1 M and 0.3 M NaCl containing the buffer. The enzyme is eluted with 0.3 M NaCl. The active fraction is concentrated and then, desalted by ultrafiltration. The concentrated solution is applied on a Sephadex G-200 column equilibrated with 0.02 M acetate buffer (pH 6.0). The enzyme elution is performed with the same buffer. The active fraction is collected and concentrated by ultrafiltration and with collodion bags in the cold until the protein concentration becomes about 3%. Saturated ammonium solution is slowly added dropwise to the concentrated enzyme solution with gentle stirring. The enzyme solution made faintly turbid with the salt is kept in the cold for several days. The precipitate, if any, is removed. Crystals of pullulanase are usually produced on allowing the solution to stand for further several days. The crystalline enzyme is washed with 70% saturated ammonium salfate solution and dissolved in a least amount of 0.02 M acetate buffer (pH 6.0). Recrystallization of the enzyme is carried out with the procedure described above. The specific activity of the intracellular pullulanase thus purified is 7,000 units per mg protein.

Assay Method : The following two methods are available for determination of pullulanase activity : one, estimating the reducing sugar produced by hydrolysis of pullulan (so-called pullulanase activity) and the other, measuring the increase of the blue color density with iodine by hydrolysis of amylopectin (so-called debranching amylase activity, isoamylase activity).

Pullulanase activity : Fifty μl of the enzyme solution is incubated with 200 μl of 0.5% pullulan buffered with 20 mM acetate buffer (pH 6.0 for *K. pneumoniae* pullulanase ; pH 5.0 for *B. acidopullulyticus* pullulanase) at 40℃ (*K. pneumoniae* pullulanase) or 55℃ (*B. acidopullulyticus*) for 5 min. The reducing sugar produced is determined by the Somogyi-Nelson method (14). One unit of the pullulanase activity is defined as the amount of the enzyme that catalyzes the hydrolysis of 1 μ mol of glycosidic bonds per min under the assay conditions.

Debranching amylase activity : The substrate solution contains 2.5 ml of 1.0% aqueous solution of glutinous rice starch and 0.5 ml of 0.5 M acetate buffer (pH 6.0 for *K. pneumoniae* pullulanase ; pH 5.0 for *B. acidopullulyticus* pullulanase). This substrate solution is incubated with 0.5 ml of enzyme at 40℃ (*K. pneumoniae* pullulanase) or 55℃ (*B. acidopullulyticus* pullulanase for one hour. An aliquot (2 ml) withdrawn from the reaction mixture is mixed with 25 ml of 0.01 N H_2SO_4 to stop the reaction. After standing at 30℃ for 30 min, 2 ml of 0.01 N iodine-potassium iodide solution is added to this solution and its volume is brought up to 50 ml with distilled water. After standing the solution for 15 min, the color density of the solution is read at 610 nm. The color value estimated immediately after the addition of the enzyme is subtracted as the reference from the value observed. One unit of debranching amylase activity is expressed as the amount of enzyme that increases the color value at E_{610nm} by 0.01 in one h under the conditions.

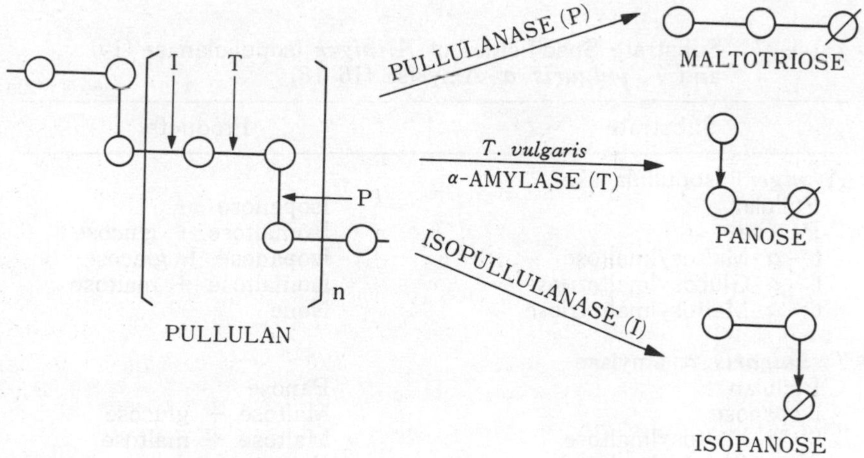

Fig. 2. Action Patterns of Pullulan-Hydrolyzing Enzymes.

Appendix : Two pullulan-hydrolyzing enzymes of other specificities (Fig. 2) are known. One is *Aspergillus niger* isopullulanase (pullulan 4-glucohydrolase) (15). This enzyme hydrolyzes the α-1, 4-glucosidic linkages in pullulan exclusively to produce isopanose. Another enzyme is *Thermoactinomyces vulgaris* α-amylase (16). This enzyme hydrolyzes α-1,4-glucosidic linkages in pullulan to produce panose. This enzyme also hydrolyzes starch to produce maltose as the main product. The enzymic and proteochemical properties of the two enzymes are shown in Tables 6 and 7, respectively.

Table 6. Enzymic and Proteochemical Properties of *A. niger* Isopullulanase (14) and *T. vulgaris* α-Amylase (15)

Property	*A. niger* Isopullulanase Extracellular	Intracellular※1	*T. vulgaris* α-amylase
Molecular weight	74,000	62,000	71,000※2
Optimum pH	3.0-3.5	3.5-4.0	5.0
Specific activity (unit/mg)	48	12.5	82
pH Stability	2.1-8.0	3.3-7.8	5.0-10.0
Thermal stability	55℃	50℃	65℃
Isoelectric point	4.65※3	4.8	5.2
Action type	Endowise	Endowise	Endowise

※1 Unpublished data (Sakano, Y. (1987)).
※2 Hiraiwa, S. (1980) the Dissertation for Master's Degree, Tokyo University of Agriculture and Technology.
※3 Tsujisaka, Y. and Hamada, N. (1971) *Proceedings of the Symposium on Amylase* (in Japanese) *6*, 17

Table 7. Substrate Specificities of *A. niger* Isopullulanase (15) and *T. vulgaris* α-Amylase (16-18)

Substrate	Products
A. niger isopullulanase	
Pullulan	Isopanose
Panose	Isomaltose + glucose
6^2-α-Maltosylmaltose	Isopanose + glucose
6^3-α-Glucosylmaltotriose	Isomaltose + maltose
6^2-α-Maltosylmaltotriose	None
T. vulgaris α-amylase	
Pullulan	Panose
Isopanose	Maltose + glucose
6^2-α-Maltosylmaltose	Maltose + maltose
6^3-α-Glucosylmaltotriose	Panose + glucose
4^G-α-Glucosylsucrose	Maltose + fructose

(Yoshiyuki Sakano)

References

1. Bender, H. &. Wallenfels, K. (1961) *Biochem. Z. 334,* 79-95
2. Ueda, S. & Nanri, N. (1967) *Appl. Microbiol. 15,* 492-496
3. Walker, G. (1968) *Biochem. J. 108,* 33-40
4. Norman B. E. (1983) *J. Jpn. Soc. Starch Sci. 30,* 200-211
5. Ueda, S., Yagisawa, M. & Sato, Y. (1971) *J. Ferment. Technol. 49,* 552-558
6. Manners, D. J. (1975) *Biochem. Soc. Trans. 3,* 49-53
7. Maeda, I., Jimi, N., Taniguchi, H. & Nakamura, M. (1979) *J. Jap. Soc. Starch Sci.* (in Japanese) *26,* 117-127
8. Ohba, R. & Ueda, S. (1975) *Agric. Biol. Chem. 39,* 967-972
9. Yokobayashi, K., Misaki, A. & Harada, T. (1970) *Biochim. Biophys. Acta 212,* 458-469
10. Kainuma, K., Kobayashi, S. & Harada, T. (1978) *Carbohydr. Res. 61,* 345-357
11. Ueda, S. & Ohba, R. (1972) *Agric. Biol. Chem. 36,* 2381-2391
12. Fujio, Y., Shiosaka, M. & Ueda, S. (1970) *J. Ferment. Technol. 48,* 8-13
13. Ohba, R. & Ueda, S. (1973) *Agric. Biol. Chem. 37,* 2821-2826
14. Somogyi, M. (1952) *J. Biol. Chem. 195,* 19-23
15. Sakano, Y., Higuchi, M. & Kobayashi, T. (1972) *Arch. Biochem. Biophys. 153,* 180-187
16. Shimizu, M., Kanno, K., Tamura, M. & Suekane, M. (1978) *Agric. Biol. Chem. 42,* 1681-1688
17. Sano, M., Sakano, Y. & Kobayashi, T. (1985) *Agric. Biol. Chem. 49,* 2843-2846
18. Sakano, Y., Sano, M. & Kobayashi, T. (1985) *Agric. Biol. Chem. 49,* 3041-3043

Ⅲ.1.c. Other Debranching Enzyme

Amylo-1,6-glucosidase/1,4-α-glucan : 1,4-α-glucan 4-α-glycosyltransferase (EC 3.2.1.33) is a multicatalytic enzyme which possesses two distinct activity sites on a single polypeptide chain (1,2). The two activities essential for debranching are of maltooligosaccharide transferase (1,4-α-D-glucan : 1,4-α-D-glucan 4-α-glycosyltransferase, EC 2.4.1.25) (abbreviated as oligotransferase), and amylo-1,6-glucosidase (dextrin 6-α-glucosidase, EC 3.2.1.33). Several evidences indicate that the activities of transferase and glucosidase are shown by separate sites, respectively, on the enzyme molecule (3,4). This enzyme is distributed in mammalian tissue (5,6) and yeast (7), and catalyzes the complete degradation of glycogen to glucose-1-phosphate and glucose in concert with phosphorylase. The enzymes from rabbit muscle (2) and *Saccharomyces cerevisiae* (baker's yeast) (8) have been purified to a homogeneous state in the test by ultracentrifugal and disc-gel electrophoretical analysis.

Reaction : The action patterns of amylo-1,6-glucosidase are schematically shown below :

(1) Action of oligotransferase

$$Gn + Gm \rightleftharpoons G_{n-2} + G_{m+2}$$
$$Gn + Gm \rightleftharpoons G_{n-3} + G_{m+3}$$

(2) Action of amylo-1,6-glucosidase

6^3-α-Glucosyl-maltotetraose ⟶ Glucose + Maltotetraose

6-α-Glucosyl-cyclodextrin ⟶ Glucose + Cyclodextrin

(3) Action of amylo-1,6-glucosidase-oligotransferase

```
   ↓
G-G-G-G              (reaction 1)                G
      |                                          | ←
G-G-G-G-G······G  ──────────────→   G-G-G-G-G-G-G-G······G
                                                     |
                                                     | (reaction 2)
      G-G-G-G-G-G-G-G······G + G  ←──────────────────┘
```

Upon incubation of the enzyme with phosphorylase limit dextrin, the action of oligotransferase is first shown and maltotriosyl groups out of the side chains of phosphorylase limit dextrin are transferred to the main chain (reaction 1). Then, the action of amylo-1,6-glucosidase appears which is specific to the stubbed glucose residues. Therefore, free molecules of glucose and maltodextrin are produced (reaction 2). Amylo-1,6-glucosidase also catalyzes the synthesis of α-1,6-glucosidic linkages by reverse reaction of the hydrolysis described above and incorporates ^{14}C-glucose into glycogen.

Specificity : In Table 1 are shown the substrate specificities of the yeast and rabbit muscle enzymes. The yeast enzyme readily debranches native glycogen and converts this molecule into a polysaccharide of a structure similar to amylopectin. On the other hand, the rabbit muscle enzyme shows only a small activity toward native glycogen (9). Table 2 shows the substrate specificity of the synthetic action of yeast and rabbit muscle enzymes. The results show that both enzymes display an inverse

Table 1. Substrate Specificity of Amylo-1,6-Glucosidases from Yeast and Rabbit Muscle (9)

Substrate	Relative rate of releasing glucose Yeast	Rabbit muscle
Rabbit liver glycogen	72	7
Amylopectin	4.5	<0.1
Phosphorylase limit dextrin from shellfish glycogen	100	100
β-Limit dextrin from shellfish glycogen	87	33
Phosphorylase limit dextin from amylopectin	29	33
β-Limit dextrin from amylopectin	43	13

specificity to that observed in the test shown in Table 1. Thus, the major differences between the specificities of the yeast and rabbit muscle enzyme are found to lie in the specificity of transferase.

Table 2. Relative Rates of Incorporation of ^{14}C-Glucose into Polysaccharide by Amylo-1,6-Glucosidases from Yeast and Rabbit Muscle (9)

Substrate	Yeast	Rabbit muscle
(100 mg/ml)		
Rabbit liver glycogen	100	100
Shellfish glycogen	94	83
Phosphorylase limit dextrin from glycogen	50	58
β-Limit dextrin from glycogen	22	24
(7 mg/ml)		
Amylopectin	67	99
Amylose	50	66
β-Limit dextrin from amylopectin	41	27
Rabbit liver glycogen	12	34

Production and Purification : The enzyme of *Saccharomyces cerevisiae* is obtained by extraction of baker's yeast cells with Eppenbach colloid mill and purified to a homogeneous statc by applying the following methods in the order : Acid treatment (pH 4.7) of the extract to inactivate other enzymes as much as possible, batchwise TEAE-Cellulose adsorption and extraction and column chromatographies by DEAE-Cellulose and Sephadex G-200. In general, five mg of purified enzyme is obtained from 8 1b. of baker's yeast.

Table 3. Enzymic and Proteochemical Properties of Amylo-1,6-Glucosidases from Yeast and Rabbit Muscle

	Yeast	Rabbit muscle
Molecular weight		
(by SDS-PAGE)	28,000	164,000
(by sedimentation equilibrium)	74,000 94,000 175,000	166,000
Optimum pH	6.0-6.5	6.0-6.5
E (1%, 280nm)	—	17.5
*K*m (glycogen phosphorylase limit dextrin)	1.6 (mM)	0.74 (mM)

The rabbit muscle enzyme is obtained by extracting rabbit muscle and purified to a homogeneous state by the ultracentrifugal and PAGE analytical tests by the methods of fractional precipitation with ammonium sulfate, DEAE-Cellulose column chromatography and hydrophobic chromatography on a column of Sepharose-$NH(CH_2)_4NH_2$.

About 200 mg of the purified enzyme is usually obtained from 1000 g muscle.

Assay Methods : The following two methods are available for assay of the enzymes. Method 1 (2). Determination of the activity to release glucose from glucosyl-α-cyclodextrin, glycogen phosphorylase limit dextrin or 6^3-α-glucosyl-maltotetraose. Two hundred μl of the reaction mixture containing 0.5% glycogen phosphorylase limit dextrin, 50 mM sodium maleate, 0.05% gelatin, 5 mM EDTA, 10 mM mercaptoethanol and 20 μl enzyme is incubated at 30℃ and at pH 6.0. Ten min later, the test tube containing the reaction mixture is transferred into a boiling-water bath for 80 sec, followed by cooling in a cold water bath. The glucose released is estimated by the method of using hexokinase-glucose 6-phosphate dehydrogenase-NADP reagents and photometrically reading the degree of reduction of NADPH at 340 nm. One unit of activity is defined as the amount of enzyme which releases 1 μmol of glucose per min under the conditions.

Method 2. Determination of the rate of ^{14}C-glucose incorporation into glycogen. One hundred μl of a solution containing 9.6 mg of glycogen and ^{14}C-glucose (20 cpm/p mol, 24 mM) is incubated with 20 μl of the enzyme in 50 mM glycylglycine, pH 7.0. After 5 and 15 min incubation at 30℃, 50 μl aliquots of the mixture are taken and adsorbed with Whatman GF/A glass fibre disc (2.1 cm diameter). The discs are washed five times with 15 ml of 66% ethanol per disc over 20 min and then, dipped into acetone and dried. The activity of ^{14}C is counted using a toluene-based scintillant. One unit of enzyme is defined as the amount of enzyme which incorporates one nanomol of glucose into glycogen per min under the conditions.

(Sumio Kitahata)

References

1. Brown, D. H. & Illingworth, B. (1964) in *Control of Glycogen Metabolism*, Whelan, W. J. & Cameron, M. P. ed., London, Churchill Ltd. p 169
2. Nelson, T. E., Kolb, E. & Larner, J. (1969) *Biochemistry 8*, 1419-1428
3. Nelson, T. E., Palmer, D. H. & Larner, J. (1970) *Biochim. Biophys. Acta 212*, 269-280
4. Bates, E. J., Heaton, G. M., Taylor, C., Kernohan, J. C. & Cohen, P. (1975) *FEBS Lett. 58*, 181-185
5. Coli, G. T. & Larner, J. (1951) *J. Biol. Chem. 188*, 17-29
6. Larner, J. & Gillespie, R. E. (1956) *J. Biol. Chem. 223*, 709-726
7. Lee, E. Y. C. & Whelan, W. J. (1966) *Arch. Biochem. Biophys. 116*, 162-167
8. Lee, E. Y. C., Carter, J. H., Nielsen, L. D. & Fisher, E. H. (1970) *Biochemistry 9*, 2347-2355
9. Lee, E. Y. C. & Carter, J. H. (1973) *Arch. Biochem. Biophys. 154*, 636-641

III.2. Branching Enzymes

Branching enzyme [1,4-α-D-glucan 6-α-D-(1,4-α-D-glucano)-transferase, EC 2.4.1.18] is a transglycosylase that splits a certain α-1,4-glucosidic linkage in α-glucans such as amylose and amylopectin (donor molecule), and transfers the chain segment produced at the non-reducing end side to another molecule or moiety of an α-1,4-glucosidic chain of α-glucan (acceptor molecule) with forming an anomalous linkage of α-1,6-bond. Namely, this enzyme converts amylose into a molecule of amylopectin type or amylopectin, and starch into a glycogen-like molecule. The transfer reaction appears to be irreversible; the reconversion of 1,6- into 1,4-linkages is not demonstrated (1). The existence of a branching enzyme was first observed in rabbit liver tissue by Coli and Coli (2), and named branching factor. On the other hand, a similar enzyme which synthesizes amylopectin-like polysaccharides was found in potato extracts by Haworth *et al.* (3) and named Q-enzyme, because it was found after phosphorylase (EC 2.4.1.1, P-enzyme).

Branching enzymes are widely distributed in plants, microorganisms and animal tissues. Plant branching enzymes convert amylose into amylopectin, but show little or no action on amylopectin. On the other hand, branching enzymes from microorganisms and animal tissues convert amylose and amylopectin into a highly ramified molecule like glycogen. The enzymes from plants are traditionally called Q-enzymes, while the enzymes from microbes and animals are called branching enzymes. These branching enzymes act in concert with starch synthase (EC 2.4.1.21) or glycogen synthase (EC 2.4.1.11) for biosynthesis of starch or glycogen.

III.2.a. Plant Branching Enzymes

Plant branching enzymes are found in potato (4-10), sweet corn (11-13), spinach leaf (14), rice plant (15), and banana leaves (16). The enzyme from sweet corn is purified to a preparation completely free from those enzyme contaminants that modify the substrates and products of the branching enzyme reaction such as α-amylase, α-glucosidase and debranching enzyme. The most extensively studied enzyme is the enzyme obtained in a crystalline state from potato.

Reaction : Amylose ⟶ Amylopectin

(1) Interchain transfer :

```
              ↓
G-G-G···G-G-G···G-G*
(donar chain)                            G-G-G···G                    G-G···G-G*
                        ⟶                        |             +
G-G-G···G-G-G···G-G*              G-G-G···G-G-G···G-G*
(acceptor chain)
```

(2) Intrachain transfer :

```
      ↓
G-G-G···G-G-G···G-G-G···G-G-G* ———→ G-G-G···G-G
                                              |
                                    G···G-G-G···G-G-G*
```

For conversion of amylose into a ramified molecule like amylopectin by the action of Q-enzyme, two mechanisms may be available, an interchain (reaction 1) or an intrachain transfer (reaction 2). The interchain transfer is evidenced by the experiments using a mixture of radioactive synthesized amylose and waxy maize amylopectin as substrate (9). But no experimental results have yet been obtained to clearly show the occurrence of intrachain transfer.

The branching action of Q-enzyme is followed by measuring the decrease in the iodine reaction color of the substrates until the converged level of color density is reached. Synthesis of branches by Q-enzyme is indicated by a decrease in the extent of β-amylolysis of the reaction mixture. Synthesis of branches is also shown by hydrolysis of the reaction mixture with isoamylase which yields various linear maltodextrins (Table 1). The absence of reducing sugars in the final reaction mixture with Q-enzyme indicates that no hydrolysis of the substrates occurs during the branching process.

Table 1. Action of Potato Branching Enzyme on Various α-Glucans (8)

	Average chain length	Degree of β-amylolysis	λ max at iodine stain(nm) before debranching	after debranching
Amylose ($\overline{DP}$ 260)	260	73	610	—
Amylose (DP 260) treated with Q-enzyme	23	48	540	565
Amylose ($\overline{DP}$ 2050)	2050	100	680	—
Amylose (DP 2050) treated with Q-enzyme	28	49	545	575
Simultaneous actions of Q-enzyme and corn phosphorylase on glucose 1-phosphate	31	50	545	575
Potato amylopectin	20	57	550	570
Potato amylopectin treated with Q-enzyme	17	56	500	540

Specificity : The Q-enzyme from banana leaf shows its action on amylose, but does not show any on amylopectin. The enzymes from potato and spinach leaf can act on both amylose and amylopectin. Two branching enzymes are obtained from sweet corn : One acts only on amylose, and the other, on both amylose and amylopectin. As shown in Fig. 1, the branches fabricated by the reaction of Q-enzyme on amylose are greatly diverse in the glucosidic chain length, though the average length becomes shorter as the reaction proceeds. This result may indicate that there is no regularity in the size of the α-1,4-glucosidic chain to be split from and transferred to the substrate by the enzyme. In the reaction at 35℃, the minimum size of amylose effective as substrate to result in branching is about 30 glucose residues (Fig. 2) (7). But, at temperature as low as 4℃, even an amylose consisting of only 10 glucose residues is effective as substrate. This result is compatible with the fact that the double helices of amylose become more stable as the temperature decreases. The average chain length ($\overline{CL}$) of the products produced by Q-enzyme on amylose ($\overline{DP}$ 260) in the presence of pullulanase (EC 3.2.1.41) is about 40. However, in the absence of pullulanase, the average chain length is nearly 20 glucose residues. This result may show that the amylose effective as substrate has a longer $\overline{CL}$ in the presence of pullulanase. French reported that amylose chains possibly exist in a state of double helices and the branch point initiates double helices (17). The above things indicate that once the first branch is formed, the following branches are formed more easily. For branching by potato enzyme, maltohexaose chain is the smallest segment to be transferred. Therefore, maltohexaose remains as an outer segment of donor molecule after the transfer reaction of a branching enzyme. The polysaccharides produced by transfer action of potato branching enzyme on native and synthetic amyloses, have a structure like amylopectin in terms of average chain length, degrees of β-amylolysis and color at iodine stain, but the profiles of the chain length of the products are different from those of native amylopectin.

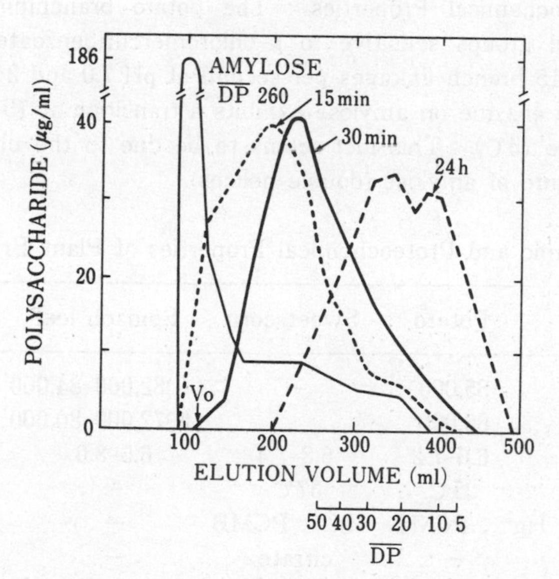

Fig. 1. Unit Chain Profiles of Products on Various Incubation Periods of Potato Q-Enzyme with Amylose ($\overline{DP}$ 260) (9).

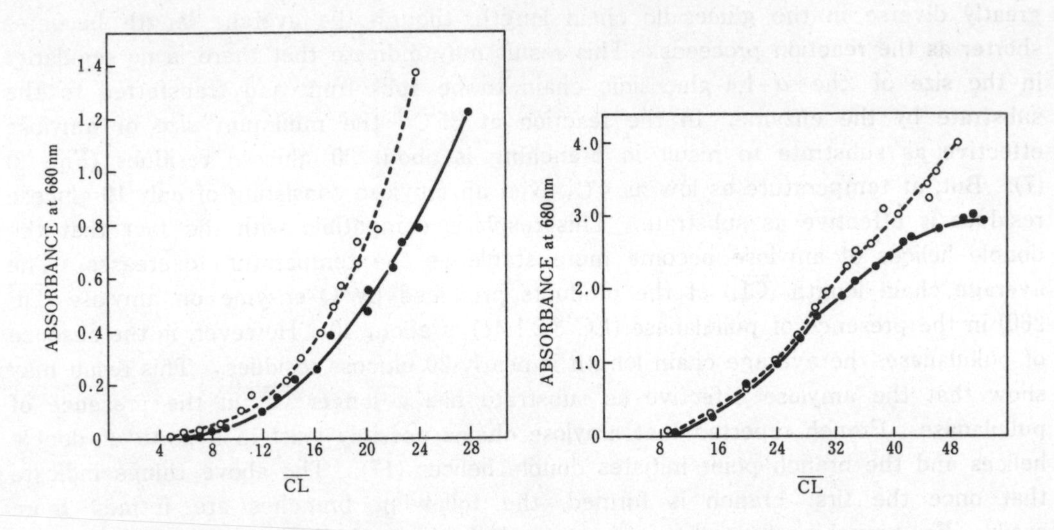

Fig. 2. Action of Potato Q-Enzyme on Growing Amylose Chains at 4℃ (left) and at 35℃ (right) (7).
The plot is of the iodine stain absorbance versus average chain length ($\overline{CL}$) of the amylose synthesized.
○…○ , Phosphorylase alone ; ●…● , Phosphorylase + Q-enzyme.

Enzymic and Proteochemical Properties : The potato branching enzyme possesses 10–12 free sulfhydryl groups sensitive to *p*-chloromercuribenzoate. One mol of the enzyme synthesizes 15 branch linkages per second at pH 7.0 and 24℃. The activation energy profile of the enzyme on amylose exhibits a transition at 15℃ (25 kJ/mol below 15℃, 63kJ/mol above 15℃). This fact seems to be due to the change in stability of the secondary structure of amylose (double helices).

Table 2. Enzymic and Proteochemical Properties of Plant Branching Enzymes

	Potato	Sweet corn	Spinach leaf	Banana leaf
Molecular weight	85,000[1] 86,000[3]	—	(I)82,000–84,000[1] (II)72,000–80,000[1]	185,000[2]
Optimum pH	6.6–7.2	6.8–7.4	6.5–8.0	6.2
Optimum temp.	25℃	37℃	—	—
Inhibitor	Hg^{2+}, PCMB	Hg^{2+}, PCMB	—	—
Activator	—	citrate	—	—

[1], [2] and [3] were determined by the method of sucrose density gradient ultracentrifugation, SDS-PAGE and gel filtration, respectively.

Production and Purification : The potato enzyme is obtained by extraction of young, red-skinned potatoes and purified to nearly a homogeneous state (Purity checked by disc-electrophoresis is over 95%.) by successive chromatographies on DEAE-Cellulose (repeated three times), hydroxylapatite, and Bio-Gel P-300 (repeated twice). A purified enzyme preparation is usually obtained in a rate of about 0.6 mg per 1.0 kg of potatoes. Two enzymes, branching enzyme I and II, are obtained by extraction of freshly harvested spinach leaves and purified to be free from α-amylase, isoamylase and disproportionating enzyme by ammonium sulfate fractionation, DEAE-Cellulose and ADP-hexanolamine-Sepharose 4B chromatographies. Enzyme I and II are usually obtained in a yield of about 3.5 and 12 mg from 1.0 kg of the raw material, respectively.

Assay Methods : Branching enzyme activity is conveniently assayed by measuring the extent in decrease of iodine reaction color of amylose and amylopectin. A mixture of 50 μl consisting of 0.1% amylose, 50 mM buffer and 5 mM mercaptoethanol is incubated with 50 μl of the enzyme at 25℃. The reaction is terminated by addition of 1 ml of 10 mM iodine solution containing 5 mM hydrogen chloride. The color density developed is measured at 660 nm and compared with that of the control containing acid-denatured enzyme. One unit of enzyme activity is defined as the amount of enzyme that decreases the absorbance at 660 nm by 1% per min under the conditions. This method, however, is not quantitative, because it is impossible to show the activity in molar basis of linkages formed.

(Sumio Kitahata & Shigetaka Okada)

References

1. Barker, S. A., Bourne, E. J., Wilkinson, I. A. & Peat, S. (1950) *J. Chem. Soc.* 84-92
2. Coli, G. F. & Coli, C. F. (1943) *J. Biol. Chem. 151,* 57–61
3. Haworth, W. H., Peat, S. & Bourne, E. J. (1944) *Nature 154,* 236-238
4. Gilbert, G. A. & Patrick, A. D. (1952) *Biochem. J. 51,* 181-186
5. Griffin, H. L. & Victor Wu, Y. (1971) *Biochemistry 10,* 4330-4335
6. Drummond, G. S., Smith, E. E. & Whelan, W. J. (1972) *Eur. J. Biochem. 26,* 168-176
7. Borovsky, D., Smith, E. E. & Whelan, W. J. (1975) *FEBS Letters 54,* 201-205
8. Borovsky, D., Smith, E. E. & Whelan, W. J. (1975) *Eur. J. Biochem. 59,* 615-625
9. Borovsky, D., Smith, E. E. & Whelan, W. J. (1976) *Eur. J. Biochem. 62,* 307-312
10. Borovsky, D., Smith, E. E., Whelan, W. J., French, D. & Kikumoto, S., (1979) *Arch. Biochem. Biophys. 198,* 627-631
11. Lavintman, N. (1966) *Arch. Biochem. Biophys. 116,* 1-8
12. Manners, D. J., Rowe, J. J. M. & Rowe, K. L. (1968) *Carbohydr. Res. 8,* 72-81
13. Boyer, C. D., Simpson, E. K. G. & Damewood, P. A. (1982) *Stärke 34,* 81-85
14. Hawker, J. S., Ozbun, J. L., Ozaki, H., Greenberg, E. & Preiss, J. (1974) *Arch. Biochem. Biophys. 160,* 530-551

15. Igaue, I. (1962) *Agric. Biol. Chem.* *26*, 424-433
16. Kumor, A. & Sanwl, G. G. (1979) *Indian J. Experimental Biology 17*, 385-387
17. French, D. (1972) *J. Jap. Soc. Starch Sci.* *19*, 8-25

Ⅲ.2.b. Mammalian Branching Enzymes

Mammalian branching enzyme converts amylose and amylopectin into glycogen-like molecules. This enzyme is found in rat liver (1-4), rabbit liver (5), and rabbit skeletal muscle (6-8). Rabbit liver has 30 times more activity than the muscle per a given weight. Glycogen branching enzyme acts in concert with glycogen synthase during synthesis of glycogen. The hereditary deficiency of glycogen branching enzyme causes a severe metabolic disease, Type Ⅳ glycogenosis (glycogen storage disease) (9). The polysaccharide in tissues of the patients is named Type Ⅳ "amylopectin" and has characteristic long, unbranched outer chains. These branching enzymes are purified to a state of SDS-disc-electrophoretic homogeneity.

Reaction : Amylose, amylopectin $\xrightarrow[\text{enzyme}]{\text{branching}}$ Glycogen-like molecule

Specificity : The mammalian branching enzyme catalyzes the transfer of α-1,4-glucosidic chains, the minimum chain length effective to be transferred is six glucose units and the maltooligosaccharide preferentially transferred is maltoheptaose (2). Maltodextrin of 16 glucose units is not available for the branching enzyme. The limit dextrin from glycogen whose outer chains are six glucose units in average is not available as substrate for more branching, while the polysaccharides whose outer chains are from 11 to 21 glucose units are available for more branching (6).

The results of the structural analysis of Type Ⅳ "amylopectin" obtained before and after treatment with branching enzyme, indicate that branching enzyme from rabbit muscle branches both A and B chains to nearly the same extent. Furthermore, the branching enzyme introduces many new outer branches at the average position where 4 glucose units are away from the nearest preexisting outer branches in the parent polysaccharide (7). The glycogen prepared *in vitro* using liver glycogen synthase, branching enzyme and UDP-glucose as glucose donor, is similar to native liver glycogen in iodine complex spectrum, sedimentation coefficient by sucrose gradient and in the effect of acid or alkali treatment. But, the glycogen obtained above differs from that prepared using phosphorylase, branching enzyme, and glucose-1-phosphate as glucosyl donor (3).

Enzymic and Proteochemical Properties : The amino acid composition is known for the enzyme from rabbit liver. Rabbit liver enzyme contains no cystine, but contains 7 residues of cysteine per mol. Two branching enzymes (BE-Ⅰ and BE-Ⅱ) are obtained from rat liver. BE-Ⅰ is associated with a RNA component, while BE-Ⅱ is not associated with this component. The specific activity of BE-Ⅱ is about half that of

BE-I. The RNA component is not essential for the activity of rat liver enzyme. The proteins of BE-I and BE-II are identical. The enzyme from rabbit muscle is also associated with 2.5 s RNA component (10). The branching activity of rabbit muscle is inhibited to 50% by 0.4 M urea, and 100% by 2 M urea. The inhibition by 2 M urea is reversed by dilution. Maltooligosaccharides consisted of 6 to 10 glucose units also inhibit the branching enzyme. Maltoheptaose is the most inhibitory among these oligosaccharides (40% inhibition at 15 mM). The activity of this enzyme is nearly doubled in the presence of 0.15 mM citrate or 50 mM glucose-1-phosphate (7).

Table 1. Action of Rat Liver Branching Enzyme on Various α-Glucans (1)

Substrates	β-Amylolysis (%)	Average chain length	Non-reducing end groups (%)	λmax* (nm)
Amylopectin	53	23	4.4	—
Amylopectin treated with branching enzyme	50	18	5.6	—
β-Limit dextrin of amylopectin	—	16.4	—	520
β-Limit dextrin of amylopectin treated with branching enzyme	8	11.7	—	460
Glycogen	45	15	6.7	—

* Determined with the iodine complex

Table 2. Enzymic and Proteochemical Properties of Mammalian Branching Enzymes

	Rabbit muscle	Rabbit liver	Rat liver
Molecular weight	92,000–103,000[1] 77,000[2], 60,000[3]	93,000[1], 71,000[2] 52,000[3]	(I)91,000[1], 82,000[2] (II)98,000[2], 82,000[2]
Optimum pH	6.8–7.8	—	6.4
Optimum temp.	30–35℃	—	—
Inhibitor	maltoheptaose, urea	—	Hg^{2+}, Mg^{2+}, pCMB
Activator	citrate, glucose-1-phosphate	—	citrate

[1], [2] and [3] were determined by the method of sucrose density gradient ultracentrifugation, SDS-PAGE and gel filtration, respectively.

Production and Purification : Both BE-I and BE-II are purified from rat liver by hydrophobic chromatography on hexylamine-Sepharose 4B, ammonium sulfate fractionation, and affinity chromatography on glycogen adipoyldihydrazide-Sepharose 4B or glycogen-ethylenediamine-Sepharose 4B. The yield of purified BE-I and BE-II are 2.8 mg and 4.1 mg in average, respectively, from 50 g of rat liver (4). Branching enzyme from rabbit liver is purified by ammonium sulfate fractionation, DEAE-Cellulose column chromatography, and chromatofocusing on Polybuffer exchanger PBE

94. The purified enzyme of 12 mg is obtained from 150 g of frozen rabbit liver (5). Branching enzyme from rabbit muscle is purified by ammonium fractionation, DEAE-Sepharose column chromatography, and Sephadex G-150 column chromatography. The purified enzyme of 0.6 mg is obtained from 4000 g of the muscle.

Assay Method : Mammalian and microbial branching enzyme activity is assayed by the following two methods besides the iodine color method ; one is the method of measuring the amount of orthophosphate formed from glucose-1-phosphate (Method 1) and the other, measuring the amount of ^{14}C-glucose incorporated into glycogen from ^{14}C-glucose-1-phosphate (Method 2). These assay methods are based on the degree of stimulation in the rate of formation of polysaccharides from glucose-1-phosphate by phophorylase caused by branching action.

Method 1 (by Satoh, K & Sato, K)(11). The reaction mixture (40 μl) containing 20 μl of 0.2 M glucose-1-phosphate and 4 mM AMP, 10 μl of branching enzyme diluted appropriately with 10 mM pH 7.4 Tris-HCl buffer containing 15 mM mercaptoethanol, and 10 μl of rabbit muscle phosphorylase b solution (1 mg/ml) is incubated at 30℃. Ten min later, the reaction is terminated by adding 0.5 ml of trichloroacetic acid and the inorganic phosphate liberated is determined by Fiske-Subbarow method. One unit of enzyme activity is defined as the amount of branching enzyme that liberates 1 μmol of inorganic phosphate from glucose-1-phosphate per min under the conditions.

Method 2 (by Boyer, C & Preiss, J) (12). The reaction mixture (0.2 ml) containing 0.1 M sodium citrate (pH 7.0), 1 mM AMP, 50 mM ^{14}C-glucose-1-phosphate (5 x 10^4 cpm/μmol) and rabbit muscle phosphorylase a solution (40 μg, 1200 units/mg) is incubated with the branching enzyme solution at 30℃. After 30, 60, and 120 min incubation, aliquots of 50 μl are removed and placed in a boiling water bath for one min. To each aliquot is added 0.1 ml of glycogen solution (10 mg/ml) and 2 ml of 75% methanol containing 1% KCl. After cooling in an ice-water bath for 5 min, the glycogen precipitated is centrifuged for 5 min in a clinical centrifuge. The glycogen precipitate is redissolved in 0.2 ml H_2O and reprecipitated with 75% methanol-KCl solution. After decantation, the precipitate is washed once with 2 ml of 75% methanol-KCl solution and then, dissolved in 1.0 ml of H_2O. A 0.5 ml aliquot is dissolved in 10 ml aquasol and counted in a scintillation spectrometer. One unit of enzyme is defined as the enzyme amount that makes transfer 1 μmol of ^{14}C-glucose transferred per min.

(Sumio Kitahata & Shigetaka Okada)

References

1. Krisman, C. R. (1962) *Biochem. Biophys. Acta 65,* 307-315
2. Verhue, W. & Hers, H. G. (1966) *Biochem. J. 99,* 222-227
3. Parodi, A. J., Mordoh, J., Krisman, C. R. & Leloir, L. F. (1969) *Arch. Biochem. Biophys. 132,* 111-117

4. Satoh, K. & Sato, K. (1982) *J. Biochem. 91,* 1129-1137
5. Zimmerman, C. P. & Gold, A. M. (1983) *Biochemistry 22,* 3387-3392
6. Brown, D. H. & Brown, B. I. (1966) *Biochim. Biophys. Acta 130,* 263-266
7. Gibson, W. B., Brown, B. I. & Brown, D. H. (1971) *Biochemistry 10,* 4253-4262
8. Caudwell, F. B. & Cohen, P. (1980) *Eur. J. Biochem. 109,* 391-394
9. Brown, B. I. & Brown, D. H. (1966) *Proc. Natl. Acad. Sci. U. S. A. 56,* 725-729
10. Korneeva, G. A., Shvedova, T. A. & Shaposhnikov. G. L. (1982) *Mol. Biol.* (in Russian) *16,* 731-738
11. Satoh, K. & Sato, K. (1980) *Anal. Biochem. 108,* 16-24
12. Boyer, C. & Preiss, J. (1977) *Biochemistry 16,* 3693-3699

Ⅲ.2.c. Microbial Branching Enzymes

Microbial branching enzyme also converts amylose and amylopectin into glycogen-like molecules. This enzyme is found in *Streptococcus mitis* (1), *Escherichia coli* (2,3), *Bacillus megaterium* (4), *B. cereus* (4), *Arthrobacter globiformis* (5), brewer's yeast (6), and *Neurospora crassa* (7,8). The enzymes from *E. coli, B. megaterium, B. cereus* and *St. mitis* are purified to a preparation free from enzyme contaminants which affect the substrates and products of the branching enzyme. The enzyme from *N. crassa* is purified to a state of disc-electrophoretic homogeneity.

Reaction : Amylose, amylopectin ⟶ Glycogen-like molecule

The branching action of this enzyme is followed by measuring the decrease in the iodine reaction color of the reaction mixture. Introduction of branching points by the enzyme into amylopectin is indicated by the decrease in wavelength of the maximum absorption of glucan-iodine complex, decrease in the extent of β-amylolysis of the reaction products, and the increase in the extent of β-amylolysis after treating the reaction products with isoamylase (Fig. 1, Table 1).

Specificity : The microbial branching enzymes react not only with amylose, but also with amylopectin, β-limit dextrin and glycogen. The enzyme from *B. megaterium* fabricates branching points on maltodextrins consisted of more than 9 glucose units and branched dextrins with α-1,4-glucosidic chains consisted of more than 12 glucose residues. The minimum oligosaccharide chain to be transferred by branching enzymes of *E. coli, N. crassa, B. megaterium* and *B. cereus* is maltohexaose.

The polysaccharides prepared from amylose and amylopectin with branching enzymes of *E. coli* and *N. crassa,* are similar to native glycogen in λmax shown by the glucan-iodine complex and average degree of polymerization of the unit chains, but the profiles of the unit chains of the products are different from those of native glycogen. In addition, unlike native glycogen, the branched glucans synthesized by the microbial enzymes are hydrolyzed by pullulanase. The glucans produced from

ADP-glucose by the combined action of *E. coli* branching enzyme and phosphorylase a or glycogen synthase in a ratio of 20 : 1 in activity are very similar to native glycogen in all the tests described above. Like native glycogen, these synthetic glucans have molecular weights greater than 10^7, and are hardly hydrolyzed by pullulanase. The profiles of unit chains after debranching with isoamylase are also similar to that of native glycogen.

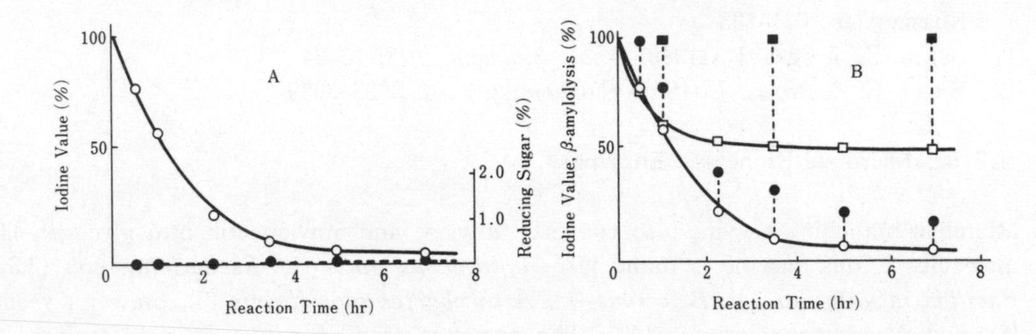

Fig.1. Action of Branching Enzyme from *B. megaterium* on Amylose (4).
A, Time course of iodine value (○) and reducing sugar formation (●); B, time course of iodine value before (○) and after (●) isoamylase treatment, and β-amylolysis before (□) and after (■) isoamylase treatment of the reaction products by branching enzyme.

Table 1. Action of *N. crassa* Branching Enzyme on Various α-Glucans (8)

Substrate	λ max of iodine complex before treatment	λ max of iodine complex after treatment	β-Amylolysis (%) before treatment	β-Amylolysis (%) after treatment
Amylopection	505	480	62	50
Amylose (DP 130)	600	570	77	48
Amylose (DP 22)	493	470	100	84
Amylose (DP 15)	437	435	100	86
Amylopectin β-limit dextrin	530	505	0	6.8
N. crassa glycogen	470	465	45	43

Enzymic and Proteochemical Properties : The amino acid composition is known for the enzymes from *N. crassa* and *E. coli* (9). The amino acid compositions of both branching enzymes are significantly similar to each other, except for the contents of proline ; the *E. coli* enzyme contains 37.5 residues of proline per mol, but the *N. crassa* enzyme contains no appreciable amount of this amino acid. The branching enzymes I and II isolated from *E. coli* have 1.16 mg and 10 μg of anhydroglucose units per mg of protein, respectively. The *N. crassa* enzyme contains 30 μg of carbohydrate as glucose per mg protein.

St. mitis branching enzyme is activated by citrate, and the optimum concentration is 50 mM. In the absence of citrate, Tris-HCl buffer (pH 7.0) activates the branching enzyme. The optimum concentration of the buffer is 0.2 M. But the activation effect of Tris-HCl buffer is only about one-third of that of citrate and no activation occurs by Tris-HCl buffer in the presence of citrate. On the contrary, the addition of 0.1 M Tris-HCl buffer in the presence of citrate inhibits the enzyme action.

Table 2. Enzymic and Proteochemical Properties of Microbial Branching Enzymes

	B. megaterium	*E. coli*	*St. mitis*	*N.crassa*	Yeast
Molecular weight	85,000[1] 51,000[3]	(I)95,000[1] (II)84,000[3],	87,000[1] 84,000[3]	—	80,000[2]	—
Optimum pH	7.6	6.8-8.0	6.8-7.0	8.0	7.0
Optimum temp.	25℃	—	30℃	27℃	20℃
Inhibitor	Hg^{2+}, Zn^{2+}, pCMB	—	Hg^{2+}, Tris	—	—
Activator	—	citrate	citrate	—	citrate

[1], [2] and [3] were determined by the method of sucrose density gradient ultracentrifugation, SDS-PAGE and gel filtration, respectively.

Production and Purification : All microbial branching enzymes are intracellular. *E. coli* B strains are cultivated in 100 ℓ of a medium containing 62.5 mM KH_2PO_4, 0.063 M K_2HPO_4, 0.6% yeast extract, and 0.4% acetate (pH 7.0) in a 130 ℓ fermentor. Two branching enzymes, I (7.41 mg) and II (1.32 mg) are obtained in a purified state from the grown cells (70 g) by extracting with 20 mM phosphate buffer (pH 7.0), followed by ammonium sulfate fractionation, heat treatment, DEAE-Cellulose and 4-aminobutyl-Sepharose chromatographies. The yeast branching enzyme is extracted from dried brewer's yeast with 0.1 M $NaHCO_3$, and purified to a preparation free from α-amylase and debranching enzyme (isoamylase and pullulanase) by fractionation with ammonium sulfate and ethanol. *N. crassa* enzyme is extracted from mycelia obtained by cultivation of wild type strain of *N. crassa* (IFO 6068) at 28℃ for 2 days in Vogel's N minimum medium with 2% glucose. The mycelial extract with 50 mM Gly-Gly buffer (pH 8.0) is purified to disc-electrophoretic homogeneity by chromatographies on DEAE-Sephacel, 6-aminohexyl-Sepharose 4B, and Toyo-pearl HW-55 S.

(Sumio Kitahata & Shigetaka Okada)

References

1. Walker G. J. & Builder J. E. (1971) *Eur. J. Biochem. 20,* 14-21
2. Sigal, N., Cattaneo, J., Chambost, J. P. & Favard, A. (1965) *Biochem. Biophys. Res. Commun. 20,* 616-620
3. Boyer, C. & Preiss, J. (1977) *Biochemistry 16,* 3693-3699
4. Okada, S., Yoshikawa, S., Taniguchi, M. & Kitahata, S. (1983) *J. Jpn. Soc. Starch Sci.* (in Japanese) *30,* 223-230

5. Zevenhuizen, L. P. T. M. (1964) *Biochim. Biophys. Acta 81,* 608-611
6. Gunja, Z. H., Manners, D. J. & Maung, K. (1960) *Biochem. J. 75,* 441-450
7. Matsumoto, A., Kamata, T. & Matsuda, K. (1983) *J. Biochem. 94,* 451-458
8. Matsumoto, A. & Matsuda, K. (1983) *J. Jpn. Soc. Starch Sci.*(in Japanese) *30,* 212-222
9. Holmes, E., Boyer, C. & Preiss, J. (1982) *J. Bacteriol. 151,* 1444-1453

III.3. Cyclomaltodextrin Glucanotransferase

Cyclomaltodextrin glucanotransferase [1,4-α-D-glucan 4-α-D-(1,4-α-D-glucano)-transferase (cyclizing), EC 2.4.1.19, CGTase] catalyzes the conversion of starch to cyclodextrin (CD) by intramolecular transglycosylation (cyclization reaction). In the presence of a suitable acceptor (such as glucose or sucrose), this enzyme also catalyzes the intermolecular transglycosylation in which glycosyl residues are transferred from an α-1,4-glucan or a CD to the acceptor (coupling reaction or disproportionation reaction). Furthermore, this enzyme catalyzes the hydrolysis of α-1,4-glucans and CDs. In 1939, Tilden and Hudson obtained a cell free enzyme preparation from cultures of *Bacillus macerans* which showed the activity to convert starch to CD (1). Since then, the enzyme has been investigated in detail (2-10), although, no other effective sources of the enzyme were found. Since Okada *et al.* in 1972 observed that *B. megaterium* strain No.5 produced CGTase, several other bacteria besides *B. macerans* were found to be producers of CGTase (11-22). At present, nine bacteria [*B. macerans* (14), *B. megaterium* (11,14), *B. circulans* (12), *Bacillus* sp. (alkalophilic) (15-17), *B. stearothermophilus* (13,18), *Klebsiella pneumoniae* (19), *B. ohbensis* (20), *B. subtilis* No.313 (21), *Bacillus* sp. AL6 (22)] are known to produce CGTase. But, no fungi or yeasts which produce CGTase have been found yet.

The enzymes described above are classified into the following three types depending on the kind of CD that is mainly produced from the α-1,4-glucans by intramolecular transglycosylation : *B. macerans* type enzyme produces mainly cyclohexaamylose (α-CD), *B. megaterium* type enzyme, mainly cycloheptaamylose (β-CD), and *Bacillus* sp. AL6 type enzyme, mainly cyclooctaamylose (γ-CD). CGTase is one of the industrially important enzymes, and utilized in production of CD (see VI. 9), maltooligosylsucrose (Coupling sugar®)(see VI.8), glycosylstevioside, etc.

Reaction : CGTase catalyzes three reactions ; intramolecular transglycosylation, intermolecular transglycosylation and hydrolysis as follows :

(1) Intramolecular transglycosylation,
Starch ⟶ α-, β-, and γ-CD ;

(2) Intermolecular transglycosylation,
Starch + sucrose (as acceptor) ⟶ Maltooligosyl-sucrose ;

(3) Hydrolysis,
Starch ⟶ Maltooligosaccharides.

(1) Intramolecular transglycosylation

CGTase produces CDs by intramolecular transglycosylation (23) attacking exowise certain α-1,4-glucosidic linkages positioned at the non-reducing end side of the α-1,4-glucans larger than maltoheptaose (G7). Table 1 shows the yields of CDs from various α-1,4-glucans by the CGTases of *B. megaterium*, *B. ohbensis* and alkalophilic *Bacillus* sp. In the case of maltose (G2) and maltotriose (G3), CGTase first catalyzes intermolecular transglycosylation to G2 and G3 to produce maltooligosaccharides larger than G7 and then, produces CDs from these maltooligosaccharides (24). The yields of CDs produced from G2 and G3 decreases with time of incubation after a certain period. This result is attributed to a decomposition of the CDs by intermolecular transglycosylation to G2 and G3 acceptors. On soluble starch, amylose or amylopectin as substrate, CGTase catalyzes only intramolecular transglycosylation and produces CDs exclusively. Therefore, the amount of CDs produced increases with time of incubation.

Table 1. Yields of CDs from Various α-1,4-Glucans by Several CGTases

Substrate	*B. megaterium*			*B. ohbensis*	*Bacillus* sp. (alkalophilic)
	Yield(%)				
	Time(min)				
	30	60	240		
Potato starch (1%)	60	62	62	58	75-80
Amylopectin (1%)	55	58	60	46	65-70
Oyster glycogen (1%)	41	41	42	26	55-60
Amylopectin β-limit dextrin (1%)	39	43	47	—	45-50
Amylose (0.2%)	72	80	82	64	85-90
Maltose (1%)	5	12	9	0	10-15
Maltotriose (1%)	18	23	20	4	20-25

The ratio of α-, β- and γ-CD produced from soluble starch differs depending on the source of CGTase. The main CDs produced are : α-CD, by CGTases from *B. macerans*, *B. stearothermophilus* (25) and *K. pneumoniae* ; β-CD , by the enzymes from *B. megaterium*, *B. circulans*, alkalophilic *Bacillus* sp. and *B. ohbensis* ; γ-CD, by the enzyme from *Bacillus* sp. strain AL6. The enzyme from *B. subtilis* No.313 characteristically produces only γ-CD from soluble starch. CGTases from *B. macerans* IFO 3490 and *B. megaterium* produce α-, β- and γ-CD in a ratio of 5.7 : 1 : 0.4 and 1 : 6.3 : 1.3, respectively, from 1% soluble starch in a relatively early stage of the reaction. The ratios of these CDs produced described above are almost constant until the production yield exceeds 30%. But at later stage of the enzyme reaction at which CD production yield goes over 30%, the ratios of α-, β-and γ-CD produced change into 2.1 : 1 : 0.3 and 1 : 2.5 : 0.5 in case of the enzymes from *B. macerans* and *B. megaterium*, respectively (Fig.1). This result may be attributed to decomposition of the α-, β- and γ-CD formed at relative early stage of the reaction followed by reproduction of CDs from the decomposition products. *B. ohbensis* CGTase belonging to the group which produces mainly β-CD, produces β- and γ-CD in the ratio of 5 : 1 and does not produce α-CD. CGTase from *Bacillus* sp. AL6 produces β- and γ-CD in the ratio of 1 : 2.7 and does not produce α-CD. CGTase of *B. subtilis* № 313 is different from other CGTases in the action on soluble starch. This enzyme produces γ-CD and various maltooligosaccharides from starch.

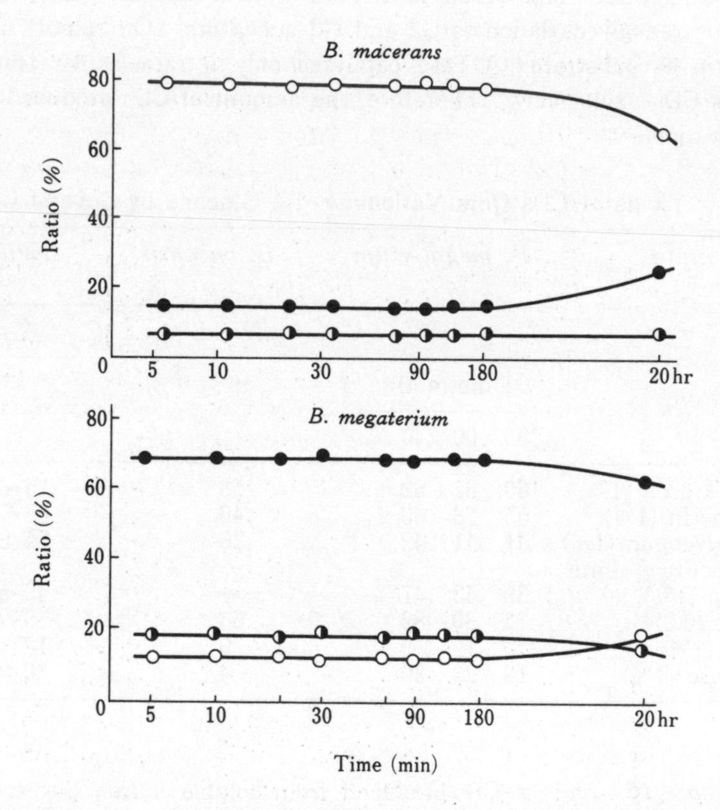

Fig. 1. Ratio of α-, β- and γ-CD Produced by CGTases from *B. macerans* and *B. megaterium*.
(○), α-CD ; (●), β-CD ; (◑), γ-CD

The production ratio of CDs is influenced, however, by the presence of surfactants which cause a helical conformational change of the substrate (26). Surfactants which have straight carbon chains as a hydrophobic moiety, such as sodium lauryl sulfate (SDS), significantly stimulate the formation of α-CD. On the other hand, surfactants such as isooctylphenylpolyoxyethylene (Triton) whose hydrophobic moieties are more bulky than straight carbon chains, are extremely effective for preferential formation of β-CD. Production of γ-CD by *K. pneumonia* CGTase is increased as much as 19% in the presence of 200 mM sodium acetate (which modifies conformation of the substrate to be favourable for formation of γ-CD) and by the addition of bromobenzene (complexing agent) after pre-incubation for 7 h (27). The addition of stevioside and glycyrrhizin as clathrate-forming compounds is greatly effective for γ-CD formation in the case of *B. ohbensis* CGTase.

(2) Intermolecular transglycosylation

The structural features of the acceptor sugars in the reaction by CGTase from *B. megaterium* and *B. macerans* are shown in Table 2 and Fig. 2 (29). The acceptor specificity of *B. megaterium* CGTase is similar to that of *B. macerans* CGTase. But, the intermolecular transglycosylation activity of *B. megaterium* CGTase is stronger than that of *B. macerans* CGTase. The effective monosaccharides and their derivatives are classified into three groups depending on the efficiency as acceptor. L-Sorbose, D-glucose, D-xylose, 6-deoxy-D-glucose, methyl-α-, and β-D-glucosides and phenyl-α- and β-D-glucosides are a group of most effective acceptors (Group A). 2-Deoxy-D-glucose and 3-O-methyl-D-glucose (Group B) are also effective as acceptors, but slightly less than those of Group A. Other sugars and sugar alcohols (Group C) are little or not effective as acceptors. These results show that the structure of an effective acceptor in the intermolecular transfer action by these CGTases is of the pyranosyl type with the same configurations as glucopyranose, namely with free C2-, C3- and C4-hydroxyl groups. D-Glucuronic acid is almost ineffective as acceptor in the case of *B. megaterium* and *B. macerans* CGTase, but it is considerably effective in the case of *B. stearothermophilus* CGTase.

These CGTases transfer glycosyl residues only to the hydroxyl group at the C4 position of D-glucose, D-xylose, 6-deoxy-D-glucose, 2-deoxy-D-glucose and 3-O-methyl-D glucose with the exception of C3-hydroxyl group of L-sorbose. D-Galactose is a very poor acceptor in the intermolecular transglycosylation of *B. megaterium* and *B. macerans* CGTase and the yield of transfer products to D-galactose is only 2~3% of those to D-glucose, L-sorbose and D-xylose (30). However, the transglycosylation to D-galactose occurs at several hydroxyl groups of the sugar ; the proportion accepted at C1-, C3- and C2-(C4-)hydroxyl groups are 26:10:1, respectively. The transglycosylation to gluco-disaccharides occurs mainly (isomaltose and nigerose) or exclusively (kojibiose) at the C4-hydroxyl group of non-reducing end glucose residue (31).

B. macerans enzyme catalyzes the intramolecular transglycosylation with glucosidic chains of mainly six glucose residues as a transfer unit, so mainly produces α-CD from soluble starch. The enzyme also catalyzes the intermolecular transglycosylation to produce radioactive G7 at the initial stage of the reaction with soluble starch or α-CD as donor and radioactive glucose as acceptor. On the other hand, the intramolecular transglycosylation catalyzed by *B. megaterium* enzyme has usually a seven glucose residue transfer unit, thus β-CD is mainly produced. However, this enzyme produces various radioactive maltooligosaccharides from the early stage of reaction in the mixture of soluble starch or α-CD and radioactive glucose (Fig. 3). The number of multiple attack calculated from the experiment using α-CD as donor and radioactive glucose as acceptor, is 0.22 for the *B. macerans* enzyme and 1.50 for the *B. megaterium* enzyme. Therefore, the difference in the products at initial intermolecular transglycosylation between the two enzymes is due to the difference in the number of multiple attack. When *B. megaterium* enzyme is incubated with soluble starch in the presence of glucose as acceptor, the intermolecular transglycosylation based on the G7 unit as well as the intramolecular transglycosylation occurs and maltooctaose (G8) is produced. But this G8 forms a new productive ES complex by "sliding" on the enzyme active site, and becomes a donor substrate (24). Therefore, no accumulation of the G8 is observed.

Table 2. Acceptor Specificity of CGTases from *B. megaterium* and *B. macerans*

Saccharide	*B. megaterium* enzyme Starch degradation[1]	*B. megaterium* enzyme Transfer products[2]	*B. macerans* enzyme Starch degradation	*B. macerans* enzyme Transfer products
No acceptor	1.0	−	1.0	−
L-Sorbose	27	++++	7.0	++++
D-Glucose	10	++++	7.0	++++
Maltose	19	++++	5.0	++++
Maltotriose	4.0	++++	1.7	+++
Sucrose	3.0	++++	1.7	+++
Cellobiose	7.3	++++	4.0	++++
Phenyl-α-glucoside	14	++++	8.5	++++
Phenyl-β-glucoside	11	++++	8.0	++++
Methyl-α-glucoside	20	++++	10	++++
Methyl-β-glucoside	16	++++	9.0	++++
6-Deoxy-D-glucose	2.8	+++	1.8	+++
D-Xylose	2.2	+++	1.6	+++
2-Deoxy-D-glucose	1.3	+	1.2	+
3-O-Methyl-D-glucose	1.2	+	1.2	+
D-Mannose	1.0	−	1.0	−
D-Ribose	1.0	−	1.0	−
D-Galactose	1.0	−	1.0	−
D-Glucuronic acid	1.0	−	1.0	−
Sorbitol[3]	1.0	−	1.0	−

[1], Rates of starch degradation in the presence of acceptors relative to those in the absence of acceptors;
[2], Intensities of the spots of transfer products on paper chromatograms (++++, very strong; +++, strong; +, weak; ±, faint; −, negative);
[3], Like sorbitol, D-fructose, D-glucosamine, N-acetyl-D-glucosamine, L-arabinose, L-rhamnose, xylitol and glycerol are negative in both tests.

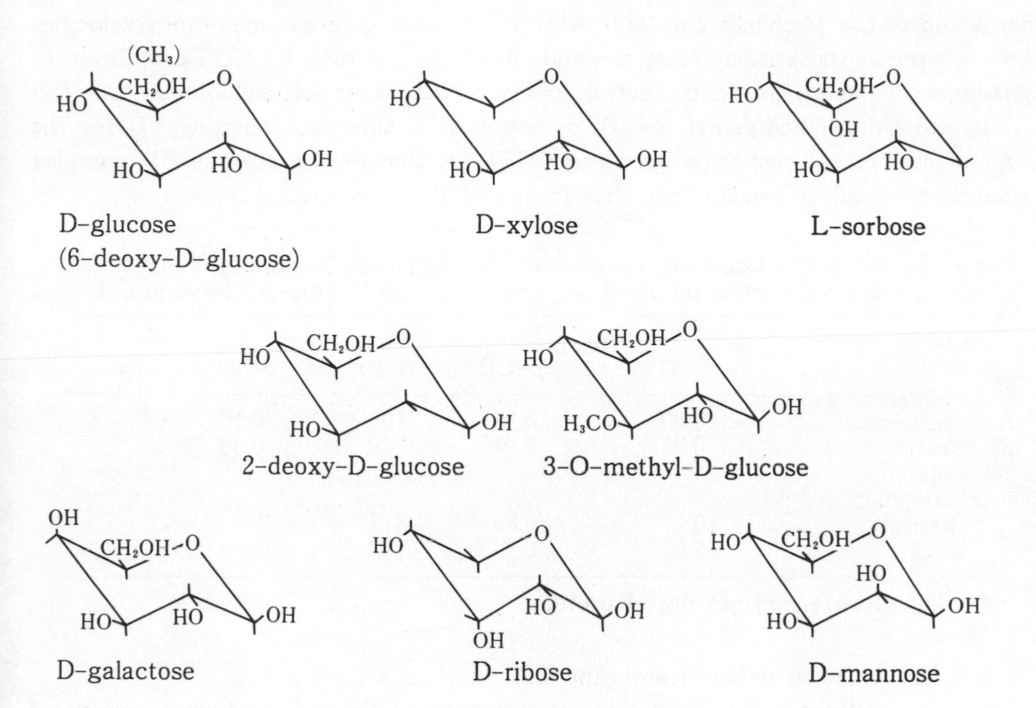

Fig. 2 The Chief Conformation of Various Monosaccharides Occupied in the Water Solution.

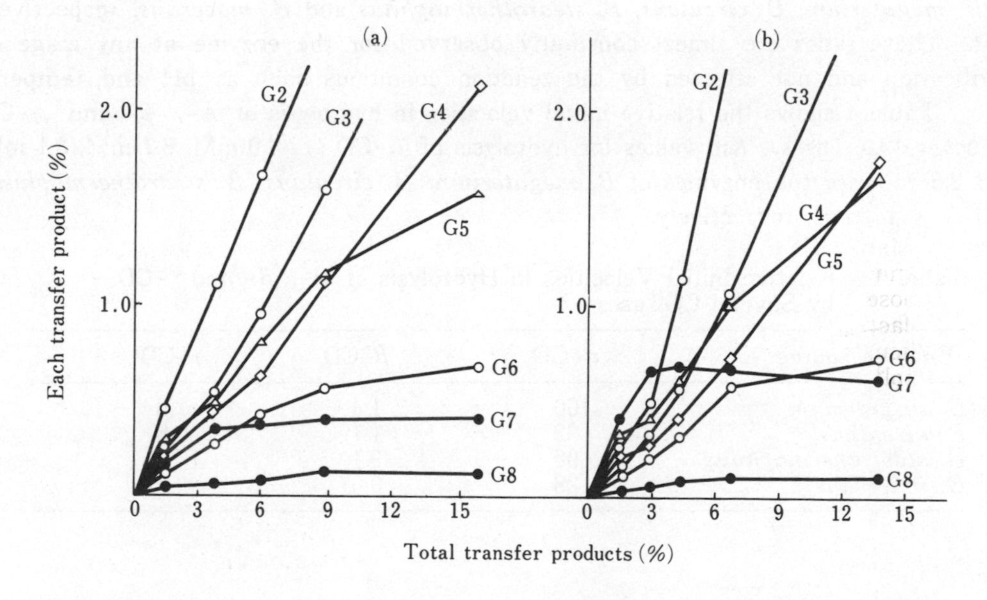

Fig. 3. Successive Change of Transfer Products from Soluble Starch to ^{14}C-Glucose by CGTase from *B. megaterium* (a) and *B. macerans* (b).

Table 3 shows the Michaelis constant (*K*m) values and apparent maximum velocities (*V*max) at transglycosylation from α- and β-CD to sucrose by CGTases from *B. megaterium, B. macerans* and neutral and acid CGTases of alkalophilic *Bacillus* sp. These results indicate that α-CD is better as a substrate than β-CD for the intermolecular transglycosylation by these CGTases, though the enzyme-CD complex formation for α-CD is smaller than that from β-CD.

Table 3. Michaelis Constant Values and the Apparent Maximum Velocity of Transglycosylation from α- and β-CD to Sucrose by Several CGTases

Enzyme source	*K*m (mM)		*V*max[1]	
	α-CD	β-CD	α-CD	β-CD
B. megaterium	0.90	0.16	106	0.19
B. macerans	0.85	0.07	79	0.03
Bacillus sp. (alkalophilic)				
neutral	10	0.83	417	70
acid	5.9	0.39	133	23

[1]: μ moles glucose/min per mg of protein

(3) Hydrolytic action on soluble starch and CDs.

CGTase also catalyzes hydrolytic action on starch and CDs and produces a series of maltooligosaccharides. The ratio of hydrolytic action to total catalytic action (sum of intramolecular transglycosylation and hydrolysis) are 1.9, 2.0, 8.3 and 2.0% for CGTases of *B. megaterium, B. circulans, B. stearothermophilus* and *B. macerans*, respectively (25). These ratios are almost constantly observed for the enzyme at any stage of purification and not affected by the reaction conditions such as pH and temperature. Table 4 shows the relative initial velocities in hydrolysis of α-, β- and γ-CD by several CGTases. *K*m values for hydrolysis of α-CD are 7.0 mM, 6.7 mM, 3.1 mM and 2.5 mM for the enzymes of *B. megaterium, B. circulans, B. stearothermophilus* and *B. macerans*, respectively.

Table 4. Relative Initial Velocities in Hydrolysis of α-, β- and γ-CD by Several CGTases

Enzyme source	α-CD	β-CD	γ-CD
B. megaterium	100	1.4	1.0
B. circulans	92	1.7	0.8
B. stearothermophilus	90	9.1	15
B. macerans	28	0.07	0.03

Table 5. Enzymic and Proteochemical Properties of Several CGTases

Enzyme source	Molecular weight	Optimum pH	pH Stability	Thermal stability	Isoelectric point
B. megaterium	75,000[1]	5.2-6.2	7.0-10.0	<55℃	6.07, 6.80
B. circulans	—	5.2-5.7	7.0-9.0	<55℃	5.80, 6.60
B. stearo-thermophilus	68,000[1]	6.0	8.0-10.0	<50℃	4.45
B. macerans	75,000[1]	5.2-5.7	8.0-10.0	<55℃	4.62
K. pneumoniae	—	5.2	5.0-7.5	—	—
Bacillus sp. (alkalophilic)					
acid	88,000[1]	4.5-4.7	6.0-10.0	<65℃	5.40
neutral	88,000[1]	7.0	6.0-8.0	<60℃	—
B. ohbensis	35,000[1]	5.5	6.5-9.5	<55℃	<4.0
Bacillus sp.AL6	45,000[2]	7.0	6.0-10.7	<55℃	—
B. subtilis No.313	64,000[1]	8.0	6.0-8.0	<50℃	7.1

[1], Determined by SDS-PAGE method;
[2], Determined by gel filtration using Sephadex G-100.

Production and Purification : CGTase is produced by submerged culture of CGTase producing bacteria. Some examples of culture conditions are as follows :

B. macerans : 1% potato starch, 1% corn steep liquor, 0.5% ammonium sulfate and 0.5% $CaCO_3$ (pH 7.0), 30℃, 65~70 h,

B. megaterium : 4% soluble starch, 1% corn steep liquor, 1% Polypepton, 1% wheat bran, 0.5% $CaCO_3$ (pH 7.0), 37℃, 70 h,

Alkalophilic *Bacillus* sp. : 1% soluble starch, 5% corn steep liquor, 0.1% K_2HPO_4, 0.02% $MgSO_4 \cdot 7H_2O$, 1% Na_2CO_3, 37℃, 2 days,

Bacillus sp. strain AL6 : 1.5% soluble starch, 2% soy bean meal, 1% soy bean oil, 0.1% $CaCl_2$, 0.9% $Na_2HPO_4 \cdot 12H_2O$, 1% Na_2CO_3, 38℃, 2 days,

Starch is the best carbon source for CGTase production and calcium or sodium carbonate is effective for maintaining the pH of the medium at a suitable value. In general, the cultured liquors of these CGTase producing bacteria do not contain starch degrading enzymes other than CGTase. Therefore, crude enzyme solutions such as cultured liquors after removing the bacterial cells, are available for the industrial applications. In the presence of ammonium sulfate, CGTases are readily adsorbed on raw corn starch at temperatures below 5℃. The enzyme adsorbed is eluted by suspending the starch into 30 mM sodium hydrogen phosphate solution at 40℃. This method is effective for purification of the enzyme. The enzymes from *B. megaterium, B. circulans, B. stearothermophilus, B. macerans, B. ohbensis,* alkalophilic *Bacillus* sp., *Bacillus* sp. strain AL6, *K. pneumoniae* and *B. subtilis* No.313 are all purified to homogeneous state. The CGTases from *B. macerans* IAM 1243 and *B. ohbensis* are crystallized (32).

Assay Methods : CGTase activity is measured by the following six methods.

(1) Dextrinogenic method (14) This method is the most convenient, if the enzyme examined is free from starch degrading enzymes other than CGTase. The reaction mixture containing 50 μl of enzyme and 450 μl of 0.55% soluble starch in 50 mM buffer (optimum pH) is incubated at 40℃ for 10 mim. The reaction is stopped with 4 ml of 10 mM I_2/250 mM KI solution containing 0.1 N hydrogen chloride. This mixture is diluted to 20 ml with deionized water and the light transmission at 660 nm is measured. One unit of the enzyme activity is defined as the amount of enzyme which causes a linear increase of 1% transmission per min.

(2) Glucoamylase method (33) This is the method that measures the amount of total CDs produced by intramolecular transglycosylation. Twenty μl of crystalline glucoamylase (4 mg/ml) and 60 μl of 20 mM acetate buffer (pH 5.2), are added to 20 μl of the reaction mixture of CGTase with 1.0% soluble starch and the mixture is incubated at 40℃ for 30 min. Then, the reducing sugar and total sugar content in the reaction mixture are determined by the Somogyi-Nelson method and the phenol-sulfuric acid method, respectively. Total CD content is estimated by subtracting the amount of reducing sugar from the total sugar content. One unit of enzyme activity is defined as the amount of enzyme that forms 1 μmol of CD per min.

(3) Coupling method (16) This is the method of determining the amount of CD degraded by transglycosylation to sucrose intermolecularly. The reaction mixture containing 5 mM α-CD, 25 mM sucrose, 0.5 mg of glucoamylase in 0.35 ml of 200 mM buffer (optimum pH of each CGTase) and 10 μl of enzyme is incubated at 40℃ for 10 min. The reaction is stopped by the addition of 1ml of 3,5-dinitrosalicylic acid (DNS) solution. The amount of reducing sugar produced is determined by the DNS-method. One unit activity of enzyme is defined as the amount of enzyme that liberates 1 μmol of glucose per min.

(4) Tilden-Hudson method (34). This is the method of determining α-CD specifically, therefore, this method is applied for assay of α-CD producing CGTase. One ml of 3% soluble starch solution is incubated with 0.5 ml of enzyme at 40℃. At certain time intervals, three drops of the digest are transferred on a microscope slide glass and mixed with a drop of 0.1 N iodine solution (2% I_2 and 4% KI in 1 N hydrogen chloride). Then the edges of the droplet are examined under a low-powered microscope for the presence of characteristic crystals of α-CD-iodine complex. Blue hexagonal plate crystals appear and later, long needles gradually spread out from the center. This stage serves as the end-point of reaction. One unit of enzyme activity is defined as the quantity which brings 1 ml of substrate to this end-point in 30 min at 40℃.

(5) Phenolphthalein method (35). β-Cyclodextrin reacts with phenolphthalein to

produce a complex. This reaction is specific for β-CD and applied for assay of β-CD producing CGTase.

(6) Bromocresol green method (36). This method is available for selective assay of γ-CD producing CGTase, because bromocresol green (BCG) produces an inclusion complex with γ-CD which shows an absorption spectrum stronger than those of the other CD's.

(Sumio Kitahata)

References

1. Tilden, E. B. & Hudson, C. S. (1939) *J. Amer. Chem. Soc. 63,* 2900-2902
2. Tilden, E. B., Adams, M. & Hudson, C. S. (1942) *J. Amer. Chem. Soc. 64,* 1432-1433
3. McClenahan, W. S., Tilden, E. B. & Hudson, C. S. (1942) *J. Amer. Chem. Soc. 64,* 2139-2144
4. Wilson, E. J., Schoch, T. J. & Hudson, C. S. (1943) *J. Amer. Chem. Soc. 65,* 1380-1383
5. French, D., Pazur, J. H., Levin, M. L. & Norberg, E. (1948) *J. Amer. Chem. Soc. 70,* 3145-3146
6. Norberg, E. & French, D. (1950) *J. Amer. Chem. Soc. 72,* 1202—1205
7. French, D., Levine, M. L., Norberg, E., Nordin, P., Pazur, J. H. & Wild, G. M. (1954) *J. Amer. Chem. Soc. 76,* 2387-2390
8. Depinto, J. A. & Campbell, L. L. (1964) *Science 146,* 1064-1066
9. Depinto, J. A. & Campbell, L. L. (1968) *Biochemistry* 7, 114-120
10. Depinto, J. A. & Campbell, L. L. (1968) *Arch. Biochem. Biophys. 125,* 253-258
11. Okada, S., Tsuyama, N. & Kitahata, S. (1972) *Proceedings of the Symposium on Amylase* (in Japanese) *7,* 61-68
12. Okada, S. & Kitahata, S. (1973) *Proceedings of the Symposium on Amylase* (in Japanese) *8,* 21-27
13. Shiosaka, M. & Bunya, H. (1973) *Proceedings of the Symposium on Amylase*(in Japanese) *8,* 43-50
14. Kitahata, S., Tsuyama, N. & Okada, S. (1974) *Agric. Biol. Chem. 38,* 387-393
15. Nakamura, N. & Horikoshi, K. (1976) *Agric. Biol. Chem. 40,* 753-757
16. Nakamura, N. & Horikoshi, K. (1976) *Agric. Biol. Chem. 40,* 935-941
17. Nakamura, N. & Horikoshi, K. (1976) *Agric. Biol. Chem. 40,* 1785-1791
18. Kitahata, S. & Okada, S. (1982) *J. Jap. Soc. Starch Sci. 29,* 7-12
19. Bender, H. (1977) *Arch. Microbiol. 111,* 271-282
20. Yagi, Y., Sato, M. & Ishikura, T. (1986) *J. Jpn. Soc. Starch Sci.* (in Japanese) *33,* 144-151
21. Kato, T. & Horikoshi, K. (1986) *J. Jpn. Soc. Starch Sci.* (in Japanese) *33,* 137-143
22. Ozaki, A. (1986) *Japan Kokai Tokkyo Koho* 274680
23. Kobayashi, S., Kainuma, K. & Suzuki, S. (1973) *Proceedings of the Symposium on Amylase* (in Japanese) *8,* 29-36

24. Kitahata, S. (1978) *Docter thesis,* Osaka University
25. Kitahata, S. & Okada, S. (1982) *J. Jap. Soc. Starch Sci. 29,* 13-18
26. Kobayashi, S., Kainuma, K. & French, D. (1983) *J. Jpn. Soc. Starch Sci. 30,* 62-68
27. Bender, H. (1983) *Carbohydr. Res. 124,* 225-233
28. Sato, M., Nagano, H., Yagi, Y. & Ishikura, T. (1985) *Japan Kokai Tokkyo Koho* 227693
29. Kitahata, S., Okada, S. & Fukui, T. (1978) *Agric. Biol. Chem. 42,* 2369-2374
30. Kitahata, S., Okada, S. & Misaki, A. (1979) *Agric. Biol. Chem. 43,* 151-154
31. Kitahata, S. & Okada, S. (1984) *Kagaku to Kogyo* (in Japanese) *58,* 50-54
32. Kobayashi, S., Kainuma, K. & Suzuki, S. (1978) *Carbohydr. Res. 61,* 229-238
33. Kobayashi, S., Kainuma, K. & Suzuki, S. (1974) *J. Jap. Soc. Starch Sci.* (in Japanese) *22,* 131-137
34. Tilden, E. B. & Hudson, C. S. (1942) *J. Bacteriol. 43,* 527-544
35. Kaneko, T., Kato, T., Nakamura, N. & Horikoshi, K. (1987) *J. Jpn. Soc. Starch Sci. 34,* 45-48
36. Kato, T. & Horikoshi, K. (1984) *Anal. Chem. 56,* 1738-1740

Ⅲ.4. Disproportionating Enzymes

Disproportionating enzyme (1,4-α-D-glucan : 1,4-α-D-glucan 4-α-glycosyltransferase, (E.C. 2.4.1.25, D-enzyme) catalyzes the transfer of a certain fragment of α-D-glucan to the C-4 position of the acceptor. The D-enzymes observed in plants and similar enzymes called amylomaltases of microorganisms are classified into this category of enzymes. D-Enzyme was at first found in potato by Peat *et al.*(1). The enzyme acts on maltooligosaccharides as substrates or donors, to split out maltosyl or larger homologous glucosidic chain units from the substrate and transfer these released glucosidic chains to another substrate molecule or glucose. Thus, the reaction results in disproportionation of maltodextrin molecules initially given as the substrate. This enzyme has been traditionally called disproportionating enzyme (D-enzyme). After this report, this enzyme has been found to be distributed in broad bean (2), carrot, tomato (3), germinated barley seed (4) and sweet potato (5).

Amylomaltase was found in *Escherichia coli* ML strain by Monod and Torriani (6). The enzyme was purified and characterized by Wiesmeyer and Cohn (7) and Palmer *et al.* (8). The same type of enzyme was observed in *Streptococcus bovis* (9), *Streptococcus mitis* (10) and *Escherichia coli* IFO 3806 (11). D-Enzyme from plants is different in the specificity from amylomaltase from microorganisms. The former enzyme does not transfer glucosyl residues, while the latter does this reaction. Schmidt and John (12) isolated amylomaltase and D-enzyme from the *Psedomonas stutzeri*. This D-enzyme was reported to use maltose as a donor unlike the plant D-enzyme.

D-Enzyme is considered to have two important physiological roles : One is to make primers and chains of about 40 glucose units to be transferred by Q-enzyme for branching. The other is to convert maltotriose into glucose and maltose efficiently by combined action with β-amylase. This maltotriose is produced from starch granules in the endosperm together with glucose and maltose by the combined action of amylolytic enzymes during the germination of seeds. The maltose thereby produced is converted by maltase into glucose to be the energy source for young shoot before it begins photosynthesis. In certain microorganisms, especially most kinds of bacteria, amylomaltase, maltodextrin permease and maltodextrin phosphorylase are induced by maltose or maltodextrins and these three enzymes form the maltodextrin utilizing system. Amylose and amylopectin are hydrolyzed to produce maltodextrins by α-amylase and debranching enzyme outside bacterial cells. The maltodextrins are then taken into the cells by maltodextrin permease, and are split into glucose and glucose-1-phosphate by amylomaltase and amylodextrin phosphorylase, respectively.

Ⅲ.4.a. Plant Disproportionating Enzymes

The plant disproportionating enzymes are classified into a minor group among various starch metabolizing enzymes, and their protein nature has not been clarified yet. However, their action patterns and specificities have been extensively investigated. The enzymes from various sources show the same action patterns and specificities.

Reaction : D-Enzyme splits off maltosyl residues or oligosaccharides larger than this from the donor molecule, and transfers them to the acceptor molecule to be bound through α-1,4-linkages. Therefore, the enzyme action results in disproportionation of maltodextrin. In the reaction, the smallest donor and acceptor are maltotriose and glucose, respectively.

$$G{\sim}G\text{-}G + G{\sim}G\text{-}G \longrightarrow G{\sim}G\text{-}G{\sim}G\text{-}G + G$$
$$G{\sim}G\text{-}G + G{\sim}G\text{-}G{\sim}G\text{-}G \longrightarrow G{\sim}G\text{-}G\text{-}G\text{-}G{\sim}G\text{-}G + G$$
$$G{\sim}G\text{-}G + G{\sim}G\text{-}G\text{-}G\text{-}G{\sim}G\text{-}G \longrightarrow G{\sim}G\text{-}G\text{-}G\text{-}G\text{-}G\text{-}G{\sim}G\text{-}G + G$$

~, Forbidden linkage ; G, glucose residue.

Specificity : (1) Reaction with maltotriose. In the reaction with maltotriose as substrate, D-enzyme produces maltopentaose and glucose in the early stages of the reaction and then, maltoheptaose, maltohexaose and maltotetraose in the subsequent stages. After a long incubation, it produces a series of maltooligosaccharides other than maltose. Jones and Whelan (13) revealed the existence of "forbidden linkages" for the action of D-enzyme from potato using ^{14}C-maltodextrins. The "forbidden linkages" are the two 1,4-linkages in a maltodextrin molecule, one being the non-reducing end linkage and the other, the linkage penultimate to the reducing end. D-Enzyme can not cleave these positional linkages. In the case of maltotriose, the maltosyl residue split from the donor by D-enzyme, is transferred to another maltotriose acceptor to produce maltopentaose. In this reaction, maltosyl residue is transferred so quickly that no accumulation of maltosyl-enzyme is observed. Maltoheptaose and glucose, once accumulated, are subjected to further enzymic action and thus, a series of maltooligosaccharides are produced as the final reaction products.

(2) Reaction with maltotetraose. D-Enzyme produces maltoheptaose and glucose in the early stages of the reaction, because maltotetraose has "forbidden linkages" in a line and only maltotriosyl is available to be transferred to another maltotetraose molecule to produce maltoheptaose. The maltoheptaose produced serves as a donor of maltosyl residue to maltotetraose to form maltohexaose. After a long incubation, a series of maltooligosaccharides other than maltose are produced.

(3) Reaction with maltopentaose. In the early stages of the reaction, maltoheptaose and maltotriose are produced. The maltosyl residue at the non-reducing end side of maltoheptaose is transferred to another maltoheptaose molecule to produce maltononaose. After a long incubation, a series of maltooligosaccharides other than maltose are produced as in the cases described above.

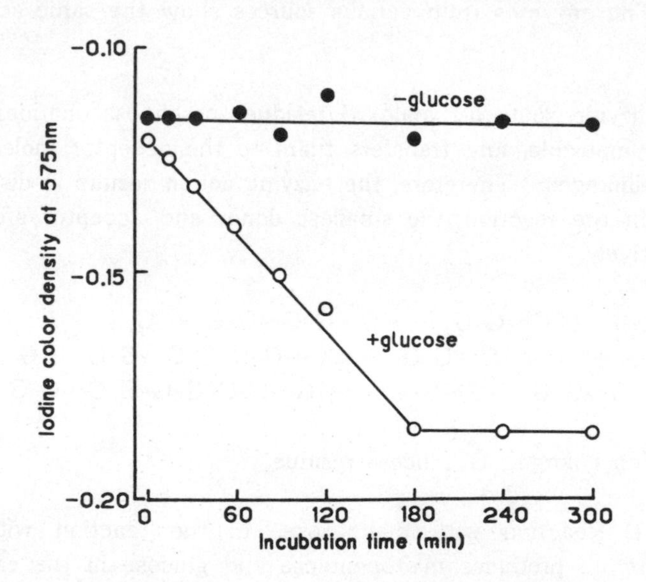

Fig. 1. Action of Barley D-Enzyme on Soluble Starch in the Presence and Absence of Glucose (4).

The reaction mixture contained 16 mU of D-enzyme and 0.4% of soluble starch in the presence and absence of 2 mM D-glucose at 30℃ and pH 6.8.

(4) Reaction with maltohexaose. In the early stage of the reaction, maltooctaose and maltotetraose are produced. Maltotriose and glucose are also produced, but slowly. A series of larger maltodextrins are also produced on a long incubation.

(5) Role of glucose. Glucose does not serve as a donor but serves as an acceptor. When soluble starch reacts with D-enzyme, the density of iodine color reaction remains unchanged. However, as shown in Fig.1, if glucose is added to the reaction mixture, the color density at 575 nm decreases linearly with time of incubation to a certain point whence no more decrease in the density occurs. In the reaction, various maltooligosaccharides are observed, which are produced by the transfer of oligosaccharides from the non-reducing end side of soluble starch to the glucose added. This reaction is equilibrated when the linear chain portions of soluble starch are consumed. Starch is decomposed in the presence of glucose. Paper chromatography reveals that maltotriose is the major reaction product.

Table 1. Some Properties of D-Enzyme from Various Sources

Source	Potato (1)	Carrot (3)	Barley (4)	Sweet potato (5)
Optimum pH	6.7	5.5	6.5	7.0
Optimum temp.	45 ℃	37 ℃	45 ℃	35 ℃
pH Stability	—	—	6.5-12.0	—
Thermal stability	—	—	25-45 ℃	—
*K*m (G3, mM)	—	—	—	7.3

The numerals indicate number of references.

Table 2. Effects of Various Metal Ions on Activity of the D-Enzyme (4)

Metal ions (1.0 mM)	Relative activity (%)
None	100
Li^+	100
Mg^{2+}	88.8
Ca^{2+}	100
Fe^{2+}	98.8
Co^{2+}	95.3
Ni^{2+}	75.3
Cu^{2+}	41.3
Zn^{2+}	90.0

Various metal ions were added as chloride salts, but Li^+ was added as sulfate salt.

Purification (4) : Malted barley (1kg) is finely powdered and added with 3 liters of 3% sodium chloride solution including 10 mM EDTA and 10 mM 2-mercaptoethanol and let stand for 3 h at 30℃ with occasional stirring. The mixture is centrifuged at 10,000×g for 30 min. The supernatant is fractionated with ammonium sulfate and the fraction precipitated between 30-60% saturation of ammonium sulfate is separated by centrifugation. This fraction is dissolved in 10 mM Tris-HCl buffer (pH 7.5) and dialyzed against the same buffer in the cold overnight. Precipitate, if it occurs, is removed by centrifugation and the supernatant is subjected to chromatography on a column of DEAE-Cellulose (4.8×18cm, Whatman DE-52) equilibrated with 10 mM Tris-HCl buffer (pH 7.5) containing 0.1 M sodium chloride. After washing the column with the same buffer, the enzyme adsorbed is eluted by a linear gradient of sodium chloride from 0.1 to 0.5 M in the same buffer. Fractions containing D-enzyme activity are combined and concentrated on an ultrafiltration apparatus (Amicon Corp.) with a PM-30 membrane. The enzyme is then dialyzed against 5 mM phosphate buffer (pH 6.8) overnight and, after centrifuging off precipitate, if it is formed, the supernatant is concentrated.

The enzyme concentrate is subjected to chromatography on a column of Sephadex G-150 (superfine, 2.6×90 cm) equilibrated with 5 mM phosphate buffer (pH 6.8). The active fractions are combined and concentrated as described above. The enzyme is then subjected to chromatography on a column (0.9×58 cm) of hydroxylapatite (DNA-Grade, BIO-RAD Laboratories) equilibrated with 5 mM phosphate buffer (pH 6.8). After washing the column with the same buffer, the enzyme adsorbed is eluted by a linear gradient of phosphate buffer from 5 to 200 mM, pH 6.8. The fraction eluted at around 90 mM usually contains D-enzyme and is concentrated. A column (1.0×31 cm) of Polybuffer exchange resin PBE-94 (Pharmacia Fine Chemicals Co.) is prepared on which Sephadex G-25 (medium, 1.0×1.8 cm) is stacked. After equilibrating the column with 25 mM Histidine-HCl buffer (pH 6.2), the enzyme obtained above is applied on it. The enzyme is eluted with Polybuffer 74 (pH 4.0 ; deaerated sufficiently) and dialyzed against 5 mM phosphate buffer (pH 6.8) before use.

Assay Method : The enzyme reaction is carried out at 30℃ for 30 min in the following mixture : 0.5% maltotriose solution, 0.1 ml; 50 mM phosphate buffer (pH 6.8), 0.1 ml; and the enzyme solution, 0.05 ml. After 30 min incubation, an aliquot (0.1 ml) of the reaction mixture is submitted to measurement of glucose produced by the glucose oxidase method of Lloyd and Whelan (14).
Reagents: 1. Tris-phosphate-glycerol buffer. Tris (hydroxymethyl) aminomethane (36.3 g) and 50 g $NaH_2PO_4 \cdot H_2O$ are dissolved in an appropriate amount of water, and 400ml glycerol and water are added to the solution to fill up to 1000 ml, adjusting the pH 7.0 with solid $NaH_2PO_4 \cdot H_2O$. 2. Glucose oxidase reagent. The following reagents are dissolved into 100 ml of Tris-phosphate-glycerol buffer : Glucose oxidase (Boehringer, catalog No.15424 EGAC), 30 mg ; horseradish peroxidase (Boehringer, catalog No.15302 EPAB), 3 mg ; o-dianisidine dihydrochloride (Sigma), 10 mg.
Procedure : To 1 ml of the test solution containing 0-75 μ g glucose is added 2 ml of glucose oxidase. The mixture is incubated at 37℃ for 30 min, and 4 ml of 5 N HCl is added to the mixture and the optical density read at 525 nm. One unit of the

D-enzyme activity is defined as the amount of enzyme that liberates 1 μmol of glucose in one min under the conditions described above. On the other hand, another aliquot of the enzymic digestion mixtures is spotted on a filter paper (Toyo No.50, 20×20 cm), and the paper is irrigated with 65% 1-propanol by the double-ascending technique at 70℃. After drying, the paper is treated with glucoamylase to hydrolyze the oligosaccharides and stained by the alkaline acetone-silver nitrate dip-method. The appearance of glucose and maltopentaose in the reaction mixture is an indication of the action of D-enzyme.

(Iwao Maeda)

References

1. Peat, S., Whelan, W. J. & Rees, W. R. (1956) *J. Chem. Soc.* 44-53
2. Rees, W. R. (1953) *Ph. D. Thesis*, University of Cambridge.
3. Manners, D. J. & Rowe, K. L. (1969) *Carbohydr. Res. 9*, 441-450
4. Yoshio, N., Maeda, I., Taniguchi, H. & Nakamura, M. (1986) *J. Jpn. Soc. Starch Sci.* (in Japanese) *33*, 244-252
5. Suganuma, T., Setoguchi, S., Fujimoto, S. & Nagahama, T. (1986) *J. Jpn. Soc. Starch Sci.* (in Japanese) *33*, 217
6. Monod, J. & Torriani, A. M. (1948) *Compt. Rend. 227*, 240
7. Wiesmeyer, H. & Cohn, M. (1960) *Biochim. Biophys. Acta 39*, 427-439
8. Palmer, T. N., Ryman, B. E. & Whelan, W. J. (1976) *Eur. J. Biochem. 69*, 105-115
9. Walker, G. J. (1965) *Biochem. J. 94*, 299-308
10. Walker, G. J. (1966) *Biochem. J. 101*, 861-872
11. Kitahata, S., Ishibashi, H. & Okada, S. (1986) *J. Jpn. Soc. Starch Sci.* (in Japanese) *33*, 217
12. Schmidt, J. & John, M. (1979) *Biochim. Biophys. Acta 566*, 100-114
13. Jones, G. & Whelan, W. J. (1969) *Carbohydr. Res. 9*, 483-490
14. Lloyd, J. B. & Whelan, W. J. (1969) *Anal. Biochem. 30*, 467-470

Ⅲ.4.b. Microbial Disproportionating Enzymes

Microbial disproportionating enzyme (D-enzyme) is generally called amylomaltase. Amylomaltase formally reported by Doudoroff *et al.* (1), is an enzyme which is inducibly formed in *Escherichia coli* ML by maltose (G2). Amylomaltase has been isolated and purified from the *E. coli* strain ML 308 (2) and IFO 3806 (3) and *Pseudomonas stutzeri* NRRL B-3389 (4). *Ps. stutzeri* produces amylomaltase as well as D-enzyme. This D-enzyme resembles the enzymes obtained from some plants. The applications of amylomaltase are still under developmental study.

Reaction : $(1,4\text{-}\alpha\text{-Glucan})_n + (1,4\text{-}\alpha\text{-Glucan})_m \rightleftharpoons (1,4\text{-}\alpha\text{-Glucan})_{n+x} + (1,4\text{-}\alpha\text{-Glucan})_{m-x}$

Amylomaltase is an enzyme which catalyzes both glucosyl-transfer and 4-α-

glucanosyl-transfer reaction. This enzyme was originally reported to catalyze the reaction shown below :

maltose + maltose ⟶ maltotriose + glucose

However, it is now known that amylomaltase does not take part in the above reaction. Therefore, it is highly likely that G2 is not effective as donor, though it is effective as acceptor (5). By the action of amylomaltase, maltooligosaccharides are disproportionated to long glucosidic chains which form iodine complexes. The addition of glucose to a reaction mixture at the equilibrium state, causes disappearance in the iodine color reaction, because the elongated chains are shortened by a disproportionate transfer to glucose as acceptor. Removal of glucose from the enzyme reaction mixture at a state of equilibrium, causes the reverse reaction and increases the degree of iodine reaction color because of elongation of the average chain length.

Specificity : Amylomaltase from *E. coli* is not specific as a 4-α-glucanotransferase, but shows a broad glycosyltransfer action. The enzyme catalyzes the transfer of both glucosyl and maltodextrinyl residues from non-reducing end of maltodextrin to acceptor molecules. When amylomaltase from *E. coli* is incubated with maltotriose (G3), maltotetraose (G4), maltopentaose (G5) or maltohexaose (G6), various 1,4-α-glucans of higher degrees of polymerization than those of the substrate are produced. In this reaction, even α-1,4-glucans of polymerization degrees higher than 25 are also formed. In reactions of amylomaltase from *Ps. stutzeri* with G3, G4, and G5, maltooligosaccharides similar to those by the action of amylomaltase from *E. coli* are produced, only that the amount of G2 and 1,4-α-glucans of polymerization degrees higher than 25 is smaller. However, in the action on G6, amylomaltase of *Ps. stutzeri* produces a series of maltooligosaccharides from G1 to G5, but no significant amount of dextrins of polymerization degrees higher than G6 is produced. D-Enzyme from *Ps. stutzeri* catalyzes mostly transfer of maltose from G3, maltotriose from G4, maltotetraose from G5 and maltopentaose from G6, respectively. Unlike other enzymes, *Ps. stutzeri* D-enzyme catalyzes glucose transfer from G2 to produce G1 and G3.

Table 1. Attack Sites of Amylomaltase from *E. coli* IFO 3806 on G3, G4 and G5

Substrate	Products from reducing end side (molar ratio)
G−G−G ↑ ↑	G1 : G2 18 : 4
G−G−G−G ↑ ↑ ↑	G1 : G2 : G3 4 : 2 : 3
G−G−G−G−G ↑ ↑ ↑ ↑	G1 : G2 : G3 : G4 1 : 1 : 2 : 3

Table 2. Acceptor Specificity of the Enzymes from Potato, *E. coli* ML 308 and *E. coli* IFO 3806

Acceptor	Potato	*E. coli* ML 308	*E. coli* IFO 3806
D-Glucose	+	+	+
Methyl α-glucoside	+	−	+
Methyl β-glucoside	+	−	+
Phenyl α-glucoside	NR	−	+
Phenyl β-glucoside	NR	NR	+
D-Mannose	+	+	+
L-Sorbose	+	NR	+
2-Deoxy-D-glucose	NR	NR	+
D-Glucosamine	−	NR	+
N-Acetyl-D-glucosamine	−	NR	+
D-Allose	NR	NR	+
D-Galactose	−	−	−
D-Xylose	+	−	+
Isomaltose	−	NR	+

NR : not reported

Table 3. Enzymic and Proteochemical Properties of Amylomaltases from Several Bacteria

	E. coli IFO 3806	*E. coli* ML 308	*Ps. stutzeri*
Molecular weight	93,000[1]	71,000[2]	47,000, 38,000[1] 74,000[2]
Optimum pH	6.5	6.9	7.7
pH Stability	6.0–8.0	−	−
Thermal stability	<45℃	−	<50℃
Isoelectric point	−	5.6	−

[1], Determined by SDS-PAGE method ;
[2], Determined by gel filtration method using Sephadex G-200.

The cleavage point by the amylomaltase from *E. coli* IFO 3806 varies depending on the kind of oligosaccharide (6) as shown in Table 1. This enzyme shows activity not only on maltooligosaccharides, but also on maltotetraitol, maltopentaitol and maltosylsucrose. This enzyme produces a series of sugar alcohols larger than maltotriitol from maltotetraitol. Also, it produces a series of maltooligosyl-sucroses larger than glucosylsucrose from maltosylsucrose. In these reactions, no reducing sugars are produced. This result indicates that amylomaltase of *E. coli* IFO 3806 does not hydrolyze maltotetraitol and maltosylsucrose. Table 2 shows the acceptor specificity of the amylomaltases. The amylomaltase of *E. coli* IFO 3806 has a wide acceptor specificity, compared with the enzymes from potato and *E. coli* ML 308. The enzyme of *E. coli* IFO 3806 transfers glycosyl residues specifically to the G4-hydroxyl groups of glucose, xylose, allose, mannose and N-acetylglucosamine.

Production and Purification : Amylomaltase is an intracellular enzyme and found in various bacteria such as *E. coli* ML 308, IFO 3806 and *Ps. stutzeri*. An example of medium composition for the bacterial culture is as follows : 1.5% Polypepton, 0.5% yeast extracts, 1% G2 (autoclaved separately), 1 mM $MgCl_2 \cdot 6H_2O$ and 0.1 M potassium phosphate buffer pH 7.0. The enzyme is inducibly produced by adding maltooligosaccharides such as G2, G3, G4 and G5 (7). The enzyme production occurs in parallel to the cell growth of bacteria. The enzyme is released by disrupting the cells by sonication. By applying the procedures of ammonium sulfate precipitation, DEAE-Sephadex, Ultrogel AcA 44, Hydroxylapatite and Butyl-Toyopearl 650M column chromatographies, the enzyme of *E. coli* IFO 3806 is purified to homogeneous state in the test of SDS-polyacrylamide gel electrophoretical analysis.

(Sumio Kitahata)

References

1 . Doudoroff, M., Hassid, W. Z., Putman, W., Potter, A. L. & Lederberg, J. (1949) *J. Biol. Chem. 179,* 921-934
2 . Wiesmeyer, H. & Cohn, M. (1960) *Biochim. Biophys. Acta 39,* 417-426
3 . Kitahata, S., Ishibashi, H. & Okada, S. (1986) *Abstract of the Annual Meeting of the Japanese Society of Starch Science* (in Japanese) p11
4 . Schmidt, J. & John, M. (1979) *Biochim. Biophys. Acta 566,* 100-114
5 . Palmer, T. N.., Ryman, B. E. & Whelan, W. J. (1976) *Eur. J. Biochem. 69,* 105-115
6 . Ishibashi, H. & Kitahata, S. (1987) *Abstract of the Annual Meeting of the Agricultural Chemical Society of Japan* (in Japanese) p79
7 . Barker, S. A., Farisi, M. A. & Hopton, J. W. (1965) *Carbohydr. Res. 1,* 97-105

Ⅲ.5 Glucose-Isomerizing Enzyme

Fructose is the sweetest of various naturally occurring sugars, and there had been a potential demand for this sugar as an alternative to other sugars such as sucrose and invert sugar. Thus, the method of isomerizing glucose to fructose has long been studied. In 1966 (1) an enzymic method of isomerizing glucose to fructose on an industrial scale was first established in Japan, and the technology is now being applied for production of fructose world-widely. Fructose production by isomerization of glucose with enzyme has the following advantages : 1) it causes no other reactions than producing fructose, 2) no loss of sugar occurs in the procedure, 3) no color reaction is introduced, 4) the product gives a fresh and gentle sweet taste, 5) the conversion of glucose to fructose is highly efficient, and 6) the reaction proceeds in solutions of high substrate concentrations.

Marshall *et al.* in 1957 first observed the fact that *Pseudomonas hydrophila* cultivated in a medium containing D-xylose produced an enzyme which isomerized glucose to fructose in the presence of arsenate (2). These findings drew the attention of many researchers to glucose-fructose conversion by enzyme. Thus, as shown in Table 1, various microorganisms including yeast, bacteria, *Actinomyces,* etc. were found to produce the enzymes which isomerize glucose to fructose. Several basic properties of the major enzymes are listed in Table 1. The enzymes produced by microorganisms may be classified into three groups as follows :

A) D-Xylose isomerases (EC 5.3.1.5, D-Xylose ketol-isomerase) ; It was demonstrated that D-xylose isomerase from *Lactobacillus brevis* which catalyzes the conversion of D-xylose to D-xylulose, also catalyzes the conversion of glucose to fructose (3). An enzyme of this type further catalyzes the conversion of D-ribose to D-ribulose. The enzyme at present under application for industry is of this type and usually called glucose isomerase.

B) Glucosephosphate isomerases (EC 5.3.1.9, D-Glucose-6-phosphate ketol-isomerase); Glucosephosphate isomerase which catalyzes the isomerization of glucose-6-phosphate to fructose-6-phosphate, isomerizes glucose to fructose in the presence of arsenate (4). This type of enzyme is produced by *Aerobacter cloacae* (5), *Escherichia intermedia* (6), etc. The enzyme of *Ps. hydrophila* (2) is also considered to be of this type, because the enzyme requires arsenate for the activity.

C) NAD-linked glucose isomerizing enzymes ; These enzymes also isomerize glucose or mannose or both of them to fructose, requiring NAD as a cofactor for the enzyme reaction. This type of enzyme is produced by *Bacillus megaterium* (7) and *Paracolobacterium aerogenoides* (8).

The reaction between glucose and fructose by the glucose-isomerizing enzyme is reversible. In Fig. 1 is shown a typical time course of this reaction. The isomerizing reaction of glucose to fructose is endothermic, and the amount of fructose produced increases with increasing reaction temperature (18~20). At about 50℃, the ratio of glucose to fructose approaches 1:1, and at 60℃, the reaction proceeds to produce 51~53 parts of fructose and 47~49 parts of glucose. For the conversion to yield more fructose, the equilibrium must be shifted by other factors than temperature. Various methods for this purpose have been suggested. They are those of using borate, water-insoluble polymer with dihydroboronyl base and oxyanion compounds of Ge, Sn, Mo and W. For example, if the reaction is carried out in the presence of borate (21) or germanium (19), the isomerization ratio is improved up to 88-94 part of fructose at the maximum. However, the most desirable method may be of using an enzyme which is thermostable and available for carrying out the reaction at temperatures as high as possible. However, for application of such a thermostable enzyme, some special cautions may be necessary, because fructose is unstable under such conditions. The glucose isomerase reaction equilibrium is an important problem that has to be studied further.

Table 1. Enzymic Properties of Various Glucose-Isomerizing Enzymes

Microorganism	Opt. pH	Opt. temp.(℃)	Cofactor	Substrate	References
Ps. hydrophila	8.5	42~43	Arsenate, Mg^{2+}	Glu, Xyl	2
Aer. cloacae	7.6	50	Arsenate, Mg^{2+}	Glu, Xyl	5
E. intermedia	7.0	40	Arsenate	Glu, G-6-P	6
B. megaterium	7.7	35	NAD	Glu	7
Par. aerogenoides	7.0	40	NAD, Mg^{2+}	Glu, Man	8
Lac. brevis[1)]	6~7	60	Mn^{2+}	Glu, Xyl, Rib	3
St. phaeochromogenes	9.3~9.5	80	Mg^{2+}	Glu, Xyl, L-Ara	9
St. albus YT-5[2)]	8.0~8.5	80	Mg^{2+}	Gly, Xyl	1
St. albus NRRL-5778	7~9	70~80	Mg^{2+}	Glu, Xyl, Rib, L-Ara	10
St. flavogriseus[3)]	7.5	70	Mg^{2+}	Glu, Xyl	11
St. griseofuscus[4)]	7.5	70	Mg^{2+}	Glu, Xyl	12
B. coagulans[5)]	7.0	70	Co^{2+}	Glu, Xyl, Rib	13
B. stearothermophilus[6)]	7.5~8.0	80	Mg^{2+}	Glu, Xyl	14
Brevibacterium sp.	8.0	70	Co^{2+}	Glu, Xyl	15
Arthrobacter sp.	8.0	60~65	Mg^{2+}	Glu, Xyl	16
Act. missouriensis	7.0~7.5	90	Mg^{2+}	Glu, Xyl, Rib	17

The molecular weights estimated are 19.1, 15.7, 17.1, 18.0, 16.0 and 13.0 x 10^4 for 1), 2), 3), 4), 5) and 6), respectively.

Reaction:

```
      CHO                        CH2OH
       |                           |
   H—C—OH                         C=O
       |                           |
  HO—C—H                     HO—C—H
       |         <====>            |
   H—C—OH                     H—C—OH
       |                           |
   H—C—OH                     H—C—OH
       |                           |
      CH2OH                       CH2OH

   D-Glucose                  D-Fructose
```

Specificity : The enzyme of Type A (D-Xylose isomerase) catalyzes the isomerization reaction between glucose and fructose and also, the isomerization between D-ribose and D-ribulose. Schray and Rose reported that the enzyme of Type A had the specificity for α-anomers of these sugars (22). Table 2 shows *K*m values of glucose isomerizing enzymes for the substrates. The *K*m for D-xylose is usually smaller than that for glucose, i. e. the enzyme has a higher affinity for D-xylose than for glucose. However, the enzymes whose affinities toward xylose and glucose are the same, are known, and the affinity for D-ribose of some of the enzymes exceeds that for either D-xylose or glucose. Furthermore, some xylose isomerases are active only on D-xylose and are inactive on glucose (23).

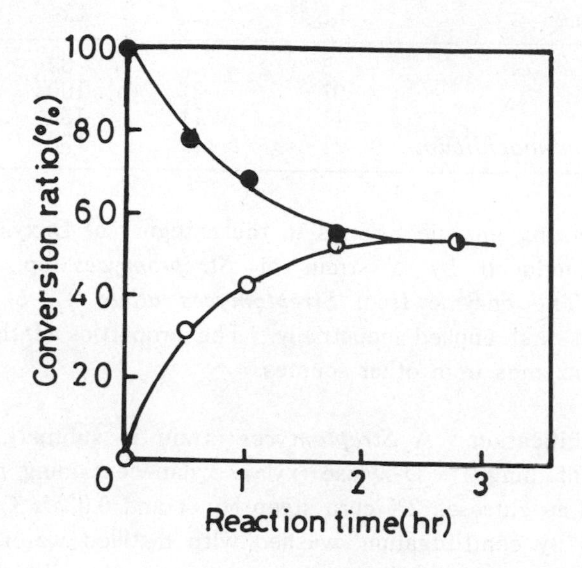

Fig.1. Time Course of Isomerization of Glucose to Fructose by Enzyme Reaction at 60℃.
Substrate concentration, 1.0×10^{-2} (%) ; phosphate buffer (pH 7.0), 4.5×10^{-2} M ; $MgSO_4$, 9.1×10^{-2}M. Substrate used : ○, glucose ; ●, fructose.

Table 2. Substrate Specificity and *K*m Values (M) of Glucose-Isomerizing Enzyme

Source of enzyme	*K*m values for			Ref.
	xylose	glucose	ribose	
Lact. brevis	0.005	0.92	0.67	3
St. albus YT-5	0.032	0.16	—	1
St. albus NRRL-5778	0.093	0.086	0.350	10
St. flavogriseus	0.078	0.249	—	11
B. coagulans	0.077	0.09	0.077	13
B. stearothermophilus	0.10	0.22	—	14

Properties : Glucose-isomerizing enzyme requires metallic ions such as Mg^{2+}, Mn^{2+} or Co^{2+} for the reaction. As shown in Table 3, the metallic ions as cofactors markedly differ depending on the source of the enzyme. The industrially applied enzyme is an Mg^{2+} requiring glucose-isomerizing enzyme.

Table 3. Metal Ions Effective for Glucose-Isomerization by Enzymes

Origin of enzyme	Mg^{2+}	Mn^{2+}	Co^{2+}	Ref.
Lact. brevis	5	100	32	3
St. albus YT-5	97	27	100	1
B. coagulans	11	11	100	13
Brev. pentosoaminoacidicum	21	11	100	15

This glucose-isomerizing enzyme belongs to the category of D-xylose isomerases and is intracellularly produced by a strain of *Streptomyces* sp., *Bacillus* sp., or *Arthrobacter* sp. The enzyme from *Streptomyces albus* TY-5 (the same as *St. rubisinosus*) (1) was first applied industrially. The properties of this enzyme are the same as those of enzymes from other sources.

Production and Purification : A *Streptomyces* strain is submergedly cultured in a medium (pH 7) containing 1% D-xylose (xylan, xylan-containing material or suitable carbon source such as glucose), 2% corn steep liquor and 0.025% $CoCl_2$ at 30℃. The cells are harvested by centrifugation, washed with distilled water and suspended in distilled water. The enzyme is liberated by autolysis of cells at 40℃ and at pH around 6.5 (24). The presence of cationic surface active agent such as cetylpyridinium chloride accelerates the release of enzyme (24). The enzyme is fractionally precipitated with cold acetone of concentrations between 44 and 67%, and then, subjected to column chromatography with DEAE-Cellulose buffered with 0.05M phosphate buffer at pH 7.0. The enzyme adsorbed on the column is eluted with 0.1M KCl in the same buffer as above and subsequently rechromatographed twice with a DEAE-Sephadex A-50 column. The crystallization is carried out by gradually increasing the concentration of acetone finally to reach about 50% (v/v) in the presence of 5×10^{-3}M $MgSO_4$ and 2×10^{-4}M $CoCl_2$ and at pH 6.5～7.0.

Assay Method : To one ml of 0.2 M glucose solution containing 0.2 M phosphate buffer (pH 7.2) and 0.01 M $MgSO_4$, one ml of the enzyme is added and incubated at 60℃. After 60 min incubation, 2 ml of 0.5 M perchloric acid are added to terminate the reaction, and the amount of fructose produced is determined by a modification of the cysteine-carbazole method*. One unit of the enzyme is defined as the amount of enzyme which converts glucose to fructose at a rate of one μ mol/min under the assay conditions.

*The reaction mixture is made up to a final volume of 50 ml with distilled water. To one ml aliquot of this solution, 0.2 ml of 1.5% L-cysteine is added followed by the addition of 6 ml of 70% H_2SO_4 and cooling with mixing. To this solution, 0.2 ml of 1.2% carbazole in ethanol is added and the mixture is kept at 60℃ for 10 min. After cooling, absorbance of the solution is measured at 560nm.

(Yoshiyuki Takasaki)

References

1. Takasaki, Y. (1966) *Agric. Biol. Chem. 30,* 1247-1251, (1969) *ibid, 33,* 1527-1534, *Fermentation Advances p.* 561-589 Academic Press (1969).
2. Marshall, R. O. & Kooi, E. R. (1957) *Science 125,* 648-649
3. Yamanaka, K. (1963) *Agric. Biol. Chem. 27,* 265-270, (1968) *Biochim. Biophys. Acta, 151,* 670-680
4. Natake, M. (1968) *Agric. Biol. Chem. 32,* 303-313
5. Tsumura, N. & Sato, T. (1961) *ibid. 25,* 616-619
6. Natake, M. (1964) *ibid. 28,* 510, (1968) *ibid. 32,* 303-313
7. Takasaki, Y. (1962) *Nippon Nogei-Kagaku Kaishi* (in Japanese) *36,* 1010-1013
8. Takasaki, Y. (1964) *Agric. Biol. Chem. 28,* 740-741
9. Tsumura, N. & Sato, T. (1965) *Agric. Biol. Chem. 29,* 1129-1134
10. Sanchez, S. & Smiley, K. L. (1975) *App. Microbiol. 29,* 745-750
11. Chen, W. P. & Anderson, A. W. (1979) *App. Environ. Microbiol. 38,* 1111-1119
12. Kasumi, T., Hayashi, K. & Tsumura, N. (1981) *Agric. Biol. Chem. 45,* 619-627
13. Danno, G. (1967) *Agric. Biol. Chem. 31,* 284-289, (1970) *ibid. 34,* 1795-1804, (1970) *ibid. 34,* 1805-1814
14. Suekane, M. (1974) U. S. Patent 3,826,714.
15. Ihimura, M., Hirose, Y., Katsuya, N. & Yamada, K. (1965) *Nippon Nogei-kagaku Kaishi* (in Japanese) *39,* 291-298
16. Lee, C. K., Hayes, L. E. & Long, M. Z. (1972) U. S. Patent 3,690,948.
17. Scallet, B. L., Shieh, K., Ehrenthal, I. & Slapshak, L. (1974) *Die Stärke 12,* 405-444
18. Takasaki, Y. (1967) *Agric. Biol. Chem. 31,* 309-313
19. Barker, S. A., Pelmore, H. & Somers, P. J. (1983) *Enz. Microb. Tech. 5,* 121-124
20. Schallet, B. L. (1974) *Die Stärke 26, 405-444*
21. Takasaki, Y. (1971) *Agric. Biol. Chem. 35,* 1371-1375
22. Schray, K. J. & Rose, I. A. (1971) *Biochemistry 10,* 1058-1062
23. Slein, M. (1955) *J. Am. Chem. Soc. 77,* 1663-1667
24. Takasaki, Y. (1970) *Rep. Ferm. Res. Inst.* (in Japanese) 23-30

Ⅳ. Clinical Assay of α-Amylase Activity in Body Fluids

The determination of α-amylase activity in serum, urine or other body fluids is widely used in the clinical laboratory for diagnosis of pancreatic disorders. Various laboratory methods of the enzyme assay are described discussing their characteristics. A recent survey of clinical laboratories revealed that 30% of the participants conducted their assays according to the amyloclastic methods and 35%, to the methods involving dye-labeled starches. Both methods are convenient, but the former is not a strictly linear reaction and the peak of color developed by starch-iodine complex shifts as the hydrolysis of starch proceeds. The latter method is simple and precise, but the substrate costs expensive. Furthermore, the results obtained by this method do not necessarily correlate with those obtained by other methods. Kinetic methods have also been accepted increasingly. These methods involve a system of coupling several enzymes or several specific substrates, and are more sensitive and better in the linearity of response to enzymic activity, compared with methods described above. More advanced techniques involving chromogenic substrates with or without a coupled enzyme system have been proposed in which the amylase activity is measured directly with negligible interference of endogeneous glucose or maltose. Many proposals have been made to standardize the method of α-amylase assay in clinics, and much effort is being devoted to provide the reference methods and reference materials as well.

Assay of Amylase Activity in Serum and Urine:
Hydrolysis of starch or glycogen by α-amylase yields mainly maltose and various polymers, including dextrins. There is a long history of the development of amylase assays, but two main kinds of methods have been used, measurement of the decrease in the amount of substrate and measurement of the increase in amounts of products. Techniques for the assay of amylase include the following :

(a) Viscometric, decrease in the viscosity of a starch solution ;
(b) Turbidimetric, decrease in the turbidity of a starch suspension ;
(c) Amyloclastic (colorimetric), decrease in starch-iodine colour ;
(d) Saccharogenic (reductometric), increase in the amount of reducing groups in a starch solution ;
(e) Use of dye-marked starches as substrate ;
(f) Use of enzyme coupled reactions.

During this decade, the amyloclastic and saccharogenic methods are most widely used in clinical laboratories. In the saccharogenic method, the increase in reducing sugars released by hydrolysis of starch is measured. Somogyi (1) used a highly alkaline copper reagent to measure titrimetrically reducing substances in the reaction mixture before and after digestion. Henry and Chiamori (2) used a less alkaline reagent and a colorimetric technique. The difference in the amount of reducing substances is expressed as glucose in mg/dl of sample for the units of amylase activity. The amylase activity of a timed urine sample can be expressed as units/h. In Henry and Chiamori's technique, buffered starch substrate is incubated with serum, diluted urine (1:2), or diluted duodenal fluid (1:25) for 34 min at 37℃, and the increase in reducing sugars is measured in terms of the reduction of Cu (II) to Cu (I). The amount of Cu (I) is assayed by its reducing effect on phosphomolybdic acid, which is converted to a blue compound, by reading absorbance against a reagent blank at 420 nm. Differences arise because of the use of different starches as the substrate. Also, the same extent of variation has been found with different amyloses. There have been questions about the reliability of amylase determinations of specimens with elevated glucose concentrations from diabetic patients. To overcome the problem, Henry and Chiamori suggested that a sample blank be measured at the same time. This helps to minimize interference even at glucose levels of 800 mg/dl (2).

Amyloclastic methods measure the decrease in the degree of blue colour developed by adding iodine to starch substrate. Wohlgemuth (3) reported on the semiquantitative method that uses serial sample dilutions, and Somogyi (4) measured the time required for the blue colour to disappear. Huggins and Russell (5) measured the decrease in the blue colour photometrically after a fixed time for digestion. With this method also, the use of different substrates gives different results.

A simple method involves the use of dye-marked (dye-coupled) starch as the substrate for amylase, and relies on the release of soluble fragments of sizes different from the substrate. Such substrates, including remazol brilliant blue starch (6), Cibachron blue F-3GA amylose (7), Reactone-red-2B-amylopectin (8), Blue starch polymer 51-A (9), and Procion brilliant red M-2BS-amylopectin (10), are insoluble in the reaction mixture. After the digestion of the substrate, the dye concentration in the supernatant is measured photometrically. What is measured are small soluble fragments coupled with dye that have been released from the substrate.

The most advanced method with a marked substrate is based on the use of *p*-nitrophenyl oligosaccharides (11). The principle of the method was reported a quarter of a century ago (12). The assay is done with the use of a synthesized *p*-nitrophenyl oligosaccharide (such as maltopentaoside, -hexaoside or -heptaoside), which is hydrolyzed by α-amylase and then catalyzed by α-glucosidase from yeast into the chromogenic product *p*-nitrophenol. The enzyme activity is assayed by a direct monitoring of the increase in absorbance of the reaction mixture at 405 nm (13).

Pierre *et al.* (14) reported a measurement of α-amylase activity using the coupling of enzyme reactions including maltose phosphorylase (MP), β-phosphoglucomutase (β-PGM), and glucose-6-phosphate dehydrogenase (G6PDH). Maltotetraose used as a synthesized substrate releases α-maltose by the catalytic activity of amylase, the maltose produced is brought into the indicator reaction by G6PDH system. To summarize the reaction sequence, maltotetraose is hydrolyzed by amylase to yield two moles of maltose per mole of substrate. MP catalyzes the phosphorolysis of each mole of maltose to one mole each of glucose and β-glucose-1-phosphate. β-PGM converts the β-glucose-1-phosphate to glucose-6-phosphate, which is oxidized to 6-phosphogluconate with the concomitant reduction of $NADP^+$ to NADPH in the reaction catalyzed by G6PDH. The high rate of production of NADPH is monitored by measurement of the increase in absorbance at 340 nm. This method gives results that may be less interfered with by the amount of glucose in the specimen than methods such as that of James *et al.* (15). James *et al.* used maltopentaose as a substrate, which is converted to glucose-6-phosphate by the coupled enzyme reactions of α-glucosidase and hexokinase. The indicator reaction they used is the same as that of Pierre *et al.*

The main sources of human serum amylase are pancreas, parotid and possibly liver. Urinary amylase is of serum amylase origin, the molecular weight of the enzyme, 45,000, allows it to filter easily through the glomeruli. An electrophoretic study has shown that there are at least two amylase isozymes with the same mobilities as β-and γ-globulin, respectively, in the serum and urine (16).

Most widely used in clinical laboratories today is the dye-amylose method (e), because of its simplicity and reproducibility, although this method is not easily adopted to automatic analytical instruments. Details of the procedure of this method are described. A test strip method for urinary amylase assays is also briefly mentioned.

Assay Method with Amylochrome (Amylochrome is the trademark of F. Hoffman-La Roche (Basel and Tokyo)) :

Reagent :

[1] Substrate tablet : Each contains 130 mg of Cibachron blue 3G-A amylose and 35 μ mol of phosphate buffer, pH 7.0

[2] Diluent solution : Each bottle contains 50 mmol of NaH_2PO_4 and 2.5 mmol of $EDTA \cdot Na_2$

[3] Standard : Cibachron blue-3G-A has 460 U/ℓ amylase activity.

Preparations of reagents :

Dilute one bottle of diluent solution [2] to 500ml with distilled water.

Assay procedure :

1. Place one substrate tablet [1] into test tubes labeled T(test) and into one test tube labeled B (blank). Only one blank is needed per each set of tests. Another test tube labeled S (standard) is prepared *without* a substrate tablet.
2. Add 1.0 ml water to each tube and mix well.
3. Place all tubes in a 37℃ water bath for 5 min to reach this temperature uniformly.
4. Add 0.05 ml serum to the tube labeled T (for urine, 0.02 ml is used).
5. Add 0.05 ml standard [3] to the tube labeled S.
6. Mix well and allow incubation for exactly 15 min.
7. At the end of the 15 min, stop the reaction by adding 4.0 ml diluted diluent solution to all tubes, and mix well.
8. Centrifuge for 10 min at 3000 rpm and the supernatant from each tube is placed into labeled T, S, and B test tubes, respectively.
9. Read the absorbance at 642 nm of the sample against water.

10. Calculation of amylase activity

$$\frac{(\text{absorbance of T}) - (\text{absorbance of B})}{\text{absorbance of S}} \times 460\,(\text{U}/\ell)$$

11. Calculation for urine

$$\frac{(\text{absorbance of T}) - (\text{absorbance of B})}{\text{absorbance of S}} \times 460 \times 2.5\,(\text{U}/\ell)$$

Interpretation :

Reference intervals of this method for serum ranges from 140 to 390 U/ℓ, with random urine samples, are from 50 to 1400 U/ℓ or from 0.1 to 1.4 U/mg creatinine; with 24-hour samples, the intervals are from 50 to 1400 U. The conversion factor to Somogyi units is 1.0 U/ℓ =0.54 Somogyi units.

Assay Method by a Test Strip (17) :

Principle :

The test strip (The test strip is distribued by Behringwerke AG (Marburg) under the name of "Rapignost" amylase) has a felt zone as the application zone, a reaction zone containing a colored starch substrate, and a paper zone for detection, all on one end of a plastic strip. The chromogenic substrate is degraded by α-amylase in urine. The soluble, colored hydrolysis products are chromatographed into the white paper zone.

The colour intensity is a measure of the amylase activity.

Reagent :

The test strip "Rapignost" amylase.

Assay procedure :

1. Pipette 0.1 ml of freshly voided urine onto the lower part of the application zone taking 3 to 5 seconds, and allow it to diffuse.
2. Place the test strip on a non-absorbent surface.
3. Compare the colouration of the detection zone with the colour scale as soon as the diffusion front has passed through the detection zone, but not before 2 min have elapsed.

Interpretation :

Pathological urinary amylase activities lead to marked reddish-violet colourations in correspondence with the positive field (+～++) of the colour comparison scale. According to the reports of several investigators, the method satisfies the condition that elevated urine amylase activity of 1000 U/ℓ by dye-coupled method gave positive results.

Clinical Significance :

Physiological Deviations — Normal values for serum and urinary amylase are shown in the assay methods. Serum amylase is first detected by ordinary techniques in the second to third month of live, rising to the low-normal adult level by one year. There is no clear difference in serum amylase levels between men and women. The levels of serum amylase in obese subjects are often higher than in lean ones, however, they are still within the normal range.

Pathological Significance :

The causes of increased levels of serum amylase (hyperamylasemia) with and without elevated urinary amylase (hyperamylasuria) include the followings :

I Pancreatic diseases

1. Acute pancreatitis
2. Chronic pancreatitis
3. Pancreatic pseudocyst or abscess
4. Pancreatic carcinoma
5. Pancreatic trauma

II Non-pancreatic diseases

(i) Hyperamylasemia without hyperamylasuria

1. Renal insufficiency
2. Salivary-type hyperamylasemia
3. Salivary gland lesions: mumps, calculus, irradiation sialadenitis, maxillofacial surgery, drugs
4. Macroamylasemia

(ii) Hyperamylasemia with hyperamylasuria

1. Biliary tract diseases
2. Intraabdominal diseases
   a. Perforated peptic ulcer
   b. Intestinal obstruction
   c. Ruptured ectopic pregnancy
   d. Dissected aneurysm
   e. Peritonitis
   f. Acute appendicitis
   g. Liver necrosis
3. Cerebral trauma
4. Burns and traumatic shock
5. Postoperative hyperamylasemia
6. Diabetic ketoacidosis
7. Drugs : morphine, codeine, corticosteroids, furosemide, methacholine, indomethacin

Pancreatic diseases :

The symptoms and sings of pancreatitis are often nonspecific, and the diagnosis of pancreatitis hinges on the serum or urinary levels of amylase. Serum amylase, when elevated, rises within 24 to 48 h of the onset of acute pancreatitis, probably the result of transperitoneal absorption of the enzyme (18,19). The peak level is usually reached within 48 h after which there is usually a large drop in the enzyme level with a return to normal values within 3 to 5 days. Normal serum amylase levels are occasionally seen in acute pancreatitis, probably following a transient rise and fall of the enzyme, extensive pancreatic necrosis, or acute exacerbation of chronic pancreatitis in which the pancreas cannot produce amylase. Persistent hyperamylasemia, on the other hand, suggests ongoing inflammation or complications, such as pancreas pseudocyst or abscess (20). An elevation in serum amylase is accompanied by elevation of urinary amylase. Elevated urinary amylase with a normal serum amylase may occur in acute pancreatitis because of the rapid clearance of amylase by the kidney. Therefore, urinary amylase may be higher and the elevation may last longer than that of serum amylase in acute pancreatitis.

Recently, the renal clearance of amylase has been studied with an attempt to improve the diagnostic accuracy of amylase assays. Several groups of investigators have accumulated observations confirming the clinical value of a high ratio of amylase to creatinine clearance (Cam/Ccr) in the diagnosis of acute pancreatitis (21). A normal Cam/Ccr ratio at an early time in the course of pancreatitis is uncommom. The sequence of the regression of parameters in pancreatitis is normalization first of serum amylase, next of urinary amylase, and finally of the Cam/Ccr ratio. Warshaw and Fuller (22) have suggested that the measurement of Cam/Ccr ratio is a means to distinguish between pancreatitis and other conditions that may cause hyperamylasemia. They found that the Cam/Ccr ratio is constantly elevated in cases of acute pancreatitis but remains normal in other conditions in which there are elevated serum levels of amylase, such as with cholelithiasis, duodenal ulcer, and intestinal diseases.

Non-pancreatic diseases :

Amylase is excreted primarily from kidneys into urine, and in the case of renal insufficiency a poor renal clearance of the enzyme takes place. High levels of serum amylase are often found in patients with renal insufficiency with oliguria or anuria. However, amylase is also metabolized via an extrarenal mechanism. Reports of the incidence and degree of elevation in hyperamylasemia are not necessarily coincident, but it is reasonable to conclude that serum amylase is seldom elevated above twice the upper limit of normal in patients with renal insufficiency (23). The salivary glands are also a rich source of amylase. Infection of the salivary glands causes high amylase levels. Studies of amylase isoenzymes have shown that there are two principal isoenzymes, the pancreatic-type (p-type) and salivary-type (s-type). Because only a small amount of the s-type isoenzyme can be excreted into the urine, hyperamylasemia of the s-type is not necessarily associated with hyperamylasuria.

Gammaglobulin binds amylase to form a macroamylase. In macroamylasemia, an unusual condition, the amylase is bound to IgA or IgG immunoglobulin. The molecular weight of these macromolecular complexes ranges from 150,000 to more than one million. The prevalence of macroamylasemia in the general population probably ranges between 1 and 2% (24). The disorder should be considered in any patient with hyperamylasemia whose urinary amylase levels are normal. A low Cam/Ccr ratio in the presence of increased serum amylase is a result that supports the diagnosis of macroamylasemia. Macroamylasemia may actually reflect dysproteinemia caused by various diseases, and may be a marker of a more fundamental pathologic aberration.

Hyperamylasemia attributed to pancreatitis has been reported in patients with cholecystitis, cholelithiasis, choledocholithiasis, and ruptured gall bladder. When a peptic ulcer perforates into the pancreas causing acute pancreatitis, the blood amylase level rises greatly. Diseases of the small or large bowel may also result in elevation of amylase because of liberation of the enzyme from the colonic or small intestinal mucosa. An elevation in serum amylase following obstruction and perforation of the small intestine is usually secondary to the pancreatitis that results, but it may result from liberation of the enzyme from the intestinal tract mucosa. Amylase is also present in the parenchymal cells of the liver. In massive hepatic necrosis, elevated serum amylase levels are seen.

Patients with cerebral trauma have hyperamylasemia. The mechanism is unclear, although it has been suggested that there is an inverse relationship between serum amylase levels and carbohydrate utilization. It has been postulated, therefore, that cortisol release after a head injury might be responsible for the hyperamylasemia because of decreased carbohydrate utilization.
Hyperamylasemia occurs frequently in diabetic ketoacidosis (25). It is most frequent when blood glucose is higher than 500 mg/dl, and the onset tends to be acute. Thus, pancreatitis is probably not usually the cause of hyperamylasemia in diabetic ketoacidosis.

(Kiyoshi Okuda, Kei-ichi Naka, Nobuo Shimojo)

References

1. Somogyi, M. (1938) *J. Biol. Chem. 125,* 399-414
2. Henry, R. J. & Chiamori, N. (1960) *Clin. Chem. 6,* 434-452
3. Wohlgemuth, J. (1908) *Biochem. Z. 9,* 1-9
4. Somogyi, M. (1960) *Clin. Chem. 6,* 23-35
5. Huggins, C. & Russell, P. S. (1948) *Ann. Surg. 128,* 668-678.
6. Rinderknecht, H., Wilding, P. & Haverback, B. (1967) *Experientia 23,* 805
7. Klein, B., Foreman, J. A. & Searcy, R. L. (1970) *Clin. Chem. 16,* 32-38
8. Bobson, A. L., Tenney, S. A. & Megraw, R. E. (1970) *Clin. Chem. 16,* 39-43
9. Ceska, M., Birath, K. & Brown, B. (1969) *Clin. Chim. Acta 26,* 437-444
10. Sax, S. M., Bridgewater, A. B. & Moore, J. J. (1971) *Clin. Chem. 17,* 311-315
11. Gillard, B. K., Markman, H. C. & Feig, S. A. (1977) *Clin. Chem. 23,* 2279-2282
12. Jansen, A. P. & Wydeveld, P. G. A. B. (1958) *Nature 182,* 525-526
13. McCroskey, R., Chang, T., David, H. & Winn, E. (1982) *Clin. Chem. 28,* 1787-1791
14. Pierre, K. J., Tung, K. K. & Naji, H. (1976) *Clin. Chem. 22,* 1219

15. James, G. P., Passey, R. B., Fuller, J. B. & Giles, M. L. (1977) *Clin. Chem. 3,* 546–550
16. Davis, J. (1972) *J. Clin. Pathol. 25,* 266–267
17. Habenstein, K. & Kohl, H. (1981) *J. Clin. Chem. Clin. Biochem. 19,* 688
18. Salt, W. B. & Schenker, S. (1976) *Medicine 55,* 269–289
19. Wolf, P. L. & Williams, D. (1973) *Practical Clinical Enzymology* pp. 129–138, John Wiley & Sons, Inc., New York
20. Brooks, F. P. (1972) *N. Engl. J. Med. 286,* 300–302
21. Levitt, M. D. (1979) *Mayo Clin. Proc. 54,* 428–431
22. Warshaw, A. L. & Fuller, A. F. (1975) *N. Engl. J. Med. 292,* 325–328
23. Levitt, M. D., Rapoport, M. & Cooperband, S. R. (1969) *Ann. Intern. Med. 71,* 919–925
24. Barrows, D., Berk, J. E. & Fridhandler, L. (1972) *N. Engl. J. Med. 286,* 1352
25. Belfiore, F. & Napoli, E. (1973) *Clin. Chem. 19,* 387–389

V. Analytical Application of Amylases and Related Enzymes

V.1. Determination of Starch Content

Starch occurs as a storage form of glucose in various plant organs and tissues. Its counterpart glycogen is found in animals and some microorganisms. These polysaccharides are polymers of D-glucopyranose linked together by α-glucosidic linkages, mainly by $(1\rightarrow4)$ and some by $(1\rightarrow6)$ linkages. The assay of starch content in plant materials, generally, consists of three steps, (i) extraction of starch, (ii) starch hydrolysis, and (iii) determination of glucose produced. The ideal method combining these steps should be applied to plant materials with a wide range of starch contents from seeds and tubers with a high starch content to leaves and sprouts with a lower one. Table 1 summarizes several couplings of the extraction with hydrolysis and glucose determination methods.

Table 1. Coupling Methods of Extraction with Hydrolysis and Glucose Determination for Assay of Starch Content (1)

Hydrolysis; Determination / Extraction	Glucoamylase; glucose oxidase-peroxidase method	Glucoamylase; hexokinase-glucose-6-phosphate dehydrogenase	Acid; chemical method
Boiling with water	MacRae (1971)[1]		
Alkali	Dekker & Richard (1971)(2)		
Perchloric acid			Pucher *et al.* (1947)[2] Hassid & Neufeld (1964)[3]
Dimethyl sulfoxide (DMSO)	Sasaki *et al.* (1986)(1)	Carpita & Kanabus (1987)(3)	Kovacs & Hill (1974)[4]

[1]MacRae, J. C. (1971) *Planta, 96,* 101-108
[2]Pucher, G. W., Leavenworth, C. S. & Vickery, H. B. (1948) *Anal. Chem. 20,* 850-854
[3]Hassid, W. Z. & Neufeld, E. F. (1964) in *Methods in Carbohydr. Chem.* Vol. 4 (Whistler, R. L., ed.) pp. 33-35, Academic Press, New York.
[4]Kovacs, M. I. P. & Hill, R. D. (1974) *Phytochemistry, 13,* 1335-1339.

Previously to the extraction and hydrolysis of starch, glucose and other soluble sugars are removed from plant materials by extraction with hot 80% ethanol (by reflux and filtration with usually 5-10 volumes of ethanol). Otherwise free glucose should be separately determined as shown in Aliquot (B) of Fig. 1 (2). Several extraction methods are known for efficient gelatinization and extraction of starch from plant materials, namely, boiling with water, heating to 130°C in an autoclave, using alkali, acid, or dimethyl sulfoxide (DMSO). The alkali (1,2) and DMSO (1,3) methods appear to be better than other methods. Starch is stable for at least 2 days in alkaline extracts and about 3 weeks in DMSO extracts (1).

For specific and sensitive assays of starch, enzymic hydrolysis using glucoamylase (EC 3.2.1.3) is preferable to acid hydrolysis and the glucose formed is specifically determined by either the glucose oxidase (EC 1.1.3.4) or hexokinase (EC 2.7.1.1) and glucose-6-phosphate dehydrogenase (EC 1.1.1.49) methods.

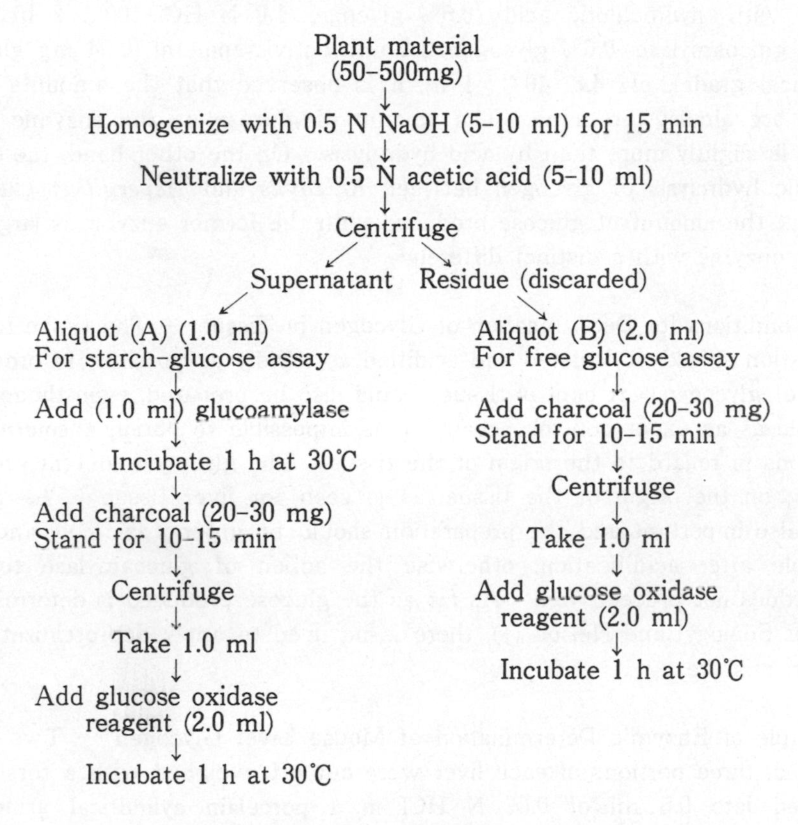

Fig. 1. Starch Analysis in Plant Material (2).

To determine glucose by the glucose oxidase method, the alkali extract is neutralized and treated with active carbon to remove inhibitory substances (2). Dilution of DMSO extracts to 20% DMSO with aqueous buffer does not affect the solubility of the starch nor interfere with the subsequent starch hydrolysis by glucoamylase of *Rhizopus niveus* (3).

(Hidetsugu Fuwa)

References

1. Sasaki, T. (1986) in *Seibutsu Kagaku Jikkenho* Vol. 19, *Denpun Kanren Toshitsu Jikkenho* (Nakamura, M. & Kainuma, K., eds.) (in Japanese), pp. 2-8, Gakukai Shutsupan Center, Tokyo.
2. Dekker, R. F. H. & Richard, G. N. (1971) *J. Sci. Food Agric.* *22*, 441-444.
3. Carpita, N. C. & Kanabus, J. (1987) *Anal. Biochem.* *161*, 132-139.

V.2. Enzymic Determination of Glycogen in Animal Tissues

Hydrolysis of Glycogen — In a comparative experiment of hydrolysis of oyster glycogen with hydrochloric acid (0.6% glycogen, 1.0 N HCl, 100℃, 2 h) and with *Rhizopus* glucoamylase [0.6% glycogen, 1.0 unit activity per ml (0.04 mg glucoamylase of analytical grade), pH 4.8, 40℃, 1 h], it is observed that the amounts of glucose produced are almost the same. The amount of glucose by the enzymic method, if anything, is slightly more than by acid hydrolysis. On the other hand, the comparison of enzymic hydrolysis of glycogen between *Rhizopus* and *Aspergillus* glucoamylases shows that the amount of glucose produced with the former enzyme is larger than by the latter enzyme with a distinct difference.

Several Conditions for Determination of Glycogen in Tissue — The tissue for glycogen determination should be isolated and acidified as rapidly as possible, in order to avoid autolysis of glycogen. A control tissue should also be prepared, even though it will be unavailable as an exact control, because it is impossible to obtain absolutely identical preparations in regard to the origin of the tissue. The glycogen content greatly varies depending on the origin of the tissue, as is seen for liver tissue. The grinding of tissue is also important and the preparation should be ground as finely and uniformly as possible after acidification, otherwise the action of glucoamylase to hydrolyze glycogen does not proceed well. As far as the glucose produced is determined by the method of Somogyi and Nelson (1), there is no need to apply deproteinization for the incubation mixture.

An Example of Enzymic Determination of Mouse Liver Glycogen — Two mice were decapitated, three portions of each liver were cut off, weighed with a torsion balance and placed into 0.5 ml of 0.05 N HCl in a porcelain cylindrical grinder vessel, respectively. The tissue pieces were ground by a high speed rotation of the cylinder to be a homogeneously cloudy state with cooling in crushed ice. The ground mixtures were transferred into small stainless steel tubes for centrifugation, washing the grinder four times, each with 0.5 ml of 0.1 M acetate buffer, pH 5.0, respectively. Then, the grinder used for test run was washed with 0.5 ml of glucoamylase dissolved in the same buffer (total activity, 1.0 unit) while the grinder for control run was similarly washed with only the buffer solution. The grinders were further washed five times, each with 0.5 ml of the buffer, and the washings transferred to combine with the former washings, each time. Each centrifugation tube containing the ground tissue and 5 ml of washings was incubated at 40℃ for 1 h, centrifuged at 2×10^4 g for 15 min, and the supernatant (0.2 - 2.0 ml) was subjected to determination of glucose by the method of Somogyi and Nelson. The results are shown in Table 1.

Table 1. Glycogen Content of Mouse Liver Determined by the Enzymic Method

Mouse	Conditions	Liver tissue (mg)	Glycogen added (mg)	Gluco-amylase added	Glucose determined (mg)	Glycogen content (%)
male	immediately	21.5	0	yes	1.82	7.54*
31 days	after	19.5	2.0	yes	3.76	7.84**
old	feeding	19.0	0	no	0.171	—
male	8 h after	27.0	0	yes	1.31	3.63*
31 days	feeding	17.0	2.0	yes	3.05	4.34**
old		22.5	0	no	0.185	—

$$* \quad \left(\frac{\text{glucose, mg, in the mixture with enzyme}}{\text{tissue weight, mg}} - \frac{\text{glucose, mg, in the mixture without enzyme}}{\text{tissue weight, mg}} \right) \times \frac{162}{180} \times 100.$$

** Glycogen content was calculated after subtracting the glycogen externally added.

The value of glycogen content of mouse liver shown in Table 1 seems to be almost correct, because the glycogen content of tissue samples is roughly the same as that examined with the control run with externally added glycogen. An experiment carried out independently, however, showed that the glycogen content of mouse liver significantly differed depending on the origin of liver tissue examined, and that the tissue around the portal vein usually showed a higher content of glycogen than other parts of liver.

(Takehiko Yamamoto)

Reference

1. Somogyi, M (1945) *J. Biol. Chem.* *160*, 69, (1952) ibid *195*, 19

V.3. Enzymic Elucidation of Structure and Homogeneity of α-1,4- : 1,6-Linked Oligoglucans (Branched Dextrins)

Cases for This Study is Necessary : A mixture of various but structurally homologous oligosaccharides, for example, a mixture of various maltooligosaccharides, can be separated into each sugar component depending on their degrees of polymerization (DP) by the technique of paper or high performance liquid-chromatography. If this mixture contains a structurally analogous saccharide, for example, a branched dextrin, the dextrin will be separated by the same techniques, because the R_F value (1) of the branched dextrin differs from those shown by a series of maltooligosaccharides. The increase in number of branching in dextrins also results in a change in the R_F value, even if the DP is the same. However, the difference of the position of branching on a maltooligosaccharide causes no effect on the R_F value. In other words, there may be the possibility that a branched dextrin isolated by the above techniques is a mixture of isomers of certian type of branched dextrin. This is the reason why the structural analysis is necessary for isolated branched dextrins.

Enzymes Available for Structural Analysis of Branched Dextrins : The method of applying enzymes may be convenient for studies of this kind, because several enzymes show their own characteristic site specificities, respectively, in the hydrolysis of α-1,4-glucosidic linkages near or around the α-1,6-glucosidic linkages present.

1) β-Amylase — This enzyme attacks α-1,4-glucosidic linkages to release maltose stepwise from the non-reducing end, but the action stops at around α-1,6-linked glucose residues and leaves the following structures as β-limit dextrins :

```
O—O                  O—O
  ↓                    ↓
O—O—O……  ,      O—O—O—O……,

O—O—O                O—O—O
    ↓                    ↓
  O—O—O……  ,      O—O—O—O……,
```

2) Pullulanase (Pullulan 6-glucanohydrolase) — The action of pullulanase against various singly branched dextrins is as follows :

```
O—O                     O—Ø        O             not attacked
  ↓     ————————→                  ↓
  O—Ø                   O—Ø   ,    O—O—Ø
O—O                     O—Ø        O—O—O          O—O—Ø
  ↓     ————————→                      ↓  ————→
O—O—Ø                   O—O—Ø,         Ø              Ø
O—O—O                   O—O—O       (a slight activity)
    ↓   ————————→
O—O—O—Os                O—O—O—Os
```

O, Glucose residue ; Ø, reducing end glucose residue or free glucose ;
O_S, Glucitol ; —, α-1,4-glucosidic linkage ;
↓, α-1,6-glucosidic linkage.

If the reducing end glucose of the dextrin has previously been reduced, the reducing oligosaccharide produced by the action of pullulanase indicates to originate to the side chain of the dextrin examined.

3) Isopullulanase (Pullulan 4-glucanohydrolase)— The followings are the specificities of isopullulanase on various branched dextrins :

```
O                O                 O—O                      O—O
↓      ——→       ↓                 ↓         ——→             ↓
O—Ø              Ø + Ø ,           O—O—Ø                     O + O—Ø,

O                O               O—O—O                      O—O—O
↓       ——→      ↓                   ↓         ——→               ↓
O—O—Ø            Ø + O—Ø,            O—O—O                       Ø + O—O + O—Ø,
                                         |                           |
                                         O—O—Ø                       Ø
```

```
O—O            O—O
  ↓              ↓
O—O—Ø   and    O—O—O—Ø   are not attacked.
```

4) *Bacillus subtilis* Saccharifying α-Amylase (BSA) — This enzyme hydrolyzes α-1,4-glucosidic linkages present at the non-reducing end side. The action specificities shown on various branched substrates are as follows :

```
O—O—O                        O—Ø  O               O
    ↓         ————→               ↓     ——→       ↓
  O—O—O—Ø                       O—O—O—Ø           O—O—Ø,

    O—O                               O—O         Ø  O
      ↓        ————→                    ↓   ——→      ↓
O—O—O—O—Ø                     O—Ø     O—O—Ø          O—O—Ø,

O—O—O                         O—Ø  O
    ↓          ————→               ↓
O—O—O—O—O—Ø                   O—Ø  O—O—Ø   Ø,

O—O—O                         O—Ø  O
    ↓          ————→               ↓
    O—Ø                            O—Ø.
```

The saccharifying α-amylase also attacks starch and glycogen hydrolyzing α-1,4-glucosidic linkages, but does not hydrolyze those linkages present near or around α-1,6-glucosidic linkages and leaves these portions as the following limit dextrin :

```
···O—O—O—O                                    Ø       O
         ↓                          ——→          +    ↓
···O—O—O—O—O—O—O······Ø                       O—Ø     O—O—Ø,

······O—O—O—O                                 Ø       O
            ↓                       ——→               ↓
······O—O—O—O—O                               O—Ø + O—O
              ↓                                       |
    O—O—O—O—O—O—O—O······Ø                            O—O—Ø,
```

```
······O—O—O—O                                      Ø          O
            ↓                                                 ↓
······O—O—O—O—O—O                        ——————→  O—Ø + O—O—O
                ↓                                               ↓
······O—O—O—O—O—O—O—O—O······Ø                                O—O—Ø,

······O—O—O—O
            ↓                                      Ø           O
······O—O—O—O—O—O—O                       ——————→         + 2 ↓
                  ↓                                O—Ø        O—O—Ø,
      ······O—O—O—O—O—O—O—O······Ø

      ······O—O—O
                |                                  Ø       O     O
······O—O—O—O   O                         ——————→          ↓     ↓
            ↓   ↓                                  O—Ø+O—O—O—O—Ø,
······O—O—O—O—O—O—O—O—O······Ø

        ······O—O—O
                  |                                Ø          O
······O—O—O—O     O                       ——————→       + 2  ↓
            ↓     ↓                                O—Ø        O—O—Ø
······O—O—O—O—O—O—O—O—O—O······Ø
```

5) Glucoamylase — Glucoamylase hydrolyzes not only α-1,4-glucosidic linkages, but also α-1,6-glucosidic linkages in starch and glycogen. However, the hydrolysis rate of α-1,4-linkages is faster than that of α-1,6-linkages, so that intermediate products are observed during the enzyme reaction as shown below :

```
O—O—O                        Ø  Ø  O                    Ø
    |           ——————→            |        ——————→
  O—O—O—Ø                       Ø  O—O—Ø                O—O—Ø ——→ 3 Ø

O—O                          Ø  O                       Ø
  |       ——————————————→       |      ——————————————→         ——————→ 2 Ø
O—O—Ø                        Ø  O—Ø                     O—Ø
```

An Example of Enzymic Analysis of Structure and Homogeneity of A Branched Dextrin (2) : α-Amylase of *Bacillus amyloliquefaciens* produces various branched dextrins in hydrolysis of starch and glycogen. The smallest molecular size branched dextrin is

```
O—O
  ↓
O—O—Ø,
```

and the next is a dextrin consisted of six glucose residues containing one α-1,6-glucosidic bond. This branched dextrin was isolated in a homogeneous state as far as examined by paperchromatography with several solvents. The degree of polymerization was determined by estimating the ratio of total sugar content to reducing sugar and the presence of α-1,6-linkage was made clear from its R_F value compared with those of several maltooligosaccharides. Upon incubation with pullulanase, the dextrin produced maltose, maltotriose and -tetraose. Also, the dextrin after reduction with sodium borohydride produced maltose, maltotriose, maltotriitol and maltotetraitol by pullulanase. These results clearly indicate that the branched dextrin was a mixture involving at least two branched dextrins : One, consisted of DP4 and DP2 and the other, DP3 and DP3, as the main- and side-chains, respectively. The possible structures that satisfy the above results are as follows :

```
O—O                O—O                   O—O                  O—O
  ↓                  ↓                     ↓                    ↓
  O—O—O—Ø,        O—O—O—Ø,          O—O—O—Ø,          O—O—O—Ø,
     a                 b                    c                    d

O—O—O              O—O—O             O—O—O
    ↓                  ↓                 ↓
    O—O—Ø,        O—O—Ø,          O—O—Ø.
     e                 f                 g
```

However, the branched dextrin was not hydrolyzed by either isopullulanase or β-amylase. Therefore, the branched dextrin of structures symbolized by a, d and e will be out of consideration. On the other hand, the branched dextrin was hydrolyzed by *B. subtilis* saccharifying α-amylase and produced glucose, maltose, 6^3-glucosylmaltotriose and 6^2-glucosylmaltose (panose).

```
                          O—Ø
                   Pul    O—O—O—Ø                  O—Ø
                 ↗                          Pul ↗   O—O—Ø
O—O               BSA     O—O
  ↓             ———→        ↓
O—O—O—Ø                   Ø  O—O—Ø          BSA    Ø  O
 (42%)                                          ↘      ↓
                 ↘ GlcA   Ø  O                         O—O—Ø
                              ↓
                          Ø  O—O—Ø→

                           O—Ø
              Pul ↗
    O—O                    O—O—O—Ø               O—Ø
      ↓
O—O—O—Ø       BSA                        Pul ↗   O—Ø
  (15%)          ↘          O—O
                              ↓
                     O—Ø   O—Ø           GlcA
                                            ↘    Ø  O
                                                    ↓      →
                                                    O—Ø
```

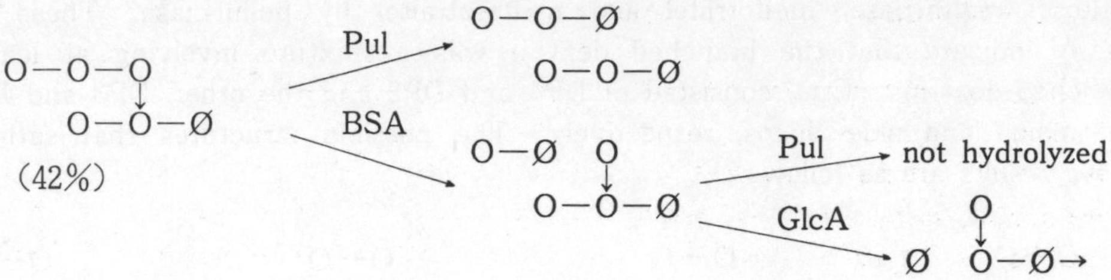

```
                        Pul         O—O—Ø
O—O—O          ─────────────→       O—O—Ø
  ↓
O—O—Ø                  BSA
                                   O—Ø   O           Pul
(42%)          ─────────────→            ↓      ─────────────→ not hydrolyzed
                                   O—O—Ø
                                                    GlcA         O
                                                ─────────────→   ↓
                                                           Ø     O—Ø→
```

(Pul, pullulanase; BSA, *B. subtilis* saccharifying α-amylase; GlcA, glucoamylase)

Also, in the digestion of the branched dextrin with BSA, three unknown branched dextrins of smaller molecular sizes were observed. Two of them were found to be O—O(↓O—Ø) (c') and O(↓O—O—Ø) (f') by the experiment of digestion first with pullulanase and then, glucoamylase or with the opposite order of the enzymes. The dextrins c' and f' are concluded to be a digestion product of c and f with BSA, respectively. Another branched dextrin newly but intermediately produced in the digestion with BSA was found to be O—O(↓O—O—Ø) (b'), the formation of this dextrin is concluded to originate to structure b dextrin. On the other hand, the digestion with BSA of the branched dextrin examined produced no isomaltose. Therefore, the presence of dextrin of structure g will be deleted. Thus the branched dextrin examined was found to be a mixture of the branched dextrins b, c, and f.

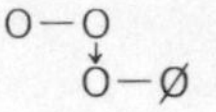

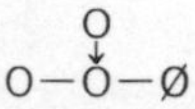

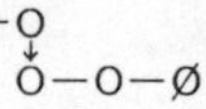

(Takehiko Yamamoto)

References

1. Martin, A. J. P. & Synge, R. L. M. (1941) *Biochem. J. 35,* 1358
2. Umeki, K. & Yamamoto, T. (1972) *J. Biochem. 72,* 101–109

VI. Application of Amylases and Related Enzymes to Industry

The history of the industrial production of enzymes dates back to the time when Dr. Jhokichi Takamine began the production of a digestive enzyme preparation by wheat bran-Koji culture of *Aspergillus oryzae* in 1894. In 1905, a textile desizing agent containing α-amylase by culture of *Asp. oryzae* was produced in Kyoto, Japan. Boiden and Effront in 1908 and 1915 patented the method of production of a desizing agent consisting of bacterial amylase (1). In 1939, industrial production of α-amylase from *Bacillus subtilis* sp. (*B. amyloliquefaciens*) started in Japan. In the early stage of the enzyme production, cultivation of the bacterium was done by wheat bran-koji culture, but in 1949 a submerged culture method was introduced. Bacterial α-amylase has been used as a desizing agent in textile industry, and also as a starch liquefying agent in starch processing industry. From around 1970, the bacterial strains for production of α-amylase were gradually replaced by those of *B. licheniformis* (2) and *B. subtilis* MN-385 (3), both of which produce more thermostable α-amylase. Glucose oxidase production from a *Penicillium* sp. started in 1957 and application of this enzyme served to facilitate the determination of glucose in sugar mixtures and thus the screening of glucose forming amylases of microbial sources was greatly advanced.

The industrial production of glucoamylase started in 1959. In the early stage of the enzyme production, wheat bran-koji culture using a strain of *Rhizopus* sp. and submerged culture of a strain of *Endomyces* sp. were applied. In 1965, the enzyme production by submerged cultures of *Asp. niger* started and has since been continued. Industrial production of dextrose powder and dextrose crystals from starch using α-amylase and glucoamylase began in 1959. The output of dextrose in Japan reached to 110,000 tons a year in 1961 and four years later, the dextrose production amounted to approximately 200,000 tons a year. Around the year 1961, isomerization of glucose to fructose by using glucose isomerase attracted attention, and the industrial production of glucose-fructose syrup started in 1969. In 1971, glucose isomerase produced by submerged cultures of *Streptomyces* became an industrial enzyme, and even an immobilized glucose isomerase was put on the market in 1976. Current industrial production of glucose-fructose syrup is being done exclusively by the continuous isomerization process using commercial preparations of immobilized glucose isomerase. Crystalline fructose is also being produced on an industrial scale. Along with production of these sugars, maltose powder, maltose crystals and syrup consisting mainly of maltotriose are also presently being produced industrially.

Recently, syrups of various homo- and/or hetero-oligosaccharides are on the market for food industry; syrup which contains branched oligosaccharides as the main component made by use of transferase, syrup containing maltotetraose as the main component by use of maltotetraohydrolase, coupling reaction products such as glucosyl-sucrose, maltosyl-sucrose, etc. from sucrose and liquefied starch by use of cyclomaltodextrin glucanotransferase, etc. Several other enzymic products such as "fructooligosaccharides" containing 1-kestose and nystose as the main components, and "galactooligosaccharides" whose main components galactosyl lactoses are also being produced by application of enzymes. Both the oligosaccharides are known to be specifically effective for growth of *Bifidobacterium* in the intestinal tract.

(Toshiaki Komaki)

References

1. A. Boiden et G. Effront, (1980) *Fr. Patent* No. 399087, (1915) Fr. Patent No. 475431, (1917) *U. S. Patent* No. 1227374, No. 1227527,
2. Shimamura, M., Amano, H. & Ohnuma, T. (1971) Japan Patent No. 623056
3. Hattori, F., Nakamura, M., Taji, N., Nojiri, M., Nakai, T. & Kusai, K. (1973) *Japan Patent* No. 923754

VI.1. Industrial Utilization of α-Amylases

1. Removal of Starch Sizer from Textile : In textile weaving, starch paste is applied for warping. The purpose is not only to give strength to the textile at weaving, but to prevent the loss of string by friction, cutting and generation of static electricity on the string by giving softness to the surface of string due to laid down nap. After weaving to cloth, the starch is removed and the cloth goes to scouring and dyeing processes. As sizer, starch or chemical sizers like polyvinyl alcohol, or their mixtures are used. The starch on cloth is usually removed by application of α-amylase.

Desizing Process

Gelatinized starch sized cloth
↓
Singeing
↓
Hot water-treatment
↓ 90℃, 10 sec.
Steeping in α-amylase solution
↓ (α-amylase 5~10 g/ℓ, nonionic surfactant 2~5 g/ℓ) 20 sec.
Steaming
↓ 70~80℃, 5~20 sec.
Hot water-washing
↓
Cold water-washing

The process mentioned above is the one being applied most at large scale works, but there are several other processes as well. For example,

(1) Dipping process, preparing an enzyme solution (5~10 g/ℓ) at 65~70℃, and dipping cloth in it (to react with the enzyme),

(2) Piling process, after dipping in the enzyme solution, the cloth is put into a piling tank to react at high temperatures in high humidity;

(3) Jig process, by letting cloth to and fro in the enzyme solution tube, to make the cloth react with the enzyme.

α-Amylase used:

At present *Bacillus subtilis* α-amylase is most widely utilized for desizing.

Activity: 3000~4000 units (IU) per g preparation. One unit activity is defined as the amount of enzyme which produces reducing sugar equivalent to 1 μmol glucose from 6 ml of 1.0% potato starch solution at pH 6.0 at 40℃ in one min.

The efficiency of desizing is examined by spraying 0.05 N I_2 onto cloth after desizing. If no color appears, the desizing is evaluated to be complete.

2. Manufacture of Dextrins: For production of various dextrin products of low molecular weights from starch, liquefaction of starch is generally necessary. Manufacture of dextrins is in general carried out by liquefaction of starch by bacterial

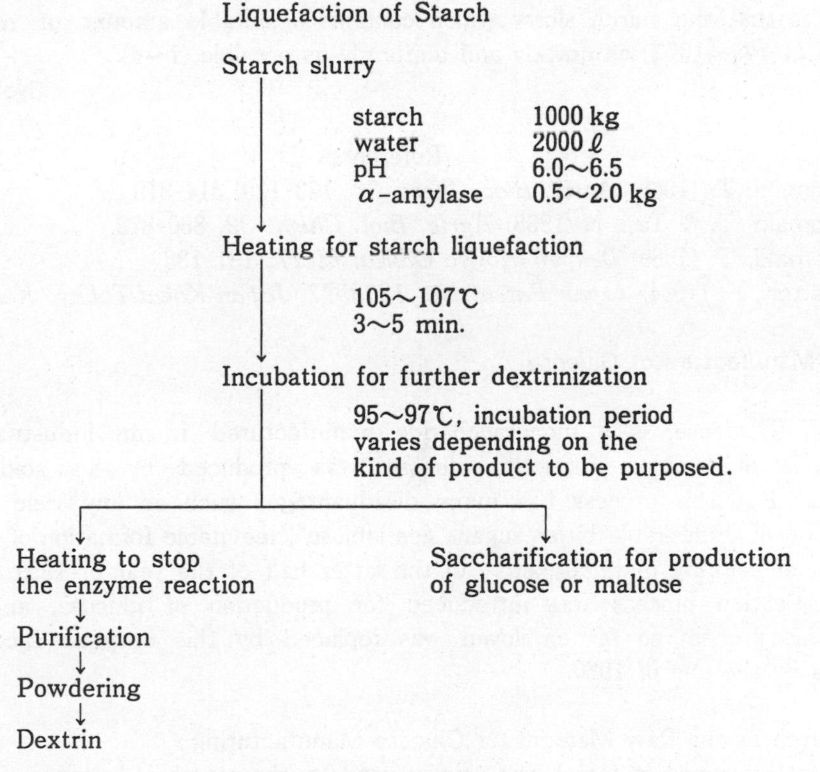

α-Amylase used: *Bacillus subtilis* α-amylase or *Bacillus licheniformis* α-amylase; activity, 10,000~24,000 IU/g.

α-amylase, followed by refining the liquefied starch solution and spray-drying. The average molecular size of the dextrin to be manufactured can be adjusted by regulating the degree of enzymic hydrolysis of starch by heating the mixture to stop the enzyme reaction. The degrees of hydrolysis expressed as dextrose equivalence (DE) of commercial dextrins are usually in a range from 3 to 25 which depends on the utilization purpose of the dextrin. Dextrin is widely used as a viscosity improver, filler or ingredient of food. For production of glucose or maltose, the dextrin prepared by liquefaction of starch with bacterial α-amylase is further incubated after adding glucoamylase or β-amylase, as shown in the flow sheet. In general, the DE of liquefied starch suitable for saccharification by these enzymes is about 10 for glucose production and 0.5 to 5 for maltose production. For production of maltose, pullulanase or isoamylase is simultaneously added with β-amylase, as described elsewhere in this book.

Liquefaction is an essential and basic process for processing of starch and if this process does not go well, various troubles such as poor filtration, turbidity of the processed solution, etc. occur. The most important factor for ideal liquefaction of starch is that the starch slurry which contains a suitable amount of α-amylase is heated at 105～107℃ as quickly and uniformly as possible (1～4).

(Noboru Taji)

References

1. Komaki, T. (1968) *Agric. Biol. Chem. 32,* 123-129, 314-319
2. Komaki, T. & Taji, N (1968) *Agric. Biol. Chem. 32,* 860-872
3. Komaki, T. (1969) *Denpun-Kogyo Gakkaishi 17,* 131-138
4. Hattori, F. (1984) *Japan Patent No. 1203977, Japan Kokai Tokkyo Koho* 51-44652

VI.2. Manufacture of Glucose

Glucose (Dextrose) is a monosaccharide manufactured in an industrial scale by hydrolysis of starch. Formerly, glucose was produced by an acid hydrolysis process. But this process has many disadvantages such as low yield of glucose, formation of undesirable bitter sugar “gentiobiose”, inevitable formation of salt, a large amount of coloring materials, etc. In the latter half of the year of 1959, an enzymic saccharification process was introduced for production of glucose, and the acid hydrolysis process so far employed was replaced by this enzymic saccharification process by the end of 1960.

1. Starch as the Raw Material for Glucose Manufacturing:

In Japan five kinds of starch are usually used by the starch industries. The amount of starch used in Japan is gradually increasing. Since 1965 domestic starch (sweet potato starch and potato starch) production has decreased and to cover the demand, import of corn to produce starch is rapidly increasing. Recently, about 62% of the total starch consumed in Japan yearly is used for production of starch sweetener. The amount of corn starch subjected to saccharification grew five fold in 1975 and thirteen fold in 1985 compared to that in 1965.

Table 1. Consumption of Starch for Starch Sweetener*

Year (Oct.~Sep.)	Starch for starch sweetener uses					Total and its ratio to the sum total starch consumed	Total starch consumed
	Sweet potato	Potato	Corn	Wheat	Imported		
1965	450	80	85	5	—	620 (51.0 %)	1215
1970	205	94	311	—	4	614 (53.1 %)	1156
1975	94	31	433	—	40	598 (53.3 %)	1123
1980	87	161	880	—	12	1140 (60.9 %)	1872
1985	121	96	1114	—	59	1390 (62.1 %)	2237

* 1000 ton

2. Manufacture of High Conversion Glucose Syrup:

High conversion glucose syrup is used for production of fine powdered glucose, crystalline dextrose or as the starting material for high fructose containing glucose syrup. The liquefaction of starch is the most important step in the enzymic production process of glucose. When the degree of dextrinization during liquefaction of starch is too high, the affinity of glucoamylase to the dextrins produced is reduced, resulting in lower rates in saccharification. If the degree of starch dextrinization is too low, the dextrins of high molecular weights form micelles which are hardly hydrolyzable with glucoamylase. Bacterial α-amylase (*B. subtilis, B. licheniformis,*) is usually used for starch liquefaction. The following phenomena may occur in cases of inefficient starch liquefaction by the enzymic process:

1) The solution of hydrolyzates is turbid,
2) Blue color reaction with iodine remains even after advanced saccharification,
3) The solution of hydrolyzates is difficult to filter and active carbon used for decoloration leaks through the filter.

Table 2. The Manufacturing Processes and Yields of Glucose

Process	Liquefaction	Saccharification	Final DE*	Glucose (%)
1	Acid	Acid	92	85
2	Acid	Glucoamylase	95	91
3	Acid/ α-Amylase	Glucoamylase	96	92
4	α-Amylase/ High pressure cooking / α-Amylase	Glucoamylase	97	93
5	α-Amylase (thermostable)	Glucoamylase	97	94
6	α-Amylase (thermostable)	Debranching enzyme + glucoamylase	97~98.5	95~97.5

* Dextrose equivalent

Manufacturing Procedure:

Starch slurry	Concn. ds* 30–40% (17–23° Bé)
↓	
Liquefaction by α-amylase	α-Amylase of *B. licheniformis* (24,000 IU/g) 0.6 kg per ds starch 1 ton, pH 6.0–6.5, 5 min at 105–110℃ and then 1–2 h at 90–95℃, final DE 5–15.
↓	
Saccharification by glucoamylase with debranching enzyme	Glucoamylase of *A. niger* (200 AGU/ml) 0.8 ℓ per ds starch 1 ton. Debranching enzyme of *Bacillus* sp. (200 PUN/g) 0.3 kg per ds starch 1 ton, pH 4.5, 60℃, 48–60 h, final DE 97–98.5, glucose content 95–97.5.
↓	
Filtration	
↓	
Decolorization	
↓	
Filtration	
↓	
Ion-exchange treatment	Cationic, anionic and mixed resin beds.
↓	
Active carbon treatment	
↓	
Filtration	*ds, dry substance.
↓	
Evaporation	
↓	
High conversion glucose syrup	

The relative yield of glucose increases with decreasing concentration of dextrin as substrate. This is because of repressing the reverse reaction by glucoamylase to polymerize glucose. However, as the concentration of glucose produced is low, the evaporation costs become expensive. Also, a large saccharification tank is necessary increasing the risk of microbial infection. On the other hand, when a debranching enzyme such as pullulanase is used together with glucoamylase (Table 2. process No.6), the degree of saccharification to glucose is further improved. In this process, the following merits are observed (1):

1) A high yield of glucose is obtaind,
2) The amounts of glucoamylase can be economized,
3) Saccharification time can be reduced,
4) Saccharification can be carried out at higher substrate concentrations.

3. Manufacture of Glucose (Dextrose):

In Japan, two kinds of glucose authorized by "Japanese Agricultural Standards" (JAS) are now being produced : Crystalline dextrose (monohydrate and anhydrous) and fine powdered glucose.

Manufacturing Procedure:

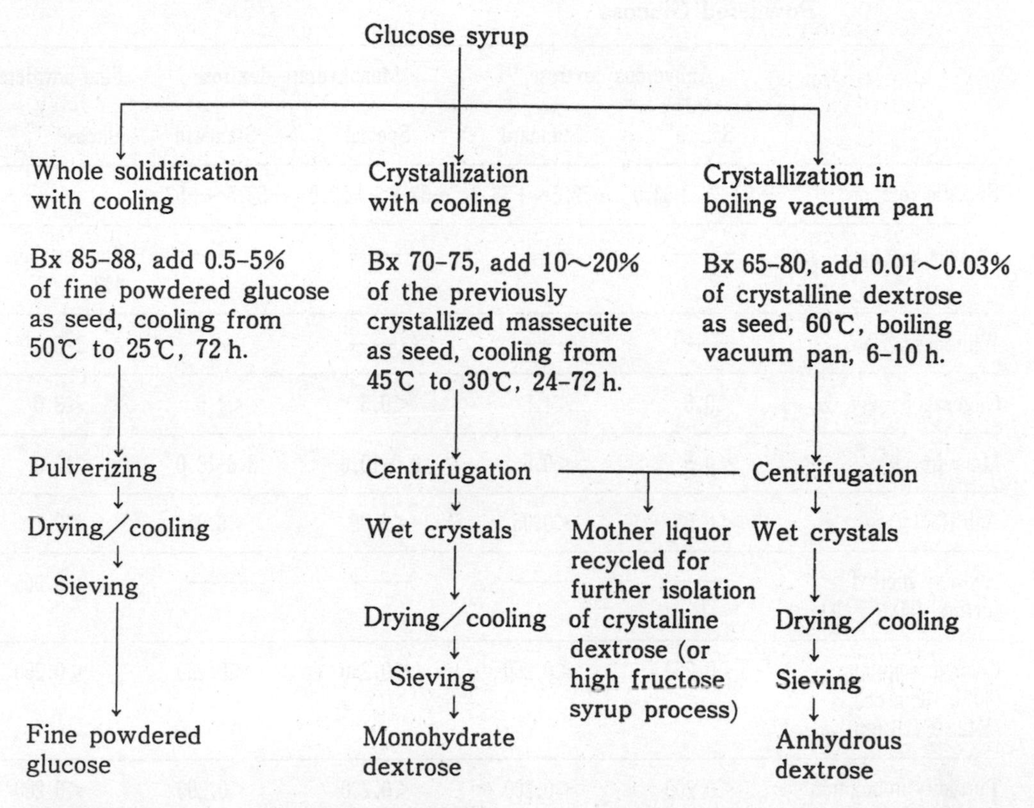

4. Specifications and Utilization of Glucose:

In Tables 3 and 4 are shown some important characteristics of JAS specifications and J.P. (Japanese Pharmacopeia) standard. The production and utilization of several grades of glucose in Japan are shown in Tables 5 and 6, respectively.

Table 3. Japanese Agricultural Standards of Crystalline Dextrose and Fine Powdered Glucose

	Anhydrous dextrose		Monohydrate dextrose		Fine powdered
	Special	Standard	Special	Standard	glucose
Specific rotation	+52.2∼+53.0	+52.5∼+53.9	+52.2∼+53.0	+52.5∼+53.9	—
Particle size (μ)	—	—	—	—	Smaller than 840
Whiteness (%)	—	—	—	—	>90
Oligosaccharides (%)	<0.5	<1.5	<0.5	<1.5	<3.0
Moisture (%)	<0.5	<0.5	8.0-10.0	8.0-10.0	<10.0
Ash (%)	<0.05	<0.05	<0.05	<0.05	<0.1
Hydroxy methyl furfural (HMF) (%)	—	—	—	—	<0.005
Color in solution (30%, 10cm cell, E420-E720 nm)	<0.250	<0.250	<0.250	<0.250	<0.250
Turbidity in solution (30%, 10cm cell, E720 nm)	<0.200	<0.200	<0.200	<0.200	<0.200

Table 4. Japanese Pharmacopeia Standards of Anhydrous Dextrose

Appearance	A white, crystalline powder, ordourless, sweet taste.
Specific rotation	+52.6 ∼ +53.2
Acidity	5.0 g/50 ml, less than 0.6 ml of 0.01 N NaOH
Chloride	<0.018 %
Sulfate	<0.024 %
Heavy metals	<4 ppm
Arsenic	<1.3 ppm
Dextrin	none
Moisture	<0.5 %
Ash	<0.1 %

Table 5. Production of Glucose in Japan*

Year (April~March)	JAS Glucose			Total of glucose including non-standard
	Crystalline dextrose		Fine powder glucose	
	Anhydrous	Monohydrate		
1970	25,337	30,347	52,253	153,707
1975	23,678	27,058	49,708	127,717
1980	23,422	27,842	51,450	130,580
1985	27,844	32,731	43,106	127,365

* ton

Table 6. Consumption of Glucose in Japan, 1985*

Use	Glucose	Liquid Glucose
Bread	7,610 (5.8)	——
Confectionary	18,427 (14.1)	5,907 (3.2)
Chewing gum	7,011 (5.4)	——
Instant powder food	13,688 (10.5)	104 (0.1)
Soft drinks	4,187 (3.2)	5,807 (3.1)
Milky drinks	215 (0.2)	827 (0.4)
Other drinks	1,577 (1.2)	303 (0.2)
Ice cream	2,012 (1.5)	31 (0.0)
Canned food	6,704 (5.1)	348 (0.2)
Jam	906 (0.7)	——
Fish jelly products	4,827 (3.7)	25 (0.0)
Sauces	12,622 (9.7)	24,119 (13.1)
Sake	5,681 (4.4)	13,237 (7.2)
Feed	9,649 (7.4)	——
Medical use	15,174 (11.7)	15,443 (8.4)
Sorbitol	——	81,293 (44.0)
Others	19,987 (15.4)	37,119 (20.1)
Total	130,279(100.0)	184,563(100.0)

* ton

(Tomoe Kanno)

Reference

1. Norman, B. E. (1983) *J. Jpn. Soc. Starch Sci. 30*, 200-211

VI.3. Manufacture of Maltose

Maltose is a naturally occurring disaccharide having the chemical structure of 4-O-α-D-glucopyranosyl-D-glucopyranose ($C_{12}H_{22}O_{11}$). It is the main component of maltosugar syrup (1). The discovery of pullulanase by Bender (2) suggested the possibility of complete conversion of starch to maltose. Since then, studies on

Manufacturing Procedure :

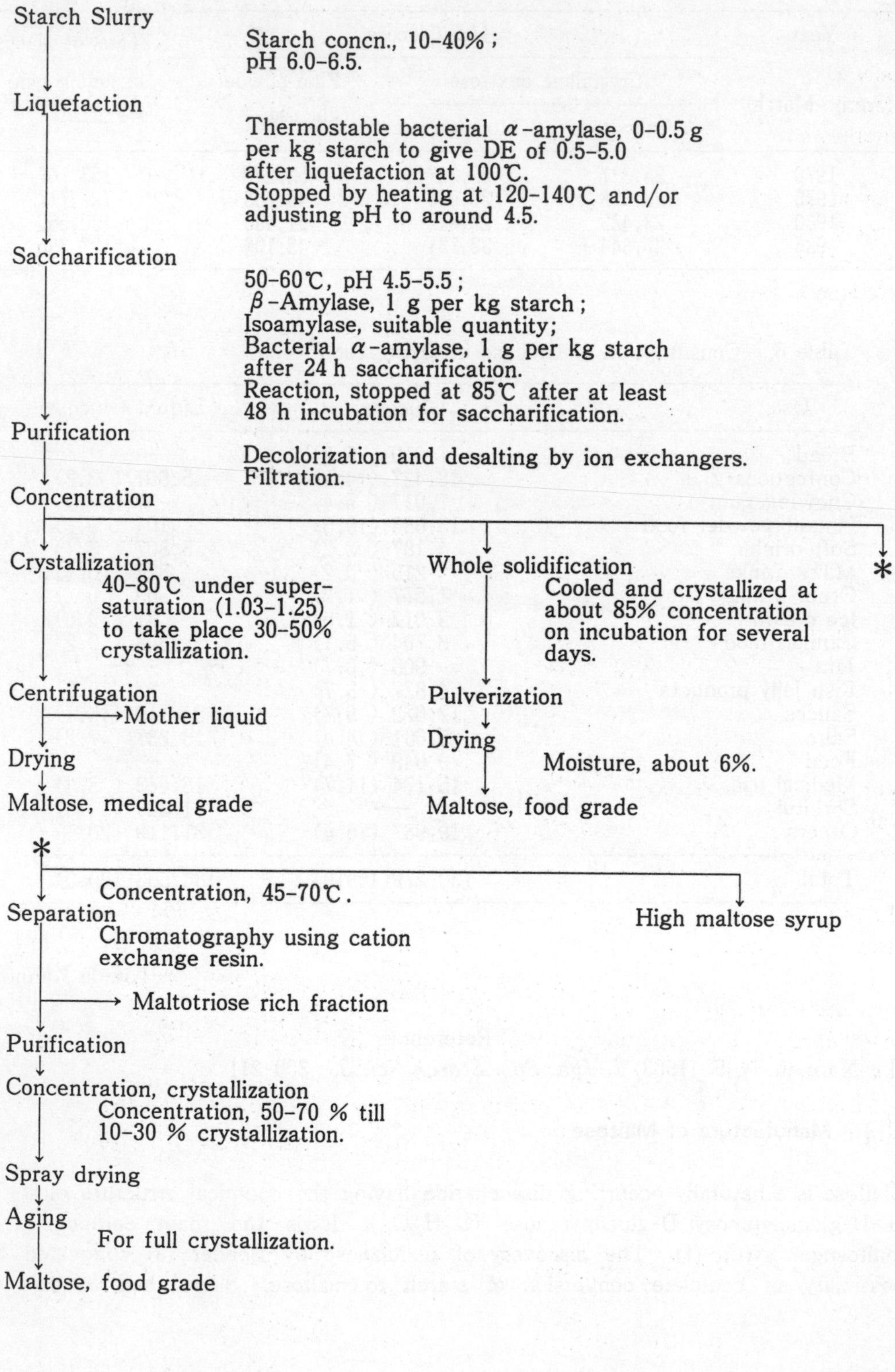

Starch Slurry
↓ Starch concn., 10–40% ; pH 6.0–6.5.

Liquefaction
↓ Thermostable bacterial α-amylase, 0–0.5 g per kg starch to give DE of 0.5–5.0 after liquefaction at 100℃. Stopped by heating at 120–140℃ and/or adjusting pH to around 4.5.

Saccharification
↓ 50–60℃, pH 4.5–5.5 ; β-Amylase, 1 g per kg starch ; Isoamylase, suitable quantity; Bacterial α-amylase, 1 g per kg starch after 24 h saccharification. Reaction, stopped at 85℃ after at least 48 h incubation for saccharification.

Purification
↓ Decolorization and desalting by ion exchangers. Filtration.

Concentration
→ (branches to Crystallization, Whole solidification, and *)

Crystallization
↓ 40–80℃ under super-saturation (1.03–1.25) to take place 30–50% crystallization.

Centrifugation
→ Mother liquid
↓

Drying
↓

Maltose, medical grade

Whole solidification
↓ Cooled and crystallized at about 85% concentration on incubation for several days.

Pulverization
↓

Drying
↓ Moisture, about 6%.

Maltose, food grade

*
→ High maltose syrup
↓ Concentration, 45–70℃.

Separation
↓ Chromatography using cation exchange resin.
→ Maltotriose rich fraction

Purification
↓

Concentration, crystallization
↓ Concentration, 50–70 % till 10–30 % crystallization.

Spray drying
↓

Aging
↓ For full crystallization.

Maltose, food grade

debranching enzymes including the discovery of isoamylase (see Ⅲ.1.a.) as well as on the conditions of liquefying starch and saccharifying the liquefied starch were performed by many reseachers to succeed in manufacturing high purity maltose powder and maltose syrup (3). Maltose is widely used as a sweetener, quality improver or as the raw material of maltitol. It is also used as an intravenous sugar supplement.

Starch as the Raw Material: Corn starch is mostly used in Japan. Potato, sweet potato and cassava starch are of course available for maltose manufacture. The concentration of starch slurry is adjusted to be 10-20% for production of medical grade maltose and 20-40% for food grade.

Enzymes: Thermostable bacterial α-amylase from *Bacillus licheniformis* (6000 JIS-units/g), bacterial α-amylase from *Bacillus amyloliquefaciens* (5000 JIS-units/g), and β-amylase from soybean (1500 JIS-units/g) to be applied are all commercial enzyme preparations. However, the utilization of isoamylase from *Pseudomomas amyloderamosa* (1250 kilounits/g) is at present limited to the patent-licencees. Pullulanase is also useful as a debranching enzyme for maltose manufacture, but the action of this enzyme is inhibitorily affected by high concentration of the substrate or products. This enzyme is known to show the reverse reaction on maltose.

α-Amylase During Saccharification: α-Amylase dextrinizes starch and causes disappearance of iodine color reaction of the hydrolyzate when the yield of maltose reaches 3-5%.

Maltose for Intravenous Injection : The maltose for intravenous injection is extensively purified by using active carbon which is effected not only for decoloration but also for removal of pyrogenic substances. The procedures after concentration of the syrup are performed in clean rooms.

Food Grade Maltose: Two methods are available; one is the method of solidifying the whole hydrolyzates and the other, a method using cation exchange resin to isolate maltotriose as a by-product (see Ⅵ.5.a). In the former method, c. 90% is the maximum of maltose content. While in the latter method, the maltose content can be increased over 90%, even from low purity hydrolyzates, though the performance may cost expensive. The commercial products of maltose are β-anomeric, monohydrate crystals. Recently, anhydrous maltose powder is being produced by crystallizing the sugar from a purified solution under high temperatures (4).

(Shuzo Sakai)

References

1. Sugimoto, K. (1977) *Denpunkagaku Handbook*, pp 450-461, Asakura, Tokyo.
2. Bender, H. & Wallenfels, K. (1961) *Biochem. Z. 334*, 79-95.

3. Sakai, S. (1981) *J. Jap. Soc. Starch Sci.* (in Japanese), *28*, 72-78.
4. Sakai, S. (1987) *Food Chemicals* (in Japanese), *29*(4), 37-42.

VI.4.a. High Fructose Containing Syrup (HFCS)

A high fructose containing syrup 42F HFCS (fructose content, 42%) is prepared by enzymic isomerization of glucose in a column reactor containing immobilized glucose isomerase. The glucose applied for isomerization is usually produced as dextrose syrup from starch whose glucose content is more than 95%. 42F HFCS is also industrially separated into a fructose rich fraction 90F HFCS (fructose content 90%) and a glucose rich fraction (glucose content, 90%) by a liquid chromatographic method. 55F HFCS (fructose content, 55%) is prepared by appropriately mixing 90F HFCS and 42F HFCS. The glucose in the glucose rich fraction can be isomerized in the above mentioned column reactors. 55F HFCS is a stable liquid sugar having a sweetness similar to that of invert sugar. It is used in many fields as a sucrose replacement. (Table 1 and 2)

Table 1. Production of HFCS

Production	Japan (1987*) (×1000 ton ds)	USA (1985) (×1000 ton ds)
42F HFCS	275	1,593
55F HFCS	420	3,297

ds, Dry substance * 1987. 4～1988. 3

Table 2. Applications of HFCS in 1985

Applications	Japan	USA
Beverages	54.9 %	67.2 %
Dairy products	12.9	4.4
Baking products	6.2	10.1
Canning	3.3	8.4
Frozen candies	6.2	—
Confectioneries	2.9	0.4
Others	13.6	9.5

Glucose Syrup: The starch as raw material is highly converted to glucose by enzymic liquefaction and saccharification. The glucose content of the hydrolyzate desirable for glucose isomerization is over 95%. Refining with ion exchangers and active carbon treatment is necessary to reduce the calcium content of the syrup to the levels below 2 ppm that do not seriously inhibit isomerase activity and to remove other undesirable substances which seriously diminish the stability of isomerase.

Manufacturing Procedure :

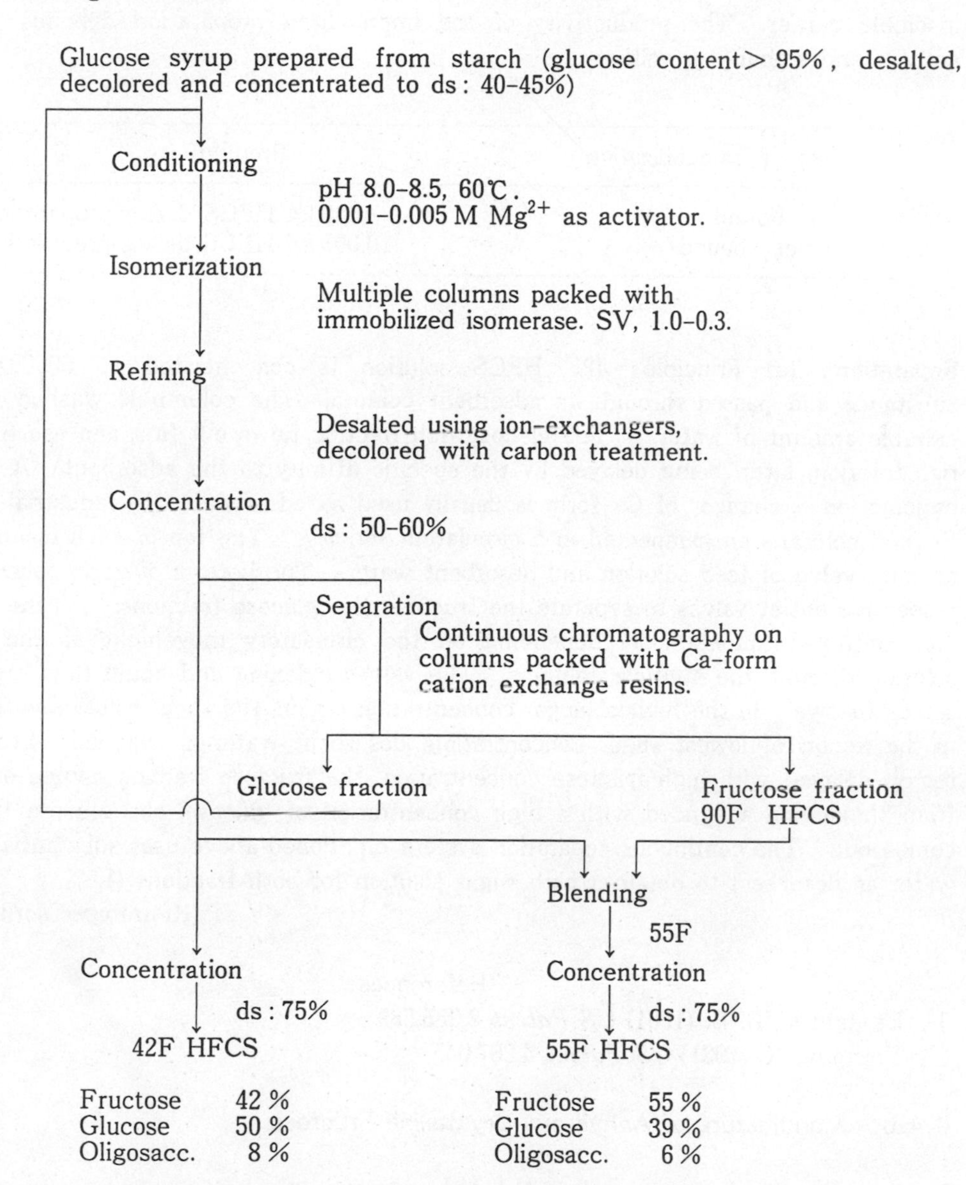

Conditioning: The presence of 0.001-0.005 M Mg^{2+} in the solution as activator is desirable to compensate the inhibitory action of Ca^{2+}. The reaction temperature, 60°C, is employed to protect from contamination of microorganisms.

Immobilized Isomerase : Commercial isomerase preparations are prepared from certain species of *Streptomyces, Bacillus, Arthrobacter* and *Actinoplanes*. For immobilization of the enzyme , two methods are being applied : (a) fixation of the enzyme as a cell bound enzyme of the microorganism, and (b) adsorption of the refined isomerase on an insoluble carrier. The productivity of the immobilized preparation depends on its activity and stability as follows:

Type of immobilization	Productivity
Cell bound	3,000 kg HFCS ds/kg preparation
Carrier bound	10,000 kg HFCS ds/kg preparation

Separation: (a) Principle; 42F HFCS solution is concentrated to 50-60% dry substance and passed through an adsorbent column. The column is washed with a suitable amount of water. The glucose rich fraction flows out first and the fructose rich fraction, later, being delayed by the specific affinity to the adsorbent. A strong cationic ion exchanger of Ca-form is usually used as adsorbent. (b) Industrial plant; Several columns are connected in a circulating series. The top of each column has an inlet valve of feed solution and desorbent water. The bottom of each column has respective outlet valves to separate the fructose and glucose fractions. All the valves are controlled automatically depending on the circulatory movement of the sugar solution through the multiple columns. The valves indexing and liquid flow rates are set as follows. In the highest sugar concentration region, the sugar solution is fed and in the region of lowest sugar concentration, desorbent water is supplied. From the region delayed with high fructose concentration, the fructose fraction comes out and from the region advanced with a high concentration of glucose, the glucose fraction comes out. The continuous separation system mentioned above uses substantially less water as desorbent to obtain a high sugar solution for both fractions (1, 2).

(Kenzaburo Yoritomi)

References

1. Broughton, D. B. (1961) *US Patent* 2,985,589
2. Yoritomi, K. (1981) *US Patent* 4,267,045

VI.4.b. Manufacture of Anhydrous Crystalline Fructose

Fructose is one of the naturally produced sugars and is often called fruit sugar or levulose. It is known as the sweetest among the various sugars and has a fresh and cool sweet taste. Fructose exists in five different structures and the product for marketing purposes is cubic anhydrous β-D-fructopyranose. It is highly soluble in water with the solubility at 50℃ reaching 86.8% (w/w). It is highly hygroscopic and tends to cake. For crystallization of fructose, it is necessary for the mother sugar to be more than 80% (w/w) pure fructose in case of ethanol (or methanol) and more than 90% (w/w) in case of aqueous solution. For crystallization of fructose from the

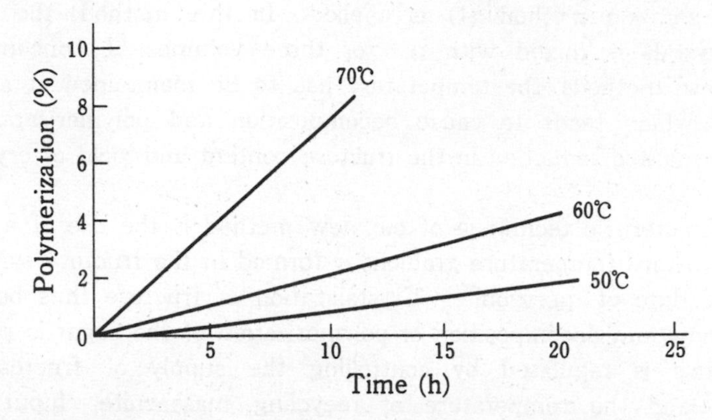

Fig. 1. Polymerization of Fructose from the Syrup [90% (w/w)].

Manufacturing Proceduce :

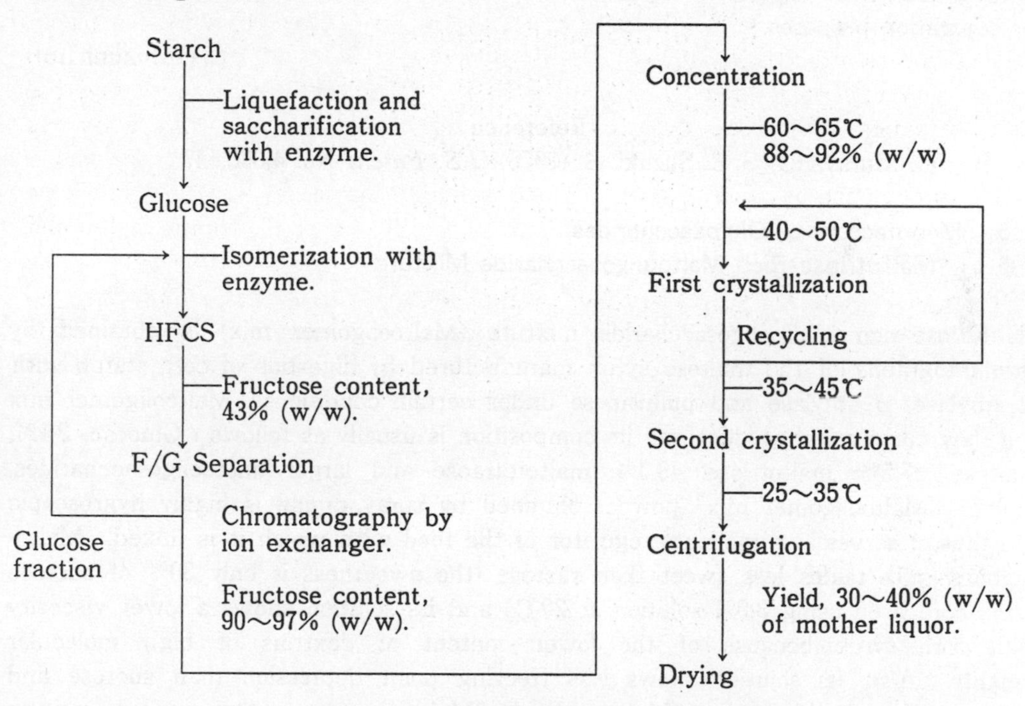

aqueous phase, the fructose solution is concentrated to 88~92 % (w/w) under reduced pressure, and the concentrate is transferred into the crystallization apparatus. Crystalline fructose powder for seed is then added to a concentration of 1~4% at 60℃. The mixture is gradually cooled to 30~35℃ at the rate of 0.1~0.3℃ per h. Then crystallization of fructose begins and continues over 100~150 h. The size of the fructose crystals obtained by this procedure ranges from 100 to 300 μm and the average yield is 30~40 %. To increase the size, a new method, "the double-staged crystallizing operation method" (1) is applied. In this method, the fructose syrup containing crystals is mixed with two or three volumes of concentrated fructose syrup. In these methods, the temperature has to be maintained at around 60℃ for long periods, which tends to cause decomposition and polymerization of fructose, resulting in a marked reduction in the fructose content and yield of crystals (Fig. 1).

The most characteristic technique of our new method is the use of a vertical cooler, so that a downward temperature gradient is formed in the fructose syrup through the recycling procedure of massecuite. Crystallization of fructose thus becomes possible at 45°C or below and decomposition or polymerization of the sugar is prevented. The size of crystals is regulated by controlling the supply of fructose solution, its concentration and the temperature of recycling massecuite. Input of the sugar crystals as seed is done only once at the starting stage and spontaneously generated crystals are utilized as seed thereafter. The size of crystals ranges from 200~1500 μm, although the bigger the crystals are, the easier is the centrifugation performed for separation purposes.

(Yoshikuni Ito)

Reference

1. Ito, Y., Murayama, S. & Suzuki, S. (1978) *U.S. Patent* No. 4,666,527

VI.5. Manufacture of Oligosaccharides

VI.5.a. Maltotriose-Rich Maltooligosaccharide Mixture

Maltotriose-rich maltooligosaccharide mixture (Maltooligomer mix) is obtained by chromatography of the maltose syrup manufactured by digestion of corn starch with α-amylase, β-amylase and pullulanase under certain conditions. Maltooligomer mix is a new commercial product and its composition is usually as follows : Glucose, 2.1%; maltose, 37.5%; maltotriose, 46.4%; maltotetraose and larger maltooligosaccharides, 14.0%. "Maltooligomer mix" powder obtained by spray drying is highly hygroscopic and thus, it serves as a moisture regulator of the food with which it is mixed. Malto-oligomer mix tastes less sweet than sucrose (the sweetness is only 30% of sucrose when compared using 3.0% solution at 20°C) and its solution shows a lower viscosity than corn syrup because of the lower content of dextrins of high molecular weights. Also, its solution shows less freezing point depression than sucrose and maltose. When added to a drink or food, "Maltooligomer mix" causes less color formation than corn syrup because of its low content of glucose. "Maltooligomer mix" is mainly used as a substitute for sucrose and other saccharides. It is also applied for preventing crystallization of sucrose in foods, and keeping a certain level of hardness of the texture during storage of foods.

Manufacturing Procedure (1):

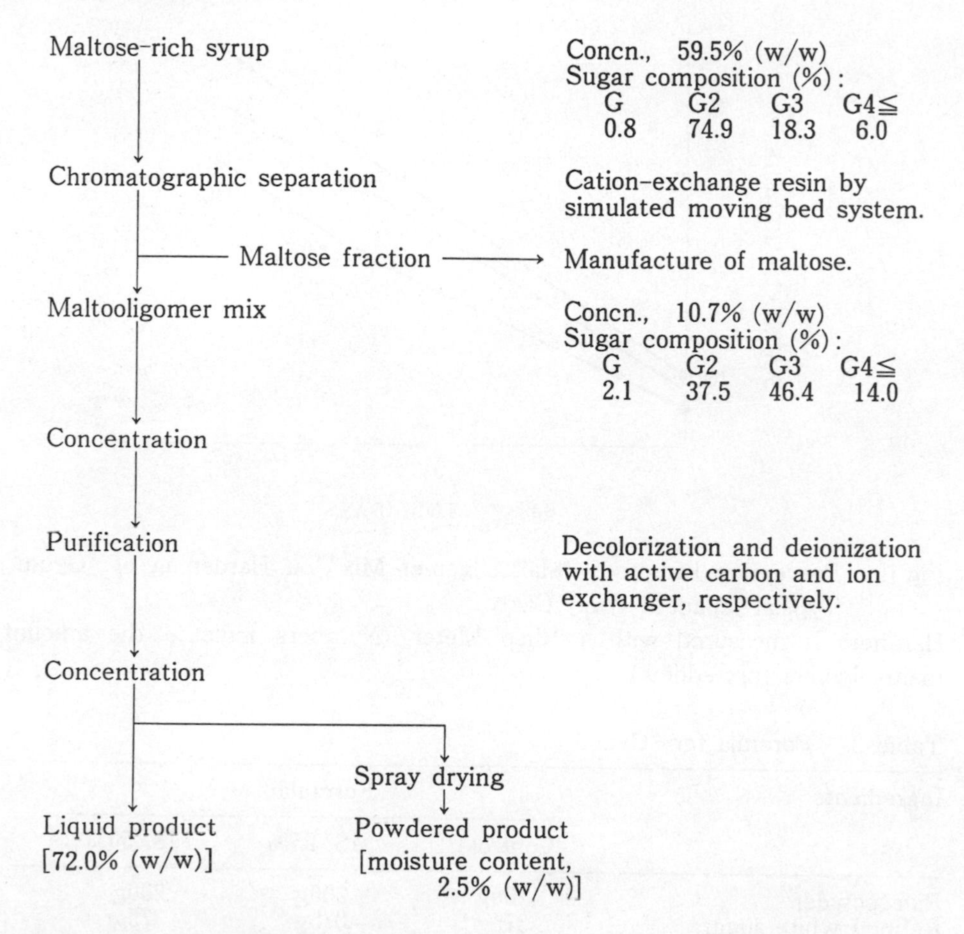

Hygroscopicity of "Maltooligomer mix" : In table 1 is shown the hygroscopicity of maltooligosaccharides (2). The hygroscopicity of "Maltooligomer mix" powder is similar to that of maltotriose.

Application test : Table 2 and Fig. 1 show the preventing effect of "Maltooligomer mix" on hardening of "Gyuhi", a typical Japanese waxy cake (3).

Table 1. Hygroscopicity of Maltooligosaccharides

24℃*	G3	>	G4	>	G5	=	G7	>	G11	>	G2		
30℃*	G3	>	G4	=	G7	>	G5	>	G6	>	G11	>	G2
38℃*	G3	>	G4	=	G5	>	G7	>	G6	>	G11	>	G2

* Examined under 90% relative humidity; Numbers neighboring G indicate degree of polymerization of glucose of the maltooligosaccharides tested.

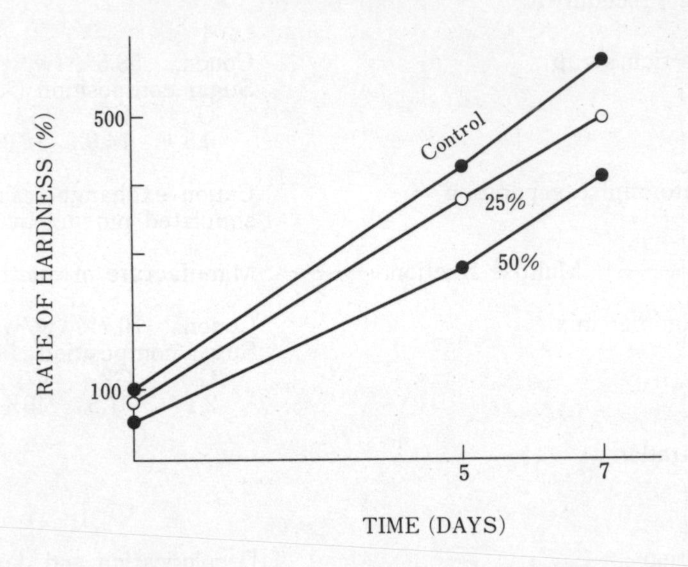

Fig.1. Preventing Effect of "Maltooligomer Mix" on Hardening of "Gyuhi", A Typical Japanese Waxy Cake.

Hardness is measured with a Rheo Meter (Numbers indicate the amount of maltooligomer mix added.).

Table 2. Formula for "Gyuhi"

Ingredients	Formula		
	Control	MS, 25%	MS, 50%
Rice powder	200g	200g	200g
Refined white sugar	375g	375g	375g
Acid corn syrup*	146g	—	—
Maltooligomer mix syrup**	—	164g	327g
Water	279g	261g	223g
Total	1000g	1000g	1000g

* Blix 85.4, ** Blix 76.4 ; MS, Maltooligomer mix syrup.

Some examples of application of "Maltooligomer mix" are as follows:

In confectionery; sweet bean-paste, sweet bean jelly, "Uiro", "Gyuhi", cakes, steamed bean jam buns, dough-cake, chewing gum, jelly, butter cream, custard cream, whipping cream, sponge cakes, frozen desserts, etc.

In foodstuffs; dumplings, frozen ground fish, boiled or roast fish paste, marine dainties, bakery products, rice cakes, crust of "Gyoza", frozen foods, jam, marmalade, etc.

In drinks; canned coffee, canned cocoa, sour drinks, fruits drinks, alcoholic beverages.

(Shigemitsu Osaki, Zenichi Yoshino)

References

1. Yoshino, Z., Shiota, K., Nakazawa, I., Murayama, M. & Kihara, T. *Japan Kokai Tokkyo Koho* 61-93192
2. Donnelly, B. J., Fruin, J. C. & Scallet, B. L. (1973) *Cereal Chem. 50*, 512-519
3. Chokusa Oligo Tou (1986) (in Japanese) Kashi Gijitsu Center, Tokyo

VI.5.b. Maltotetraose Syrup

Maltotetraose syrup (G4 syrup) is produced by subjecting starch to the action of a maltotetraose-forming amylase (G4 amylase), obtained from a strain of *Pseudomonas stutzeri* (cf. II.1.b).

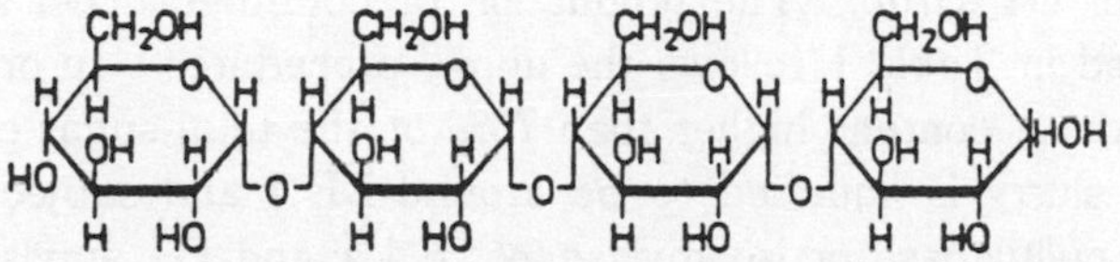

Structure of Maltotetraose

Table 1. Comparison of Sugar Composition of Several Syrups

	G4 syrup	Acid-conversion syrup	Enzyme-conversion syrup
DE	32.4	32.6	47.0
Glucose (%)	1.0	10.3	3.0
Maltose (%)	6.6	8.3	53.5
Maltotriose (%)	9.5	8.4	15.8
Maltotetraose (%)	50.2	8.1	3.2
Maltopentaose (%)	1.5	6.9	2.0
Dextrins, G6≦ (%)	30.3	57.8	23.0

As shown in Table 1, G4 syrup contains 50% or more maltotetraose and is markedly different in its sugar composition from acid- or α-amylase-conversion starch syrups in a comparison based on the same dextrose equivalent.

Properties and Applications of G4 Syrup: As can be seen from the sugar composition, G4 syrup has the following properties and application fields:

1) Less sweet taste; The sweetness of the syrup is as low as only about 20% of sucrose's. Therefore, a partial replacement of sucrose with G4 syrup reduces the sweetness of foods without affecting their inherent taste and flavor.

2) High moisture retention; Because of the high moisture retention power of the predominant component, maltotetraose, G4 syrup serves to prevent retrogradation of starch ingredient and makes retain a suitable moisture in foods.

3) Less coloration; The contents of glucose and maltose of G4 syrup are much less than those of other syrups. Therefore, G4 syrup shows a less Maillard reaction.

4) High viscosity: G4 syrup shows a viscosity of about 2.5 times higher than that of sucrose, and is useful in improving the texture of foods.
5) A mild freezing point; Based on the same concentration of solid matter, G4 syrup depresses the freezing point of water more moderately than sucrose or high-fructose syrup. So, G4 syrup can be used to control the freezing points of frozen foods.
6) Low osmotic pressure; G4 syrup is a low osmotic pressure supplier as well as a gloss imparter. G4 syrup can be advantageously used for processing foods, including bakery products, confectionery, jams, jellies, frozen desserts and soft drinks, and for industrial purposes, such as a paper sizer.

Manufacturing Procedure: The flow chart is an example of an industrial process for the production of G4 syrup. The production of commercial G4 syrup with the sugar composition listed in Table 1 follows the above procedures. In order to obtain a syrup with a maltotetraose content higher than 75% of the total sugar content, potato starch is used and the slurry is liquefied to be around DE 1 and subjected to the combined action of either pullulanase or isoamylase (cf. Ⅲ.1.a) and G4 amylase. Fractionation of the resulting high maltotetraose containing solution is done through a column of a special ion exchange resin which yields a maltotetraose syrup of a purity exceeding 97%.

Manufacturing Procedure:

Starch slurry preparation
↓ One part corn starch, 3 parts tap water and thermostable α-amylase, 1.0g/kg starch.
Liquefaction
↓ Heated to 100℃ at pH 6.5 to obtain DE 4.5.
Inactivation
↓ Heated to 135℃ for 5 min to inactivate α-amylase.
Saccharification
↓ Added with G4 amylase 4,000 units/kg starch at 65℃ and incubated at 55℃, pH 6.5 for 46 h.
Filtration
↓
Purification
↓ Purified with active carbon and ion exchangers.
Concentration
↓
G4 syrup

Enzymes Used: Commercial thermostable α-amylase of *B. licheniformis* or *B. subtilis* is used. Since *Ps. stutzeri* NRRL 3389 was reported in 1971 as a G4 amylase-forming bacterium (1), many researchers investigated the enzyme source originating to bacteria and their enzyme properties (2,3). The original strain of *Ps. stutzeri* NRRL 3389, reported as the enzyme producer, however, was found to be too low in the enzyme productivity and unfavorable for G4 syrup production. The enzyme used in the practical production of G4 syrup is a strain isolated by screening various natural sources as potent G4 amylase producers. This strain also belongs to *Ps. stutzeri*, but its G4 amylase productivity is 50~100 times greater than those so far known. The optimum temperature of the G4 amylase is 50℃, higher by about 5℃ and its pH stability, more or less wider than the enzymes of other strains. One unit of G4 amylase is defined as the amount of enzyme that liberates reducing sugar equivalent to one μmol of glucose from soluble starch at pH 6.5, 40℃ in one min.

(Yoshio Tsujisaka)

References

1. Robyt, J. F. & Ackerman, R. J. (1971) *Arch. Biochem. Biophys. 145*, 105-114
2. Schmidt, J. & John, M. (1979) *Biochim. Biophys. Acta 566*, 88-99
3. Sakano, Y., Kashiyama, E. & Kobayashi, T. (1983) *Agric. Biol. Chem. 47*, 1761-1768

Ⅵ.5.c. Anomalously Linked Oligosaccharides Mixture ("Alo mixture")

"Alo mixture" named here indicates a mixture containing isomaltose, panose, isomaltotriose (6^2-α-glucosylisomaltose) and several other so called branched oligosaccharides composed of four and five glucose residues. The "Alo mixture" practically produced usually contains glucose and maltosc other than those anomalously linked oligosaccharides described above (Table 1).

Table 1. Sugar Composition of "Alo Mixture" Syrup on Marketing

DP*	Sugar	Content (%)
1	Glucose	40.5
2	Maltose	6.7
	Isomaltose	16.9
	Others**	4.7
3	Maltotriose	0.8
	Panose	12.5
	Isomaltotriose	3.4
	Others**	2.3
4	Anomalously linked tetraoligomers**	8.9
5	Anomalously linked pentaoligomers**	3.3

* Degree of polymerization.
** The structure(s), not determined.

"Alo mixture" syrup has many favorable properties for application to food industry. It is mildly sweet and shows a low viscosity, high moisture retaining capacity and low water activity convenient in controlling microbial contamination. "Alo mixture" syrup is used at present in bakery, confectionery, soft drinks and Sake-making, and several other food processing industries.

Manufacturing Procedure of "Alo mixture": Starch is dextrinized by using thermostable bacterial α-amylase. The degree of hydrolysis (DE) of starch is kept between 6.0 to 10.0. The simultaneous reaction of saccharification and transglucosidation of the dextrin is carried out by applying soybean β-amylase and *Aspergillus niger* transglucosidase, a kind of α-glucosidase. The reaction mixture is finally purified and concentrated to 25.0% moisture.

Reactions Occuring in Making "Alo mixture": The reactions of hydrolysis and transglucosidation occur simultaneously. Maltose formation is catalyzes by β-amylase and isomaltose or other anomalously linked oligosaccharide formations are catalyzed by transglucosidase (an α-glucosidase), as shown in Fig. 1.

Manufacturing Procedure:

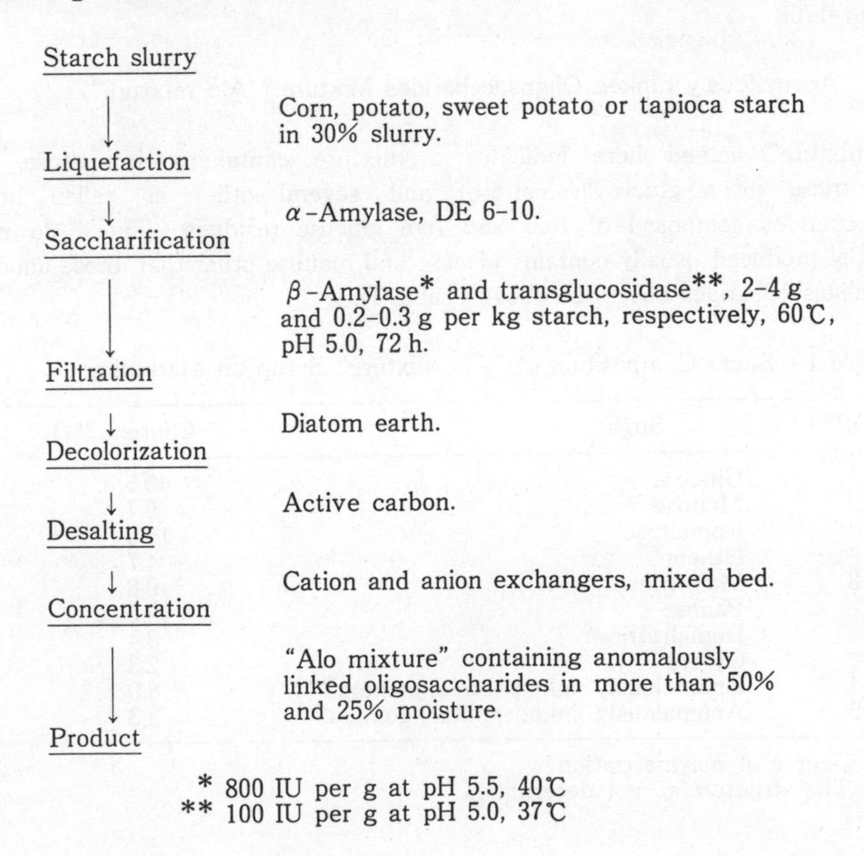

* 800 IU per g at pH 5.5, 40℃
** 100 IU per g at pH 5.0, 37℃

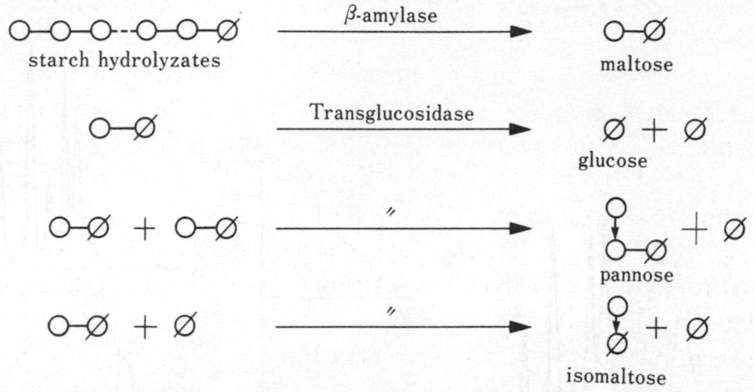

O–O–O--O–O–Ø —β-amylase→ O–Ø
starch hydrolyzates — maltose

O–Ø —Transglucosidase→ Ø + Ø
glucose

O–Ø + O–Ø —〃→ pannose + Ø

O–Ø + Ø —〃→ isomaltose + Ø

Fig. 1. Simultaneous Reactions.

(Hajime Takaku)

References

1. Hamada, S. (1984) *J. Jpn, Soc. Starch Sci. 31*, 83-91
2. *Bunki Oligo Tou* (1987) (in Japanese) Kashi Gijitsu Center, Tokyo

VI.5.d.1. High Molecular Weight Branched Dextrins from Corn Starch

Branched dextrins of high molecular weights are prepared by hydrolysis of corn starch with α-amylase followed by chromatography to separate them from various linear dextrins. The extent of the starch degradation to be carried out depends on the variety of starch and the physical properties desired (such as viscosity and insuring easy fractionation conditions carried out thereafter). The branched dextrins finally obtained as powder by spray drying generally show the following properties:
(1) An aqueous solution causes no retrogradation on a long standing even at a low temperature regardless of the concentration.
(2) It shows a viscosity considerably lower than the starch digests of corresponding DE.
(3) It gives a luster film which is not hygroscopic.
(4) The spray dried product is not hygroscopic and retains its powdery state under humid conditions.
(5) It adsorbs hydrophobic substances in aqueous phase possibly due to the helical structure of the side chains of the molecule.
The branched dextrin mixture has favorable effects on processing of foods. It is used as an extender and a glazing agent for production of powdery foods and rice cakes, respectively.

Liquefaction of Starch Slurry: In the case of corn starch as raw meterial, it is preferable to use two kinds of α-amylases. One should be a highly heat stable (105 ℃) α-amylase and the other, a less thermostable bacterial α-amylase.

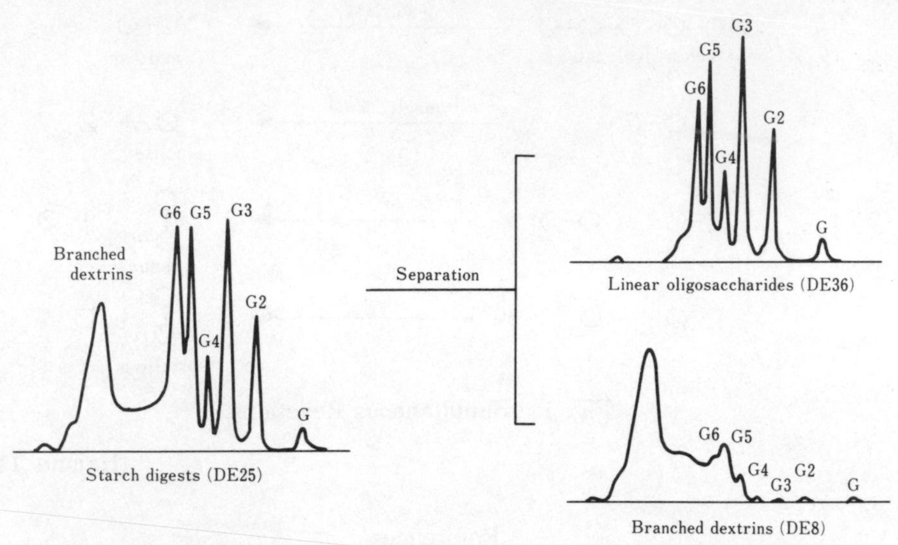

Fig. 1. Separation Patterns of Branched Dextrins by HPLC.

Manufacturing Procedure:

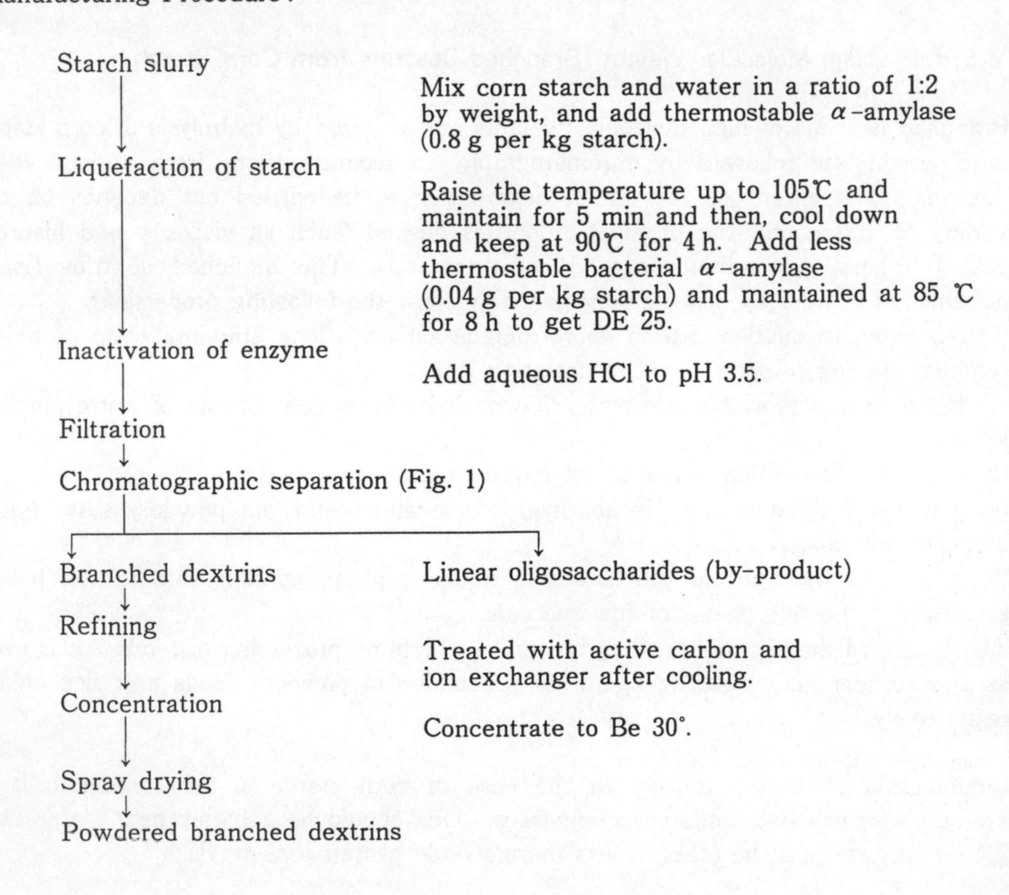

Fractionation: As the gel-type filtering agent, a sodium type cation exchange resin with cross-linkage degrees of a range from 4 to 6 is most practical. It is necessary for the particle size of the resin to be in a range from 40 to 60 mesh and uniform in the shape. To fractionate branched dextrins and linear oligosaccharides on an industrial scale, the simulated moving bed system is applied to the continuous flow of the saccharified solution of starch digests with α-amylases.

(Tsukasa Yoshida)

VI.5.d.2. High Molecular Weight Branched Dextrins from Waxy Corn Starch

The preparation of high molecular weight branched dextrin mixture is done under extremely specified conditions (1). The degree of hydrolysis is maintained between 2 and 5 and the viscosity of hydrolyzate solution between 50 and 150 cps. (30% at 30℃). Also, the solution shows little retrogradation and keeps its clarity on a long storage. The powder obtained by spray-drying the solution must not be hygroscopic or deliquescent. The branched dextrin product is used almost exclusively as a mixing or dilution powder in foods and pharmaceuticals where non-sweet, non-retrograde and non-deliquescent properties are desired.

Waxy Corn Starch : Commercial waxy corn starch is availabe.

α-Amylase : Thermostable bacterial α-amylase.

Manufacturing Procedure :

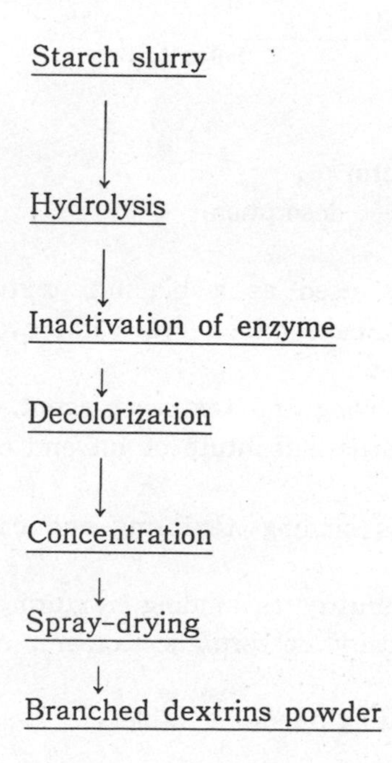

Mix waxy corn starch with two weights of deionized water and add thermostable α-amylase 1.0 g per kg starch.

Elevate the temperature up to and maintain at 90℃ for 10 min.

Heat at 110℃ for 5 min.

Treat with active carbon and ion-exchanger after cooling.

Concentrate to 35%.

Estimation of Enzymic Hydrolysis : The progress of the enzymic reaction is followed by measuring the change in viscosity which occurs with hydrolysis of starch.

Composition of the Branched Dextrin Mixture: The composition of the branched dextrin mixture industrially produced is shown in Table 1. Branched oligosaccharides smaller than six glucose residues comprise generally less than 5.0%. The average molecular weight of the dextrins is estimated to be around 4,000 from the DE of the mixture.

Table 1. DE and Sugar Composition of Branched Dextrins

DE	G	G 2	G 3	G 4	G 5	G 6	Higher
4.1 6	0. 2	0. 7	0. 7	0. 6	0. 6	1. 7	9 5. 5

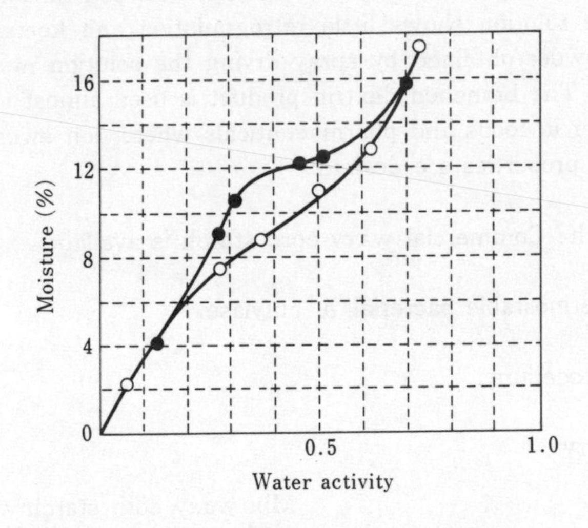

Fig. 1. Water Sorption Isotherm.
—○—, Sorption; —●—, desorption.

Applications : 1) The branched dextrins mixture is used as a binding, texturing, stabilizing and glazing agent, e.g. in custard powders, cake mixes, bread improvers, a partial substitute of fat and oil in low-calorie foods, etc.

2) It is also applied to canned foods as binding, bodying and texturing agent, taste and flavour carrier, e.g. sterilized canned soups, a partial substitute of fat and oil for low-calorie dressings, etc.

3) In confectionery is used the dextrins mixture as binding, texturing and coating agents.

4) Also, the dextrins mixture is applied in dairy industry as binding, texturing and stabilizing agents, taste and flavour carriers, e.g. dairy desserts, ice cream, coffee whiteners, etc.

5) The dextrin mixture is used as a binding and coating agent for tablets.

The spray-dried powder of the dextrins mixture is also utilized as an additive to spray-dried soy sauce, hydrolyzed proteins, fruit juices, various other hygroscopic food products, etc.

(Yoshio Hanno)

Reference

1. Hanno, Y., Ueno, F., Ogura, M., Takayanagi, Y. & Mochizuki, Y. (1972) *Japan Patent* No. 1090725

VI.6. Manufacture of Pullulan

Pullulan is an α-glucan extracellularly produced by a strain of *Aureobasidium pullulans* (1,2,3,4). Pullulan has the structure shown below and predominantly comprises maltotriose units which are repeatedly polymerized at their ends by α-1,6-glucosidic linkages. However, various investigations revealed that a few portions are occupied by maltotetraose instead of maltotriose (3,4). The molecular weight of pullulan is adjustable in the range of 50,000 to 3,000,000 by varying the cultivation conditions, i.e. medium composition and duration (5,6,7).

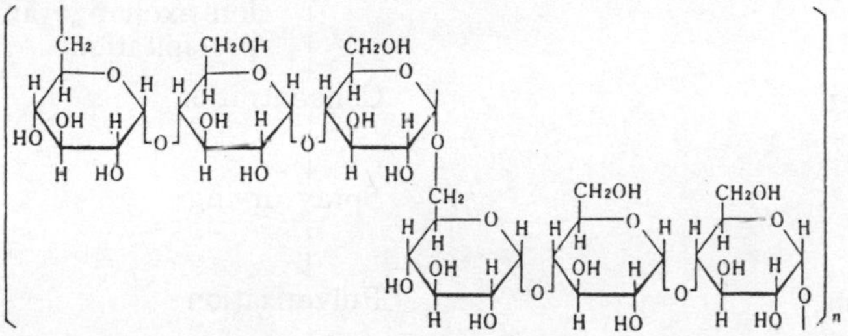

Structure of Pullulan

Manufacture of Pullulan (5,8,9,10) : In manufacturing pullulan a strain of *Aureobasidium pullulans* is cultivated by submerged culture in a culture medium comprising 15% acid-conversion starch syrup (DE 40-50), 0.2% peptone, 0.2% KH_2PO_4, 0.04% $MgSO_4 \cdot 7H_2O$ and 0.001% $FeSO_4$. Initially, the medium is adjusted to pH 6.5. During cultivation, aeration at a flow rate of about a half of the medium by volume is effected. The oxygen absorbance rate, OAR, of 2 mmol O_2/min/liter is constantly maintained by adequate aeration and stirring. Cultivation is continued for about 100 h, when pullulan with a molecular weight of 200,000 is obtained at a yield of 75%. As schematically shown below, production of pullulan involves two methods. Product 1 is used mostly for foodstuff or processing, while Product 2 is directed to film production and pharmaceutical uses.

Table 1. Analytical Data of Product-1 and -2

Composition	Product 1 (%)	Product 2 (%)
Moisture	6	6
Ash	4	0.1
Water insolubles	10	0～ 4
Pullulan	80	90～94

Manufacturing Procedure:

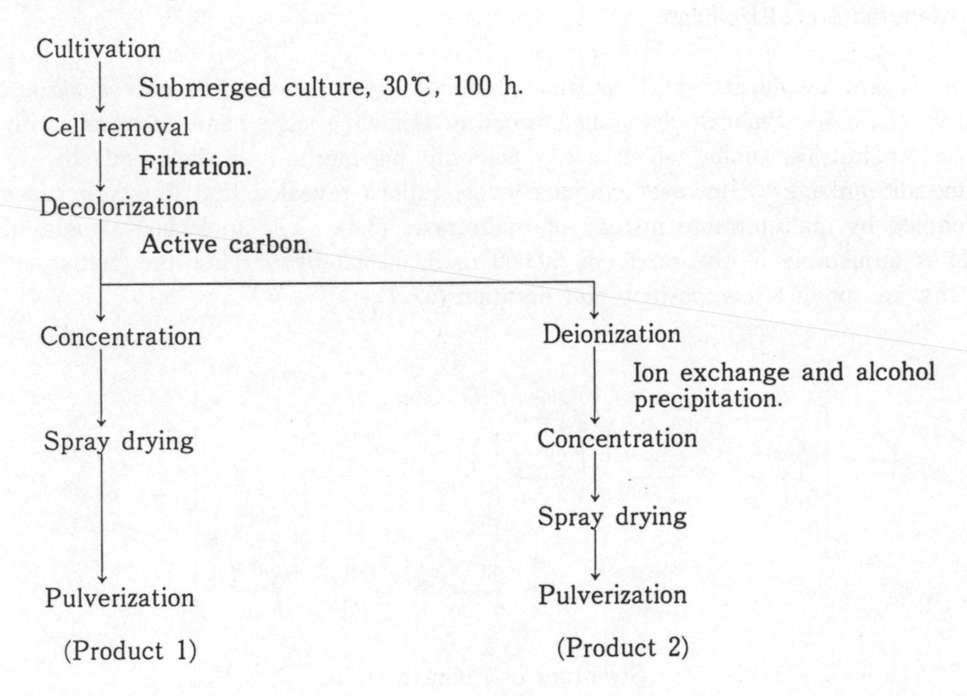

Properties of Pullulan : The following properties of pullulan are distinguished from those of other polysaccharides.

1) Pullulan readily dissolves in cold water and the aqueous solution is low in viscosity and does not gel.

2) Aqueous pullulan has a high surface tension comparable to that of water.

3) Aqueous pullulan is stable in a wide pH range, from 2 to 12, and its viscosity remains unchanged in the presence of various metal ions.

4) Aqueous pullulan has an intensive adhesion.

5) Pullulan can be prepared into shaped articles by compression molding or extrusion. Transparent films are also preparable. The films are characterized in being cold water soluble and nearly gas inpermeable.

Applications of Pullulan : Pullulan is used in many fields including food, pharmaceutical and chemical industries. The following examples are some of the uses:

1) Food industry ; Since pullulan is hardly decomposed by digestive enzymes it is proposed as a low caloric food ingredient. Pullulan is used to bind fragments of cooked fish meat, beef, pork and poultry to shape them into snacks or fast-foods of various forms. Pullulan is also used as a viscosity enhancer in beverages, creams and sauces (11). Since pullulan films exhibit a negligible gas permeability and are oil resistant, the films are ideal for wrapping and packaging oily foods that are liable to oxidation during storage. In addition, sheet foods incorporated with spices, seasonings, colorings and flavorings are prepared with pullulan (12).

2) Pharmaceutical industry; Pullulan is used as a bulking agent or bulk imparter in sustained release tablets (13). Development of contact lenses compatible to eyes is proceeding encouragingly (14). In addition, diphtheria and tetanus vaccines prepared by conjugating the toxin and pullulan are under study (15). Pullulan is useful as a plasma expander (16).

3) Others; Fertilizer sticks and rods prepared by using pullulan as a binder are welcomed by horticulturists and gardeners (17). In the production of coated paper, the addition of pullulan provides papers excellent in gloss and ink absorption (18). Fractionated pullulan products of various molecular weights, but with a narrow range of distributions, are used as standards in high performance liquid chromatography (19).

(Yoshio Tsujisaka, Masakazu Mitsuhashi)

References

1. Bender, H., Lehman, J. & Wallenfels, K. (1959) *Biochem. Biophys. Acta. 36*, 309-316
2. Ueda, S. (1964) *Kogyokagaku-zasshi* (1964) (in Japanese) *67*, 757-760
3. Walenfels, K., Keilich, G., Bechtler, G. & Freudenberger, D. (1965) *Biochem. Z. 341*, 433-450
4. Catley, B. J. & Whelan, W. T. (1971) *Arch. Biochem. Biophys. 143*, 138-142
5. Sugimoto, K. (1978) *Hakko-to-Kogyo* (in Japanese) *36*, 98-108
6. Kawahara, K., Ohta, K., Miyamoto, H. & Nakamura, S. (1984) *Carbohydrate Polymer 4*, 335-356
7. Catley, B. J. (1970) *FEBS Letters 10*, 190-193
8. Kato, K. & Shiosaka, M. (1977) *Japan Patent* 863550, Filed Oct. 11 (1971)
9. Kato, K. & Shiosaka, M. (1977) *Japan Patent* 866889, Filed June 23 (1972)
10. Kato, K. & Nomura, T. (1979) *Japan Patent* 981540, Filed Feb. 23 (1974)
11. Kato, K. & Shiosaka, M. (1976) *Japan Patent* 812139, Filed Jan. 13 (1972)
12. Hijiya, H. & Shiosaka, M. (1977) *Japan Patent* 907576, Filed Dec. 22 (1972)
13. Nakamura, S., Ohta, K., Miyamoto, H. & Kawahara, K. (1982) *Polymer Preprints Japan* p 31
14. Himi, T., Watanabe, Y., Fukuda, T., Mizutani, Y., Kuriaki, M. & Sato, T. (1980) *J. Japan Contact Lens Soc. 22*, 233-238
15. Matsuhashi, T., Yamamoto, A., Sadahiro, S. & Ikegami, H. (1981) *Naturwiss. 68*,

S. 49
16. Igarashi, S., Nomura, K., Naito K. & Yoshida, M. (1986) *Japan Patent* 1321752, Filed May 25 (1979)
17. Matsunaga, H., Fujiwara, S., Namioka, H., Tsuji, K. & Watanabe, M. (1981) *Japan Patent* 1043596, Filed Feb. 24 (1975)
18. Nomura, T. (1981) *Japan Patent* 1063600, Filed Sept. 5 (1973)
19. Yoshida, M. (1982) *Japan Kokai Tokkyo Koho* 57141401, Filed Oct. 15 (1980)

VI.7. Alcohol Fermentation of Uncooked Cereal Grains by Using Glucoamylase

Alcohol production by fermentation is increasing year by year, with increase in the demand not only for beverages, but also for utilization as solvent, fuel, raw materials in chemical industries, etc. Industrial production of alcohol from starchy materials such as cereal grains has been done in general by cooking the raw materials prior to saccharification and fermentation. Many researchers have investigated the techniques of alcohol fermentation of starchy materials without the preliminary cooking (1-7). Recently, the above idea has been applied to fermentation on an industrial scale. The procedure of alcohol fermentation of uncooked maize grains by using saccharifying enzyme (a preparation from *Rhizopus*) in the industrial scale is shown in Fig. 1 (5).

Mashing Process : Pulverized maize is mixed with two weights of water in a mash tank equipped with an agitator. The enzyme for saccharifying raw material is added to the above mixture and the mash is immediately transferred to a fermentor containing the yeast seed.
Pulverizing of Maize : Dry milling is quite enough for obtaining maize powder suitable for non-cooking alcohol fermentation. Table 1 shows the typical size distribution of pulverized maize for the above purpose.

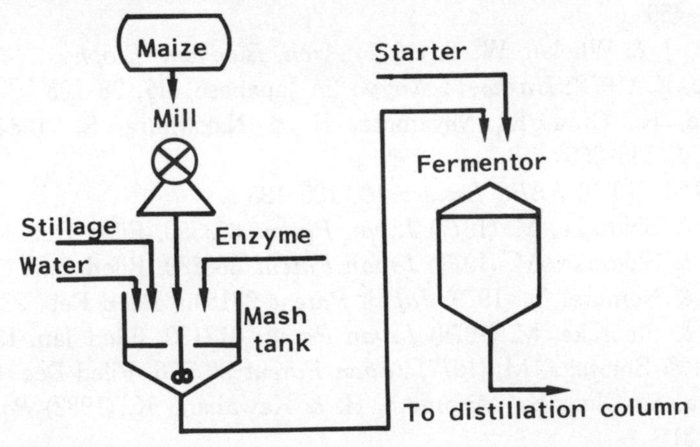

Fig. 1. Procedure of Alcohol Fermentation of Uncooked Maize Grains by Use of Enzyme.

Saccharifying Enzyme : A saccharifying enzyme from *Rhizopus* which is the glucoamylase first observed, is the most suitable for the non-cooking alcohol fermentation. The enzyme composition of the enzyme preparation is shown in Table 2, indicating that the preparation contains several other enzymes besides the saccharifying ones, such as liquefying enzymes as well as acid, neutral and alkaline proteinases, cellulase and pectinase.

Table 1. Size Distribution of Pulverized Maize

Average diameter (μ m)	Distribution (%)
>3360	0
3360-840	24.5
840-350	35.6
350-250	7.6
250-177	13.6
< 177	18.7

Table 2. Several Enzymes in A Saccharifying Enzyme Preparation

Enzymes	Activity (units/g)
Liquefying	2000
Dextrinizing	5450
Saccharifying	2147
Proteinase	
Acid	3920
Neutral	1340
Alkaline	1130
Cellulase	580
Pectinase	95

Table 3. Analytical Result of Fermented Mash by the Non-Cooking Method

Volume (kl)	112.5
pH	4.8
Total acidity (ml)	3.3
Alcohol (v/v%)	14.5
Residual total sugar content as glucose (%)	1.57
Residual reducing sugars as glucose (w/v%)	0.18
Fermentation efficiency (%)	88.5

Water for Mashing: Stillage is used in a range of 20 to 30% of the total water to be added. The stillage is added while it is hot and its addition increases the buffering capacity, prevents bacterial contamination, and has a favorable effect on the quality of the alcohol to be distilled.

Starter : The starter is used in a proportion of about 7% of the final mash volume.
Fermentation: The fermentation temperature and period usually employed are 26-32°C and about 96 h, respectively. Table 3 shows the results of alcohol fermentation by the non-cooking method using a fermentor of about 120 kl.

(Nobuya Matsumoto)

References

1. Yamasaki, I., Ueda, S. & Shimada, T. (1963) *J. Ferm. Ass. Jpn. 21*, 83-86
2. Ueda, S. & Koba, Y. (1980) *J. Ferment. Technol. 58*, 237-242
3. Svendsby, O., Kakutani, K., Matsumura, Y., Iizuka, M. & Yamamoto, T. (1981) *J. Ferment. Technol. 59*, 485-487
4. Kumagai, C., Miyairi, M., Huang, C. T., Suzuki, I., Tanaka, T. & Akiyama, H. (1982) *Hakkokogaku Kaishi* (in Japanese) *60, 77-86*
5. Matsumoto, N., Fukushi, O., Miyanaga, M., Kakihara, K., Nakajima, E. & Yoshizumi, H. (1982) *Agric. Biol. Chem. 46, 1549-1558*
6. Hayashida, S., Ohta, K., Flor, P. Q., Nanri, N. & Miyahara, I. (1982) *Agric. Biol. Chem. 46*, 1947-1950
7. Yoshizumi, H., Ashikari, T., Nakamura, N., Kunisaki, S., Tanaka, Y., Kiuchi, N. & Shibano, Y. (1987) *J. Jpn. Soc. Starch Sci. 34*, 148-154

VI.8. Enzymic Synthesis of Maltooligosylsucrose Syrup, Fructooligosaccharides, Palatinose and Galactooligosaccharides

VI.8.a. Maltooligosylsucrose Syrup

Cyclomaltodextrin glucanotransferase (CGTase) is a unique enzyme which has the ability to convert starch to cyclodextrin. This enzyme also catalyzes a transglycosylation reaction whereby glycosyl moieties are transferred from starch to acceptors (see III.3.). When CGTase is incubated with the mixture of starch and sucrose, various maltooligosylsucroses ("Coupling sugar®", glucosyl residues of starch are transferred to conjugate with glucose residues of sucrose by α-1,4-linkage), are produced. The sugar composition of "Coupling sugar" varies depending on the degree of starch liquefaction and ratio of sucrose to be added. Table 1 shows an example of sugar composition of "Coupling sugar". The sweetness of "Coupling sugar" is about 60% of sucrose's and it has many favorable physical and physicochemical properties (1). Sucorse is an excellent sweetner, enchantingly palatable and high calorie. However, it is a potential factor in the formation of dental caries. Development of new sweetners with a low cariogenicity but still retaining the desirable features of sucrose has been in every increasing demand. "Coupling sugar" is a unique starch-sucrose based sweetener developed with the attempt to fulfil such demands from dental professionals and those concerned over juvenile dental hygiene.

Dental caries is caused by oral microbials. A bacterium, *Streptococcus mutans*, produces a water-insoluble glucan from sucrose and this glucan covers the dental surface forming a caries conductive plaque to promote adherence of oral bacteria on the surface. The presence of a saccharide such as glucose, fructose or sucrose during plaque formation leads to bacterial acid production in the plaque, subsequently initiating the occurrence of dental caries. The cariogenicity of "Coupling sugar" was studied by a group of the Department of Dental Research, Japanese National Institute of Health and other institutes and universities. The followings are the results (2).
(1) No formation of water-insoluble glucan—— "Coupling sugar" is not only unavailable for the synthesis of insoluble glucan, but also inhibits the synthesis of insoluble glucan from sucrose by *St. mutans*.
(2) Low acidogenecity —— The acid production from the principal constituent of "Coupling Sugar" by *St. mutans* is low compared to that from sucrose.
The above findings prove that "Coupling sugar" is a sweetener which hardly induces dental caries. This "Coupling sugar" is introduced in various foods such as cookies, candies and jams, as a substitutional sweetener for sucrose.

Manufacturing Procedure : The following flow chart illustrates the outline of the practical process for production of "Coupling sugar".

Starch slurry

↓ Mix one part of corn starch and two parts of tap water by weight, thermostable α-amylase, 2.0 g/kg starch.

Liquefaction

↓ Liquefy the slurry (pH 6.5) by heating at 100℃ to reach DE 5.0.

Inactivation

↓ Inactivate the enzyme by heating at 130℃ for 5 min.

Cooling

↓ Cool the hydrolyzate to 65℃ and add 0.5 parts sucrose and thermostable CGTase in a rate of 3,000 units/kg starch.

Transglycosylation

↓ Incubate at pH 5.6 and at 65℃ for 24 h.

Purification

↓ Purify with active carbon and ion exchangers.

Concentration

↓

Maltooligosylsucrose Syrup
"Coupling Sugar"

Table 1. Sugar Composition of "Coupling Sugar"

Fructose	0.5 ～ 1.5%
Glucose	5 ～ 7 %
Sucrose	11 ～ 15 %
Maltose	10 ～ 12 %
Glucosylsucrose	11 ～ 15 %
Maltotriose	7 ～ 9 %
Maltosylsucrose	7 ～ 11 %
Higher oligosaccharides	35 ～ 41 %

(Shigetaka Okada)

References

1. Okada, S. & Kitahata, S. (1975) *Nippon Shokuhin Kogyo Gakkaishi* (in Japanese) *22,* 420-424
2. Araya, S. (ed.) (1980) *Proceedings of the 5th Conference on Dental Caries and Coupling Sugar* (in Japanese) National Institute of Health, Tokyo, Japan

VI.8.b. Fructooligosaccharides

In this section the following 1^F-(1-β-fructofuranosyl)$_{n-1}$-sucrose oligomers (n = 2～4) are described : 1-kestose (GF_2), nystose (GF_3), 1^F-β-fructofuranosylnystose (GF_4). These saccharides occur naturally in plants such as onion, asparagus, Jerusalem artichoke, etc. Mixtures of them, "Neosugar" G and -P, are now commercially produced from sucrose by the action of fungal β-fructofuranosidase, as shown in Fig. 1 (1). Of the enzymes of various microorganisms examined, *Aspergillus niger* enzyme is the most potent and suitable for production of the fructooligosaccharides.

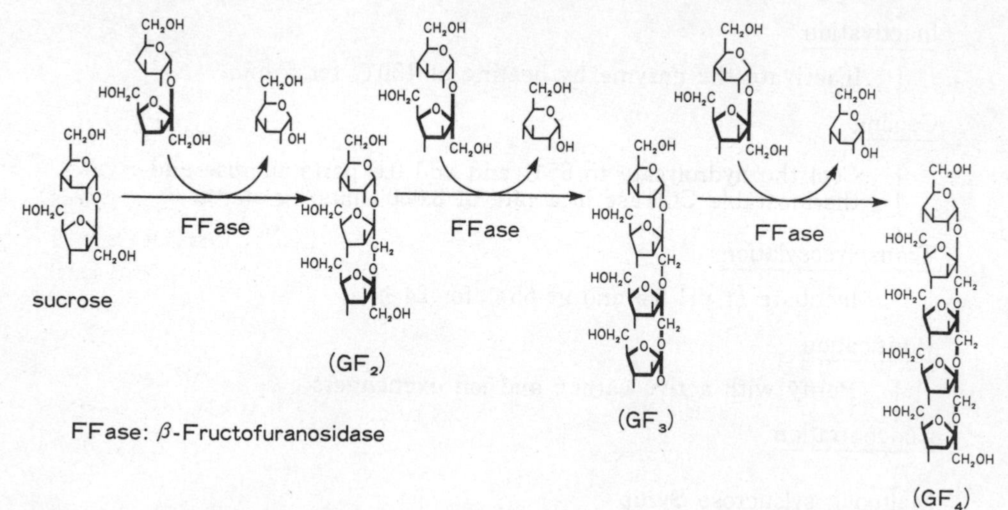

Fig. 1. Structures and Enzymic Synthesis of Fructooligosaccharides.

Table 1. Sugar Composition (wt %) of "Neosugar" -G and -P

Product	G(F)	GF	Fructooligosaccharides GF_2	GF_3	GF_4	(GF_{2-4})
Neosugar G	< 33	< 12	25	25	11	> 56
Neosugar P		<5	35	50	11	> 95

Manufacturing Procedure :

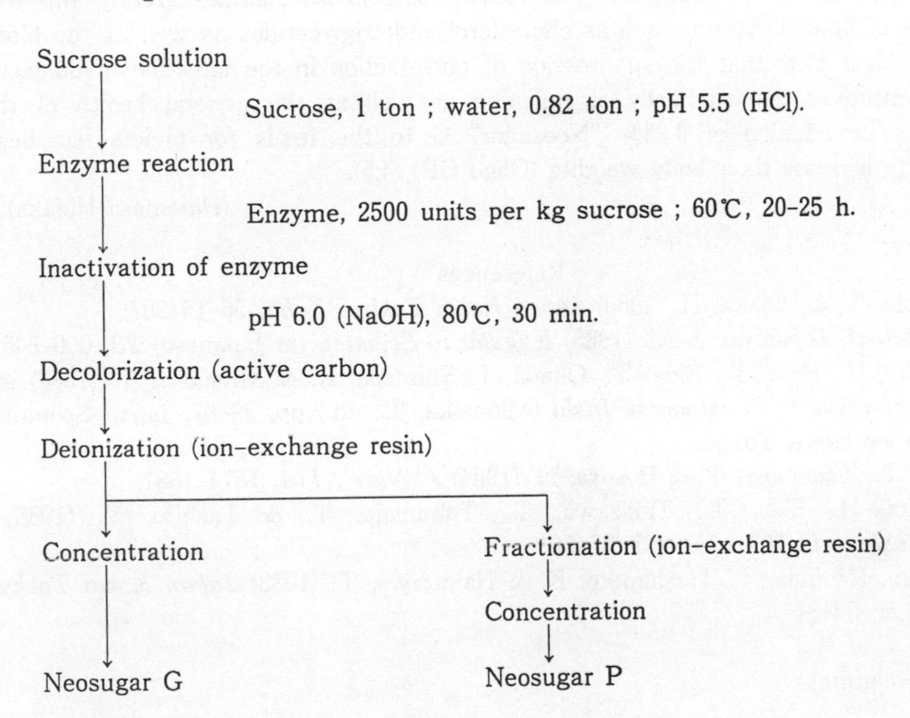

The recommended amount of enzyme to be applied is 2–3 activity units to one g sucrose. One unit of enzyme activity is defined as the amount of enzyme which produces 1 μ mol of 1–kestose per min under the following conditions: 10% (w/ v) sucrose, 0.04 M McIlvain buffer, pH 5.0, 40℃. Fructooligosaccharides have useful physical and physiological properties and are widely applicable to foodstuffs and animal feedstuffs.

Physical Properties : Fructooligosaccharides are colorless and odorless, giving a sweetness of noble quality. The sweetness of "Neosugar" G and -P is about 60% and 30% of that of sucrose, respectively. Both "Neosugars" are stable at neutral pH and at temperatures up to 140℃.

Physiological Properties : Fructooligosaccharides are nondigestible (2,3), but they are selectively utilized by bifidobacteria in the human intestinal canal (2,3). Studies from the medical point of view revealed that administration of the saccharides results in improvement of intestinal bacterial flora followed by the improvement of constipation and blood lipid level of hyperlipemia patients. The administration of these fructooligosaccharides has been reported to be effective in suppression of the production of intestinal putrefactive substances in both man and animals. The administration of "Neosugar" P of 8 g daily to man for over 2 weeks results in 10 fold increase in the number of bifidobacteria in the stool. In case of patients with hyperlipemia, the administration of the above fructooligosaccharides greatly improves the levels of lipids in serum such as cholesterol and triglycerides as well as the blood pressure. It is clear that the suppression of putrefaction in the intestine of domestic animals improves the efficiency of nutrition as well as the general health of the animals. The addition of 0.25% "Neosugar" G to the feeds for piglets has been reported to increase their body weights (Oligo GP) (4,5).

(Hidemasa Hidaka)

References

1. Adachi, T. & Hidaka, H. (1980) *Japan Kokai Tokkyo Koho,* 56-154967; Hidaka, H. & Hirayama, M. (1985) *Kagaku to Seibutsu* (in Japanese) *23,* 600-605
2. Hidaka, H., Hara, T., Eida, T., Okada. J., Shimada. K. & Mitsuoka, T. (1983) in *Chonai Flora to Shokumotsu Inshi* (Mitsuoka, T., ed.) pp, 39-67, Japan Scientific Societies Press, Tokyo
3. Oku, T., Tokunaga, T. & Hosoya, N. (1984) *J. Nutr., 114,* 1574-1581.
4. Hidaka, H., Eida, T., Takizawa, T., Tokunaga, T. & Tashiro, Y. (1986) *Bifidobacteria Microflora 5,* 37-50
5. Hidaka, H., Eida, T., Hashimoto, K. & Nakazawa, T. (1983) *Japan Kokai Tokkyo Koho* 60-34134

VI.8.c. Palatinose

6-0-α-D-Glucopyranosyl-D-fructofuranose is conventionally called isomaltulose, or more popularly palatinose. It is produced in a crystalline form for food use. It tastes sweet, and is used mainly in confectionery as a sugar substitute (1). Palatinose is non-cariogenic (2,3,4), and slowly digestible in the small intestine (5,6).

Sources of α-Glucosyltransferase : Three bacterial strains, e.g. *Protaminobacter rubrum, Serratia plymuthica* and *Erwinia rapontici* are known to show a marked increase of the enzyme activity in the cells when they are grown on sucrose (7).
Immobilized α-Glucosyltransferase : Whole bacteria cells of the above bacterial strains are immobilized by the gel entrapment and/or cross linking methods, and used as immobilized enzyme (8).
Products of the Enzymic Reaction: The column effluent contains palatinose, α-glucosyl-1,1-fructose (trehalulose), isomaltose, isomelezitose, glucose, fructose and

sucrose in an approximate ratio of 85.7, 8.7, 0.4, 0.2, 1.8, 2.4 and 1.1, respectively (8).
Palatinose Syrup : Palatinose syrup contains palatinose and trehalulose as the major components. Their contents are about 20 and 40% of the total solid, respectively. Palatinose syrup as well as crystals are used in confectionery as a sugar substitute, because palatinose is sweet (8), low cariogenic (9) and digestible in the small intestine (10).
Palatinose Crystals : Palatinose crystals contain water in a molar ratio of 1:1 (11).

Manufacturing Procedure :

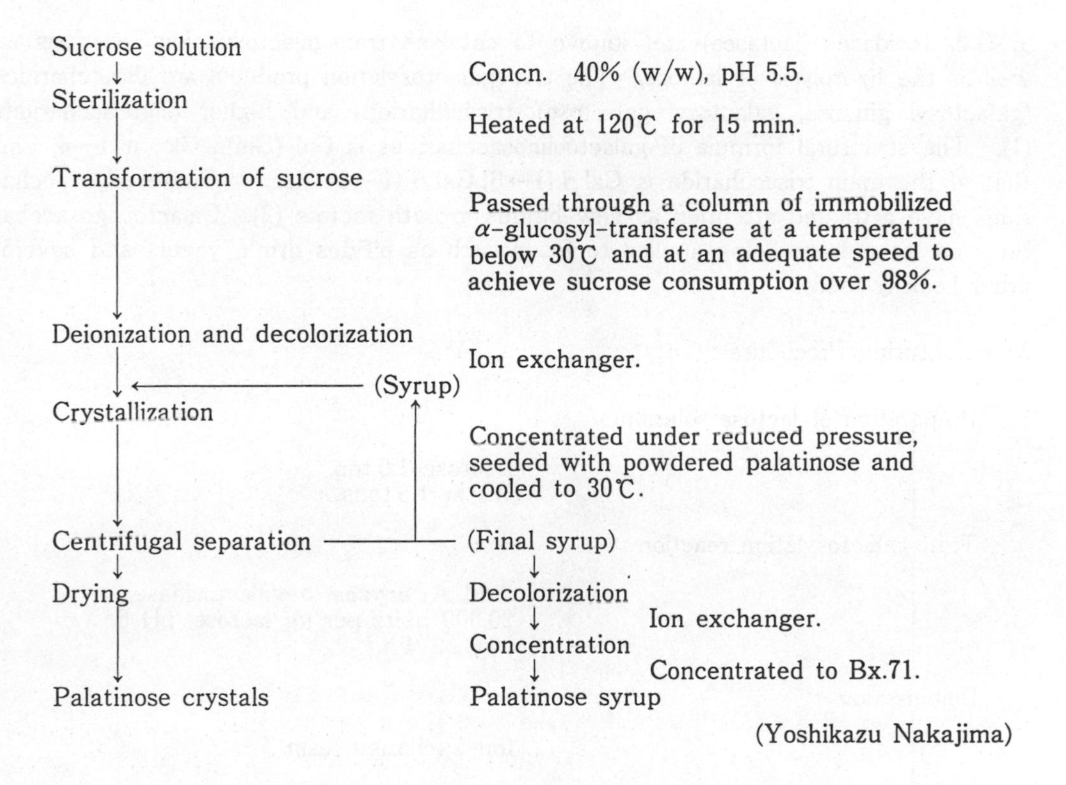

(Yoshikazu Nakajima)

References

1 . Kaga, T. & Mizutani, T. (1985) *Seitogijutu Kenkyu Kaishi* (in Japanese) *34,* 45–57
2 . Ohta, K. & Takazoe, I. (1983) *Bull. Tokyo Dent. Coll. 24,* 1–11
3 . Ohshima, T., Izumitani, A., Sobue, S., Okahashi, N. & Hamada, S. (1983) *Infect. Immun. 39,* 43–49
4 . Ohshima, T., Izumitani, A., Sobue, S. & Hamada, S. (1983) *Japan J. Med. Sci. Biol. 36,* 219–223
5 . Goda, T. & Hosoya, K. (1983) *J. Jpn. Nutr. Food Sci.* (in Japanese) *36,* 169–173
6 . Kawai, K., Okuda, Y. & Yamashita, K. (1985) *Endocrinol. Japon 32,* 933–936
7 . Weidenhageen, R. & Lorenz (1957) Ger. Patent 1,049,800

8 . Nakajima, Y. (1984) *Seitogijutu Kenkyu Kaishi* (in Japanese) *33*, 55-62.
9 . Izumitani, A., Sumi, N., Ohshima, T. & Sobue, S. (1985) *Shouni-Shikagaku Zasshi* (in Japanese) *23*, 592-599
10. Yamada, K., Shinohara, H. & Hosoya, N. (1985) *Nutrition Reports International 32*, 1211-1220
11. Fujii, S., Kishihara, S., Komoto, M. & Shimizu, J. (1983) *Nippon Shokuhin Kogyo Gakkaishi (*in Japanese*) 30*, 339-334

VI.8.d. Galactooligosaccharides

β-Galactosidases (lactases) are known to catalyze transgalactosylation reactions as well as the hydrolysis of lactose. The transgalactosylation products are disaccharides (galactosyl glucose, galactosyl galactose), trisaccharides and higher oligosaccharides (1). The structural formula of galactooligosaccharides is Gal-(Gal)n-Glc, n=1~4, and that of the main trisaccharide is Galβ (1→6) Galβ (1→4) Glc. These oligosaccharides have attracted attention as new bifidus growth factors (2). Galactooligosaccharides are at present being applied to foods such as bifidus drink, yogurt and several other healthy foods.

Manufacturing Procedure :

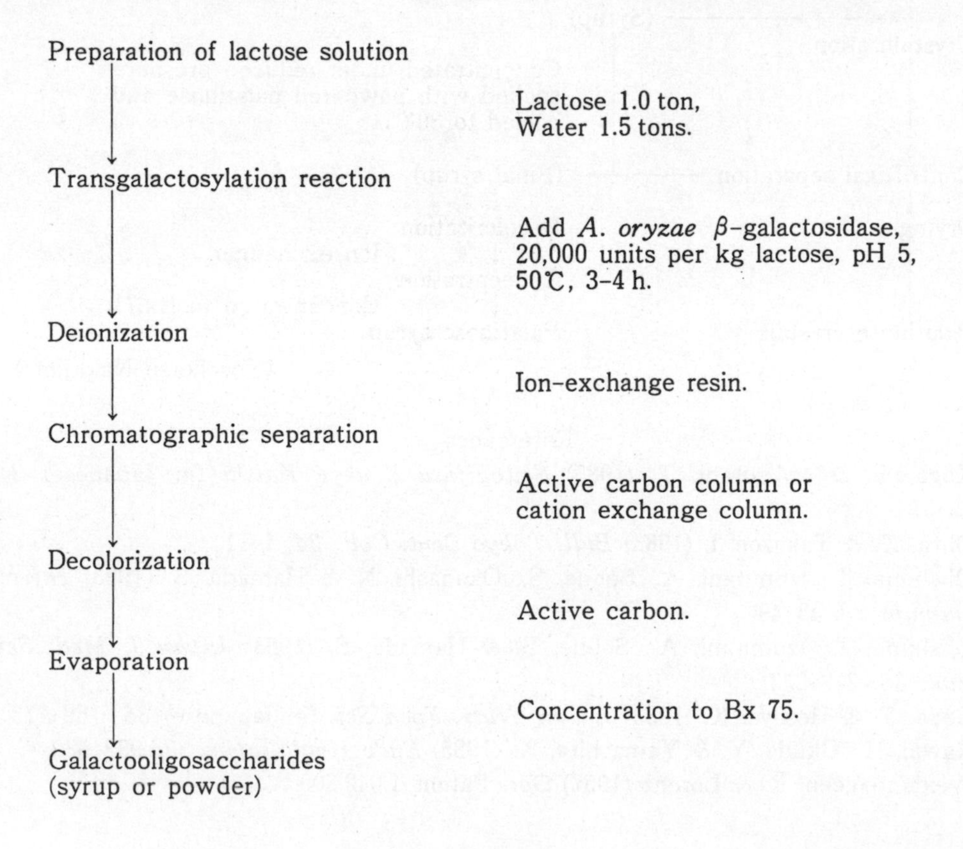

β-Galactosidase : The galactooligosaccharide formation has been reported for the enzymes from *Kluyveromyces fragilis, K. lactis, Bacillus circulans* and *Aspergillus oryzae* (1). Recently, the production of galactooligosaccharides from lactose has been achieved by β-galactosidase from *A. oryzae* (3).

Assay Method: β-Galactosidase activity is assayed using 5 mM *o*-nitrophenyl-β-D-galactopyranoside (ONPG) as a substrate, pH 4.5 at 30℃. One unit of the enzyme activity is defined as the amount of enzyme which liberates 1 μmol of *o*-nitrophenol per min under the above conditions.

(Keisuke Matsumoto, Akio Kuroda)

References

1. Toba, T. (1985) *Japn. J. Dairy Food Sci. 34,* A169-A182
2. Tanaka, R., Takayama, H., Morotomi, M., Kuroshima, T., Ueyama, S., Matsumoto, K., Kuroda, A. & Mutai, M. (1983) *Bifidobact. Microflora 2,* 17-24
3. Mutai, M., Terashima, I., Takahashi, T., Tanaka, R., Kuroda, A., Ueyama, S. & Matsumoto, K. *Japan Patent* No.1367596

VI.9. Production and Application of Cyclodextrins

VI.9.a. Production of Cyclodextrins

Cyclodextrins (CDs) are cyclic, non-reducing oligosaccharides composed of six, seven, or eight glucopyranose units and called α-CD, β-CD and γ-CD, respectively. These CDs are manufactured on an industrial scale. In addition to these 1,4-linked cyclic CDs, various branched CDs have been reported, but only maltosyl CDs (maltose conjugated CDs by α-1,6-linkage) are also produced industrially. In the reaction of cyclodextrin glucanotransferase (CGTase) on liquefied starch, α-, β-, γ-CD and hydrolyzed starch are produced. The CD producing ratios are different depending on the source of the CGTase and the reaction conditions (See III.3.). Many methods for production of a specific CD have been suggested so far. In 1980s, some industrial processes for manufacturing CDs were successfully established. For example, a method without using any organic precipitants is adopted on an industrial scale (1). Another unique manufacturing process is the method developed by the present author group in collaboration with the National Food Research Institute (Japan) (2). This method uses membrane technologies such as ultrafiltration membrane (UF) and reverse osmosis membrane (RD).

Manufacturing CDs Using Membrane Methods : The CD formation is carried out mainly in a reactor, and the resulting reaction mixture is filtered by means of a UF system. CDs and linear or branched oligosaccharides of small molecular weights permeate through the UF membrane being separated from the reaction mixture. CGTase and starch digests of larger molecular weights remain and are returned to the reactor. During the circulation through the UF and reactor system, acyclic dextrins are converted into CDs. The yield of CDs by the batch system is below 40% of the starch applied, but it goes up to 70% by the UF membrane system. In this method the CD forming reaction is carried out on a substrate of extremely low concentration, and after heat inactivation of the enzyme, the reaction mixture is concentrated by using a RO membrane. The spray-dried product of a mixture of CDs and dextrins is used for foods. The addition of glucoamylase produced a mixture of CDs and glucose, and this glucose is removed by UF membrane filtration. For separation and fractionation of each CD, β-CD is first separated by crystallization, because the solubility of β-CD is far less than that of the other two CDs. For separation of α-CD and γ-CD at present, Taka amylase and glucoamylase are added to the mixture whereby γ-CD is hydrolyzed into glucose while α-CD remains as it is. The mixture is then filtered through a UF membrane to isolate α-CD.

Manufacturing of Branched CDs : α-, β- and γ-CD are hardly soluble (The solubility of β-CD is very small). To improve this defect, several ideas were proposed (3-6) in which maltosyl CD preparation method was industrialized. The maltosyl CD is at present on sale (commercial name, ISOELEAT).

Manufacturing Procedure of Maltosyl CD : The synthetic method and industrial production of maltosyl CD were achieved by collaboration of researchers of the National Food Research Institute of the Ministry of Agriculture, Forestry and Fishery (7), Tokyo Noko University, Faculty of Agriculture and Technology (8), Nikken Chemicals Co., Ltd. (9), and Ensuiko Sugar Refining Co., Ltd. Maltosyl CD is manufactured by using CDs and maltose as starting material, and thermostable pullulanase as the enzyme. The maltose is conjugated with CD by the reverse reaction of hydrolysis shown by pullulanase.

Table 1. Composition of Cyclic and Acyclic Oligosaccharides of a Commercial Maltosyl CD "ISOELEAT"*

G2	3.3 (%)
G3	0.7
G4	8.3
α-CD	17.3
β-CD	1.3
γ-CD	1.0
G2-α-CD	30.5
G2-β-CD	10.6
G2-γ-CD and $(G2)_2$-γ-CD	14.6
$(G2)_2$-β-CD	8.9
$(G2)_2$-γ-CD	0.3
unknown-1	1.1
unknown-2	1.2

*, Analyzed by HPLC

Reaction Conditions : Concentration of substrate, 70 to 80% in which the amount of maltose is 4 to 5 times as much as that of the CD ; reaction temperature, 60 to 70°C, and reaction period, 72 h. The unreacted maltose is recycled after separation by chromatography.

Manufacturing Procedure:

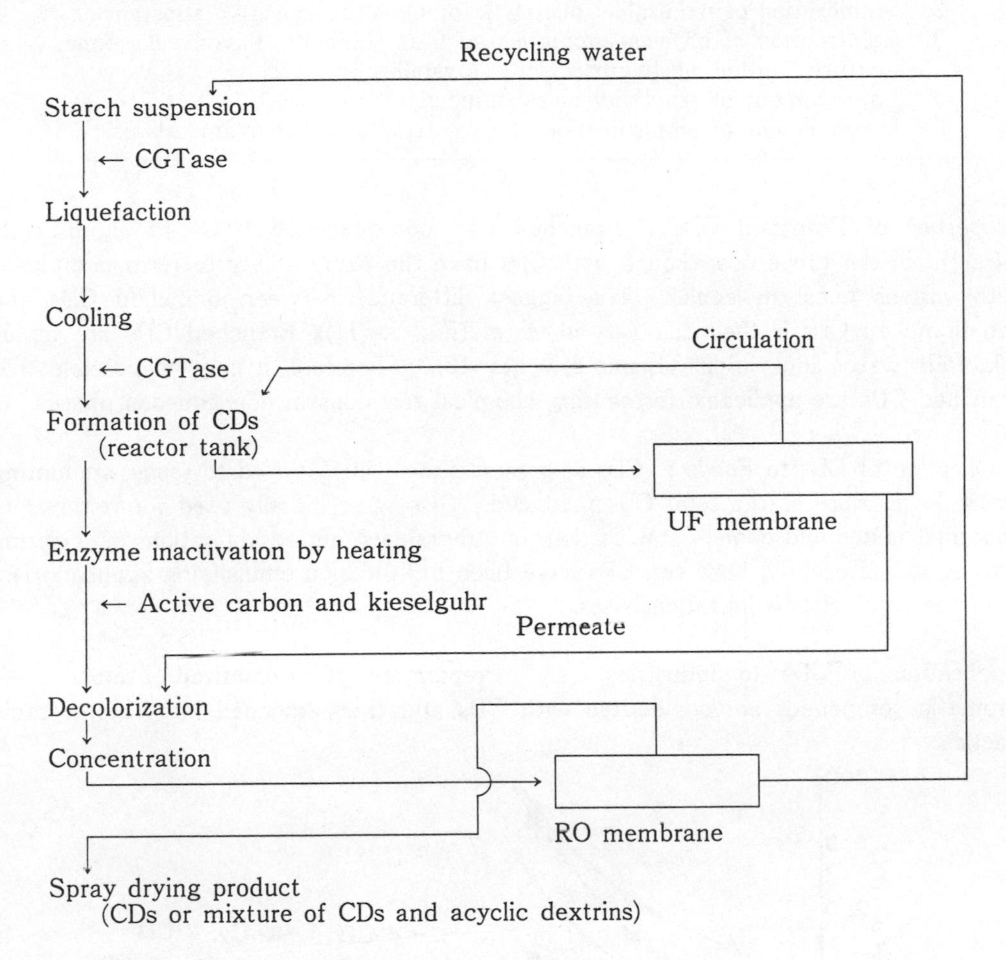

VI.9.b. Application of Cyclodextrins

The most characteristic function of a CD is the formation of inclusion complexes by taking various substances (guests) into its hydrophobic cavity. CDs are being used in various industries because of this ability to form inclusion complexes. Recently, the use of CDs is rapidly increasing. Substances available as the guest molecule greatly vary from polar substances such as acids, amines, and various kinds of ions to highly non-polar aliphatic and aromatic hydrocarbons, further to even rare gases. It is known that the stability of the inclusion complexes depends on the cubic fitness between the host and guest molecules, and that Van der Waals, hydrogen bonding and charge transferring forces do work individually or cooperatively as the complexing power (10). CDs have been used in various fields including foods and medicines (See VI 10).

Table 2. Function of CDs

1. Regulation of volatility of flavor or smell masking volatile agents.
2. Stabilization of oxidizable, photolytic or moisture sensitive substances.
3. Reformation of physical properties such as solubility, flavor, color tone, texture, hardening, hygroscopicity, crystallization, etc.
4. Improvement of reactivity of substances.
5. Improvement of emulsification of fats, fatty acids or hydrocarbons.

Properties of Branched CDs : Branched and non-branched CDs are significantly different in the properties, though both CDs have the same ability to form complexes with various guest molecules. The biggest difference between branched CDs and non-branched CDs is their solubility in water (Fig. 1, (11)). Branched CDs are highly soluble in water and various organic solvents (12). Therefore, it may be possible that branched CDs are applicable for causing chemical reactions in non-aqueous phases.

Application of CDs to Foods : CDs at present are mostly applied to foods, amounting for 90 % or more of the total CDs produced. CDs were initially used for removal of abnormal tastes and odors, stabilization of fragrances and stabilization of coloring substances. Recently, however, CDs have been utilized as a emulsifying agent carriers or in the form of powderization bases.

Application of CDs to Industries : (a) Preparation of aromatized plastics —— Aromatic compounds are powderized with CDs and then, kneaded into thermostable plastics.

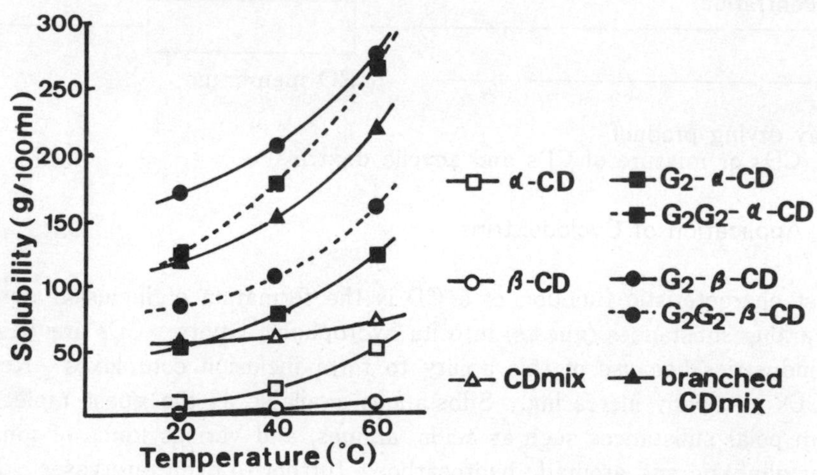

Fig. 1 Solubility of CD and Maltosyl-CD

(b) Aromatized insence sticks —— An aromatic compound is mixed with CD, and modeled to give a special odor at burning.

(c) Ink —— To control the surface tension.
(d) Agricultural chemicals —— The inclusion complex of "Pyrethroid" an insecticide and β-CD, has been patented for its manufacture. Application of CDs to herbicides, pheromones, plant growth regulators and repellents are also under investigation. In addition, application of CDs to moth proofing sheets is under development by kneading organic phosphoric camphor into plastics.

Other Application of CDs : Excellent reviews about the properties of CDs as catalysts in hydrolysis of esters etc. have been published by Bender and Komiyama (13). Also the hydrophobic cavities of CDs are regarded to incorporate oxygen. CD has catalytic ability of hydrolysis which is due to hydroxyl groups, but the activity is not so strong and the cavity is limited in size. Hattori and Toda (14) synthesized an enzyme model by introducing nicotinamide groups into CDs producing three kinds of CD derivatives. They seem to be oxido-reductase models. Hattori and Toda (14) observed that in a simple non-enzymic system containing CD, adenosine triphosphate (ATP) is formed from adenosine diphosphate (ADP) by reaction with creatine phosphate. CDs and modified CDs are thus an attractive research subjects.

Harada and Morimoto prepared a CD polymer by allowing the CD to react with epichlorohydrin, and investigated the properties (15). Two types of polymer were obtained depending on the reaction conditions. The water-soluble polymer serves to increase the solubility of sparingly soluble substances. On the other hand, water isoluble polymers of bead-form have been developed into exchangers by introducing carboxyl or dimethyl amino groups (15). Shaw and Wilson reported on a study of removing the bitter taste of orange and grapefruit by employing a β-CD polymer (16).

Armstrong *et al.* (17) reported on a novel liquid chromatography to separate isomers of various chemicals by employing CD-bonded silica-gel columns. The facts described above show that CDs are prospective material for chromatography beds to separate or fractionate various substances.

(Hitoshi Hashimoto)

References

1. Horikoshi, K. & Nakamura, N. (1981) in *Proceedings of the First International Symposium on Cyclodextrins* (Szetli, J., ed.) pp 25-39
2. Kobayashi, S., Otani, T., Watanabe, A., Kainuma, K. & Umeda, K (1985) *Japan Patent* 1290643
3. Taylor, P. M. & Whelan, W. J. (1966) *Arch. Biochem. Biophys. 113*, 500-502
4. French, D. & Abdullah, M. (1970) *Arch. Biochem. Biophys. 137*, 483-493
5. Hizukuri, S., Abe, J., Mizowaki, N., Koizumi, K. & Utamura, T. (1986) *J. Jpn. Soc. Starch Sci.* (in Japanese) *33*, 119-126
6. Okada, S., Yoshimura, Y. & Kitahata, S. (1986) *J. Jpn. Soc. Starch Sci.* (in

Japanese) *33*, 127-132
7. Kobayashi, S. & Kainuma, K. (1986) *Japan Kokai Tokkyo Koho* 86-92592
8. Sakano, Y. (1986) *Japan Kokai Tokkyo Koho* 86-70996
9. Sakano, Y., Shiraishi, T. & Niwa, H. (1986) *Japan Kokai Tokkyo Koho* 86-197602
10. Matsui, Y. (1986) *Hyomen* (in Japanese) *24*, 332-344
11. Shiraishi, T. & Sakano, Y. (1987) *Abstract of the Annual Meeting of the Agricultural Chemical Society of Japan* (in Japanese) p82
12. Oda, T., Hara, K., Takaku, H. & Oku, S. (1987) *Food Chemical* (in Japanese) *7*, 20-26
13. Bender, M. L. & Komiyama, M. (1987) *Cyclodextrin chemistry (Reactivity and Structure concepts in Organic chemistry 6) Springer Verlag*
14. Hattori, K. & Toda, F. (1982) *Yukigoseikagaku-Kyokaishi* (in Japanese) *40*, 1180-1188
15. Harada, K. & Morimoto, S. (1985) *Japan Patent* 1287106
16. Shaw, P. E. & Wilson Ⅲ, C. W. (1983) *J. Food Science 48*, 646-647
17. Armstrong, D. W., DeMond, W., Alak, A., Hinze, W. L., Riehl, E. & Khank, H. B. (1985) *Anal. Chem. 57*, 234-237

Ⅵ.10. Modified Cyclodextrins as Drug Carrier

α-, β- and γ-Cyclodextrins are the most popular cyclodextrins (CDs). They consist of six, seven and eight D-glucopyranose residues, respectively, linked by α-1,4 bonds. Because of the difference in the cavity diameter, each CD has its own degree of forming inclusion complexes with guest molecules of specific sizes. CDs have been widely utilized for improvements of the physical, chemical and biological properties of drug molecules (1,2), but they have several unfavorable properties as drug carriers. Various CD derivatives have thus been prepared to improve these unfavorable properties (3). This section deals with recent advances in the utilization of chemically modified CDs as drug carriers, classifying them into hydrophilic and hydrophobic CD derivatives.

Hydrophilic CD Derivatives : Hydrophilic CD derivatives have better aqueous solubilities than the original CDs. The aqueous solubility of β-CD is only 1.8% (w/v) at 25℃. The hydroxyl groups of CDs can most easily be structurally modified, and thus, incorporations of various functional groups have been carried out.

Methylated CDs : The low aqueous solubility of β-CD is ascribable partly to its stable crystal-lattice, being due to the widespreading hydrogen bonding network between the hydroxyl groups of β-CD and water molecules in the crystal (4). One of the methods proposed to increase the aqueous solubility is the introduction of substituents with less hydrogen bonding ability onto the hydroxyl groups of β-CD. Underlying this idea, methylated CDs such as heptakis (2,6-di-O-methyl)-β-CD (DM-β-CD) and heptakis

(2,3,6-tri-O-methyl)-β-CD (TM-β-CD) were prepared and their pharmaceutical applications, examined (5). DM- and TM-β-CDs are readily soluble in water (more than 30% (w/v) at 25°C) and organic solvents, less hygroscopic and highly surface active, compared with parent β-CDs. The inclusion ability of β-CDs is in most cases in the order to DM-β-CD > β-CD > TM-β-CD. The methylated β-CDs markedly improved the solubility, dissolution rate in vitro, and the oral, rectal and percutaneous absorptions in vivo of poorly water-soluble drugs such as fat-soluble vitamins (6), steroid hormones and cardiac glycosides (3). For example, the low aqueous solubility ($<1 \times 10^{-7}$ M) of vitamin E esters is increased by a factor of 10^5 by DM-β-CD complexation, leading to a markedly enhanced bioavailability (about 70 fold) after oral administration to dogs (Fig.1), though such enhancement is hardly achieved by the parent β-CD (7). Methylated CDs are inhibitory to various types of reactions, since the hydroxyl groups of the CDs, which act as nucleophilic agents and general base catalysts (8), are blocked. In addition, the less hygroscopic nature of the methylated CDs is an advantage, because of the moisture sorption that initiates hydrolytic decomposition of drugs in the solid state (9).

Hydroxypropylated CDs : 2-Hydroxypropyl CD derivatives are prepared by condensation with propylene oxide in alkaline conditions (10). The hydroxypropylation occurs statistically at many hydroxyl groups in the CD molecule, thus consisting of the amorphous mixture of chemically related components with different degrees of substitution. This multi-component character prevents crystallization of hydroxypropyl CDs from water, owing to a higher aqueous solubility, compared with those of the parent CDs. Upon inclusion complexation with hydroxypropyl CDs, crystalline drug molecules are converted to an amorphous solid-state. The products show a higher aqueous solubility and a faster dissolution rate, thus resulting in rapid absorption in vivo of poorly water-soluble drugs on oral and sublingual administrations (11-13). Hydroxypropyl CDs in the pharmaceutical field are superior in their bioadaptability, i.e. lower toxicity. As shown in Table 1 (14), the hemolytic activity of hydroxypropyl β-CD on human erythrocytes is significantly lower than that of the parent and methylated β-CDs, due to the diminished interaction with membrane components such as cholesterol, proteins and phospholipids to be removed from. Furthermore, the hemolytic activity decreases with increasing average degree of substitution in the hydroxypropyl-β-CD molecule. Also, the hydroxylpropyl CDs irritate the mucous membranes less than the parent or methylated CDs. Hydroxypropyl CDs are thus useful as drug carriers particularly for injections and liquid preparations for mucous membranes. Also, hydroxypropyl CDs with various carrier characteristics can be prepared by controlling the degree of substitution.

Branched CDs : Glucosyl (G1), maltosyl (G2) and dimaltosyl $(G2)_2$-CDs are so-called branched CDs in which one or two primary hydroxyl groups of the CDs are substituted by the mono- and di-saccharides through α-1,6 linkages, respectively (15,16). These branched CDs have recently received considerable attention in the pharmaceutical field, due to their high aqueous solubility (solubilities of G1- and G2-β-CDs at 25°C are about 97 and 147 w/v %, respectively). The solubilization effects of the branched β-CDs on poorly water-soluble drugs, particularly steroid hormones, are remarkable, which is related to their intrinsic high solubility as well as inclusion ability (14). The hemolytic activity against human erythrocytes decreases in the order of β-CD > G1-β-CD > G2-β-CD > $(G2)_2$-β-CD (Table 1). Thus, the inclusion complexations with these branched CDs as well as hydroxypropyl CDs are promising parental carriers enhancing the solubility of poorly watersoluble drugs and reducing the local toxicity such as hemolysis induced with various membrane-perturbing drugs.

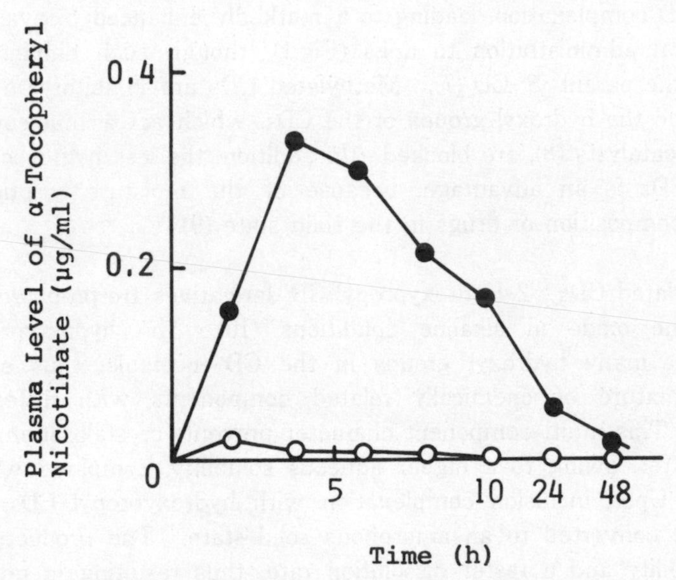

Fig. 1 Plasma Levels of α-Tocopheryl Nicotinate Following Oral Administration of Capsules Containing α-Tocopheryl Nicotinate or Its DM-β-CD Complex (equivalent to 100 mg drug) to Dogs.
○ : α-Tocopheryl nicotinate, ● : complex.

Epichlorohydrin-CD Polymers : A number of hydrophobic CD Polymers have been synthesized in the form of powders, films, gels, or beads. Among them, an O-alkylated CD polymer cross-linked with epichlorohydrin is the most popular (17). Various epichlorohydrin-CD polymers with different degrees of polymerization are available depending upon the preparation procedure. For example, β-CD polymers of relatively low molecular weights (<10,000) are considerably more soluble in water than the parent β-CDs and exist in an amorphous state having a rapid dissolution rate. These polymerization products effectively solubilize various poorly water-soluble drugs such as phenytoin, indomethacin, steroids and furosemide, leading to enhance oral or sublingual absorptions in vivo (11, 12, 18, 19).

On the other hand, CD polymers with high molecular weights are reported to be effective as disintegrating agents in tablets made by direct compression (20). Generally, the CD polymers show no untoward or toxic effects when administered chronically to mice p.o. for a 16-week period (11). They are not absorbed and may be eliminated into feces, because of the high molecular weight. The CD polymers with their superior abilities and apparent lack of oral toxicity are highly likely to serve as pharmaceutical additives for the preparation of various dosage forms.

Table 1. Hemolytic Activities[a] of β-CD Derivatives on Human Erythrocytes

β-CD	0.6
DM-β-CD	0.1
TM-β-CD	0.4
HP-β-CD (D.S.[b] = 2.5)	1.3
HP-β-CD (D.S. = 5.6)	1.4
HP-β-CD (D.S. = 6.8)	1.7
HP-β-CD (D.S. = 8.0)	1.9
HE-β-CD[c]	5.5
G1-β-CD	1.1
G2-β-CD	1.4
$(G2)_2$-β-CD	2.5
Epichlorohydrin-β-CD Polymer	0.5

[a] Expressed as concentration (w/v %) of CDs to induce 50% hemolysis in isotonic phosphate buffer at 37℃.
[b] Degree of substitution.
[c] 2-Hydroxyethyl β-CD.

Hydrophobic CD Derivatives :
There is little information concerning hydrophobic CD derivatives. However, they may be useful for controlling the release rates of water-soluble drugs. The following are our recent results on pharmaceutical applications of hydrophobic CD derivatives.

Ethylated CDs : When ethyl groups are introduced into the hydroxyl groups of β-CD, its physicochemical properties are markedly changed depending upon the degree of substitution. For example, heptakis (2,6-di-O-ethyl)-β-CD (DE-β-CD) and heptakis (2,3,6-tri-O-ethyl)-β-CD (TE-β-CD) are slightly soluble in water (DE-β-CD > TE-β-CD), less hygroscopic and more surface active than the parent β-CD (21). The dissolution and release rates of water-soluble drugs with short biological half-lives, such as diltiazem hydrochloride, isosorbide dinitrate, 5-fluorouracil, cimetidine and theophylline, are significantly retarded by complexation with the ethylated β-CDs. In addition, the hydrophobic CD complexes show pH-independent dissolution behaviors because of unionization at physiological pH regions (14). Table 2 shows the pharmacokinetic parameters obtained after oral administration of a single dose of tablets containing diltiazem or its ethylated β-CD complexes to rats (21). The mean residence time (MRT) in the systemic circulation, variance of residence time (VRT) and the time (t_{max}) required to reach the maximum plasma level (C_{max}) are markedly increased by administration of the drug as complexed form, showing the complex to be an excellent sustained-release preparation. Of interest is that the area under the plasma concentration curve (AUC) is also increased in the case of DE-β-CD complex. These data indicate that ethylated CDs are capable of controlling the release rate of water-soluble drugs. They may be applicable for hydrophobic drug carriers of sustained-release-type preparations.

Carboxymethylethyl CDs : Introduction of ionizable groups into hydroxyl groups of β-CD molecule causes an aqueous solubility change depending on environmental pHs. Carboxymethylethyl β-CD derivative (CME-β-CD) in which hydroxyl groups of DE-β-CD were substituted with carboxymethyl groups was first prepared by us (14). This derivative is less soluble in water at low pH regions, but freely soluble in alkaline regions due to the ionization of the carboxyl group (pKa about 4-5).

Table 2. Pharmacokinetic Parameters of Diltiazem Following Oral Administration of Tablets Containing Diltiazem or Its Ethylated β-CD Complexes (equivalent to 6 mg drug) to Rats

System	C_{max} (ng/ml)	t_{max} (min)	AUC[a] (h·ng/ml)	MRT (h)	VRT (h)
Diltiazem alone	411.1	30.0	1190	7.7	50.0
DE-β-CD complex	161.5	120.0	4200	20.4	164.3
TE-β-CD complex	33.1	30.0	971	20.6	118.1

[a]UP to 48 h post-administration.

Thus, the CME-β-CD complexes with various drugs, such as theophylline, 5-fluorouracil, cimetidine and propranolol hydrochloride, release the drugs very slowly in a low pH region of the gastric tract (pH about 1.2), while rapidly in the intestinal tract (pH about 6.8), the main absorption site. These results indicate that CME-β-CD may serve as an enteric type drug carrier, preventing the degradation of the drug in gastric juice and/or irritation of the gastric mucous membrane.

In the near future, new improved CD derivatives will be prepared and the inclusion complexations become more effective and valuable for enhancing the efficacy of drug activity.

(Kaneto Uekama and Fumitoshi Hirayama)

References

1. Szejtli, J. (1982) *Cyclodextrins and Their Inclusion Complexes*, Akadémiai Kiadó, Budapest
2. Duchêne, D., Glomot, F. & Vaution, C. (1987) *in Cyclodextrins and Their Industrial Uses* (Duchene, D., eds.) pp. 211-257, Editions de Sante, Paris
3. Uekama, K. & Otagiri, M. (1987) *in Critical Reviews in Therapeutic Drug Carrier Systems* (Bruck S.D., eds.) vol. 3, pp. 1-40, CRC Press, Boca Raton, FL

4. Saenger, W. (1984) *in Inclusion Compounds* (Atwood, J. L., Davies, J. E. D. & MacNicol, D. D., eds.) vol. 2, pp. 231-259, Academic Press, New York
5. Uekama, K. & Irie, T. (1987) *in Cyclodextrins and Their Industrial Uses* (Duchêne, D., eds.) pp. 393-439, Editions de Sante, Paris
6. Pitha, J. (1981) *Life Sci. 29,* 307-311
7. Uekama, K., Horiuchi, Y., Kikuchi, M., Hirayama, F., Ijitsu, M. & Ueno, M. (1988) *Incl. Phenom.* in press
8. Bender, M. L. & Komiyama, M. (1978) *Cyclodextrin Chemistry,* Springer-Verlag, Berlin
9. Kikuchi, M., Hirayama, F. & Uekama, K. (1987) *Int. J. Pharm. 38,* 191-198
10. Pitha, J., Milecki, J., Fales, H., Pannell, L. & Uekama, K. (1986) *Int. J. Pharm. 29,* 73-82
11. Pitha, J. & Pitha, J. (1985) *J. Pharm. Sci. 74,* 987-990
12. Pitha, J., Mitchell, H. & Michel, M. E. (1986) *J. Pharm. Sci. 75,* 165-167
13. Müller, B. M. & Brauns, U. (1986) *J. Pharm. Sci. 75,* 571-572
14. Uekama, K., unpublished data
15. Kobayashi, S., Maruyama, K. & Kainuma, K. (1983) *J. Jpn. Soc. Starch Sci.* (in Japanese) *30,* 231-239
16. Koizumi, K., Utamura, T., Sato, M. & Yagi, Y. (1986) *Carbohydr. Res. 153,* 55-67
17. Hoffman, J. L. (1973) *J. Macromol. Sci. Chem.* A7, 1147-1157
18. Uekama, K., Otagiri, M., Irie, T., Seo, H. & Tsuruoka, M. (1985) *Int. J. Pharm. 23,* 35-42
19. Uekama, K., Udo, K., Irie, T., Yoshida, A., Otagiri, M., Seo, H. & Tsuruoka, M. (1987) *Acta Pharm. Suec. 24,* 27-36
20. Fenyvesi, E., Takayama, K., Szejtli, J. & Nagai, T. (1984) *Chem. Pharm. Bull. 32,* 670-677
21. Uekama, K., Hirashima, N., Horiuchi, Y., Hirayama, F., Ijitsu, T. & Ueno, M. (1987) *J. Pharm. Sci. 76,* 660-661

VII. Amylolytic Preparations as Digestives in Japan

The following three sources for preparation of amylolytic digestives are at present approved in Japan: malt, swine pancreas and certain microorganisms, as shown in Table 1 (1). The amylolytic action and activity are examined and standardized estimating either the dextrinization activity or the saccharification activity. The digestives for pharmaceutical purposes are qualitatively and quantitatively standardized according to the Japan Pharmacopoeia. However, in this section only the manufacturing and assay methods of the amylolytic enzyme preparations from which the digestives are prepared for pharmaceutical purposes, are described.

Table 1. Sources and Types of Amylolytic Enzymes for Digestives in Japan

Source	Type	Opt. pH	Active pH range**
Bacillus subtilis	Amyloclastic	5.8	4.5-8.0
B. amyloliquefaciens	"	6.0	4.5-8.0
B. amylosolvens	"	5.0-6.0	4.5-8.0
Aspergillus oryzae	Mixed*	4.8	3.5-6.5
Asp. aureus	"	4.5-5.0	3.5-6.0
Asp. awamori	"	5.0	3.0-7.0
Asp. niger	"	4.0-4.5	2.0-8.0
Rhizopus chinensis	Saccharogenic	5.5	5.5-9.0
Malt	"	4.8	4.0-6.5
Swine pancreas	"	6.5-7.2	5.5-9.0

* The amylolytic enzymes produced by these microorganisms consist of both amyloclastic and saccharogenic types.

** pH-Range showing activites of more than 50% of that at the opt. pH.

VII.1. Swine Pancreatic Preparation

Swine pancreatic amylase is classified as α-amylase (1,4-α-D-Glucan glucanohydrolase). It produces glucose, maltose and various oligodextrins from starch. The amylase is protected by Ca^{++} and activated by Cl^-. Pancreatin J.P. (Japan Pharmacopoeia) is a digestive preparation consisted of swine pancreatic amylase, proteinase and lipase. The preparation is manufactured by processing swine pancreas with solvents to remove lipids and water followed by pulverizing the defatted and dried pancreatic tissue. Dextrin, lactose or crystalline cellulose are used as diluent and the preparation, standardized to meet the J.P. specifications. A brief manufacturing procedure is shown below :

Manufacturing Procedure of Pancreatin J.P :

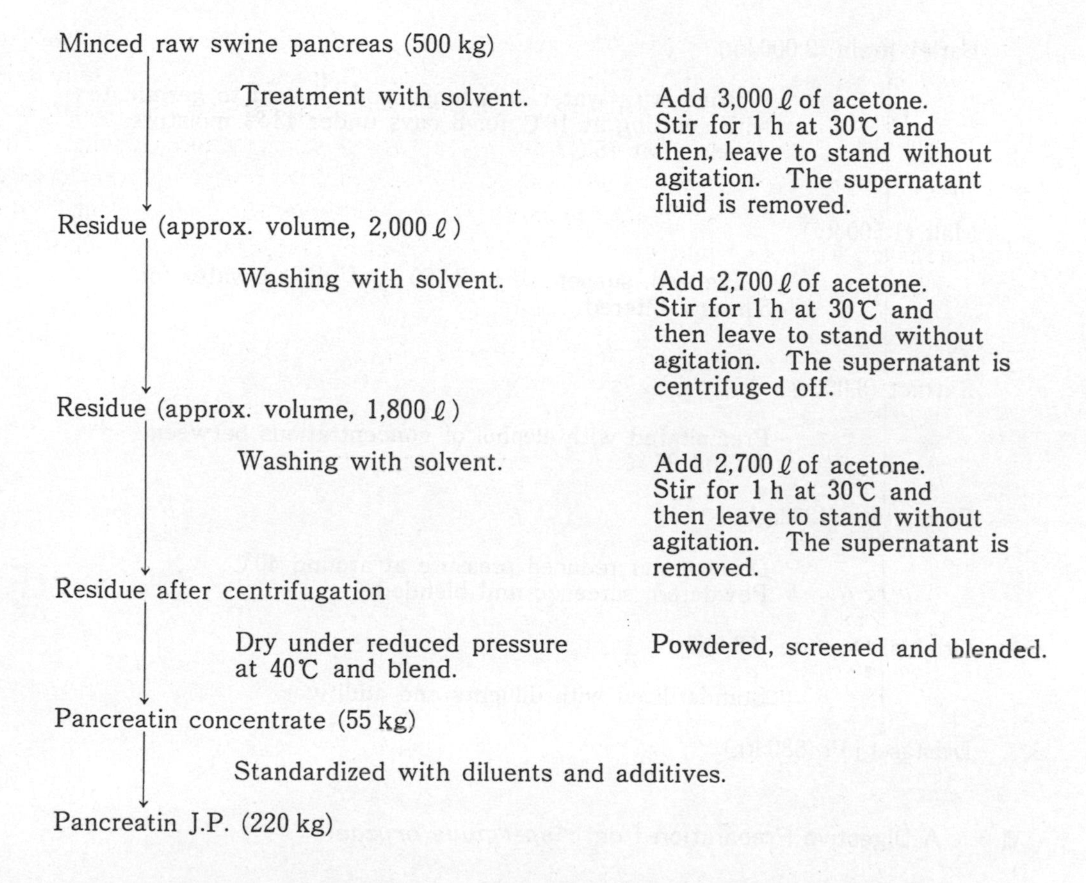

The recovery yield of amylase activity depends on the freshness of the raw pancreas to be processed, and generally the more fresh the pancreas, the higher the recovery of activity. Pancreatin has a tendency to absorb moisture and is generally sensitive even at room temperature.

Ⅶ.2. Malt Preparation

Malt contains both α-amylase (1,4-α-D-Glucan glucanohydrolase) and β-amylase (1,4-α-D-Glucan maltohydrolase), and the two activities work together on starch. The final reaction products are glucose, maltose, maltotriose and several oligosaccharides. The enzyme activity is protected by Ca^{++} like in the case of swine pancreatic amylase, but the activation by Cl^- is not observed. Diastase J.P. is a digestive which consists mainly of malt diastase standardized to meet the J.P. specifications with a diluent of dextrin or lactose. The malt diastase preparation is also hygroscopic and it is necessary to store the preparation in a sealed container at temperatures lower than 30℃. Malt diastase is produced through several steps of processing as shown below:

Manufacturing Procedure:

Barley grain (2,000 kg)
↓ Washed with water, moistened and allowed to germinate with aerating at 16℃ for 8 days under 43 % moisture. Air-dried at 75℃.
Malt (1,500 kg)
↓ Pulverized, suspended in 3,750 ℓ of chilled water for 5 h and filtered.
Extract (3,000 ℓ)
↓ Precipitated with alcohol of concentrations between 38 and 66 %.
Precipitate (120 kg)
↓ Dried under reduced pressure at around 40℃. Powdered, screened and blended.
Original Diastase (58 kg)
↓ Standardized with diluents and additives.
Diastase J.P. (580 kg)

VII.3. A Digestive Preparation from *Aspergillus oryzae*

The amylase preparation derived from *Asp. oryzae* has one of the highest α-amylase activity. The enzyme reaction products from starch consist mainly of maltose. Small amounts of glucose and several oligosaccharides are also produced. The α-amylase activity is protected by Ca^{++} but is not activated by Cl^{-}. A selected strain of *Asp. oryzae* is cultured on a wheat bran medium and the enzyme concentrate is obtained through several steps including fractionation with alcohol and finally dried in a powder form as shown below:

Manufacturing Procedure:

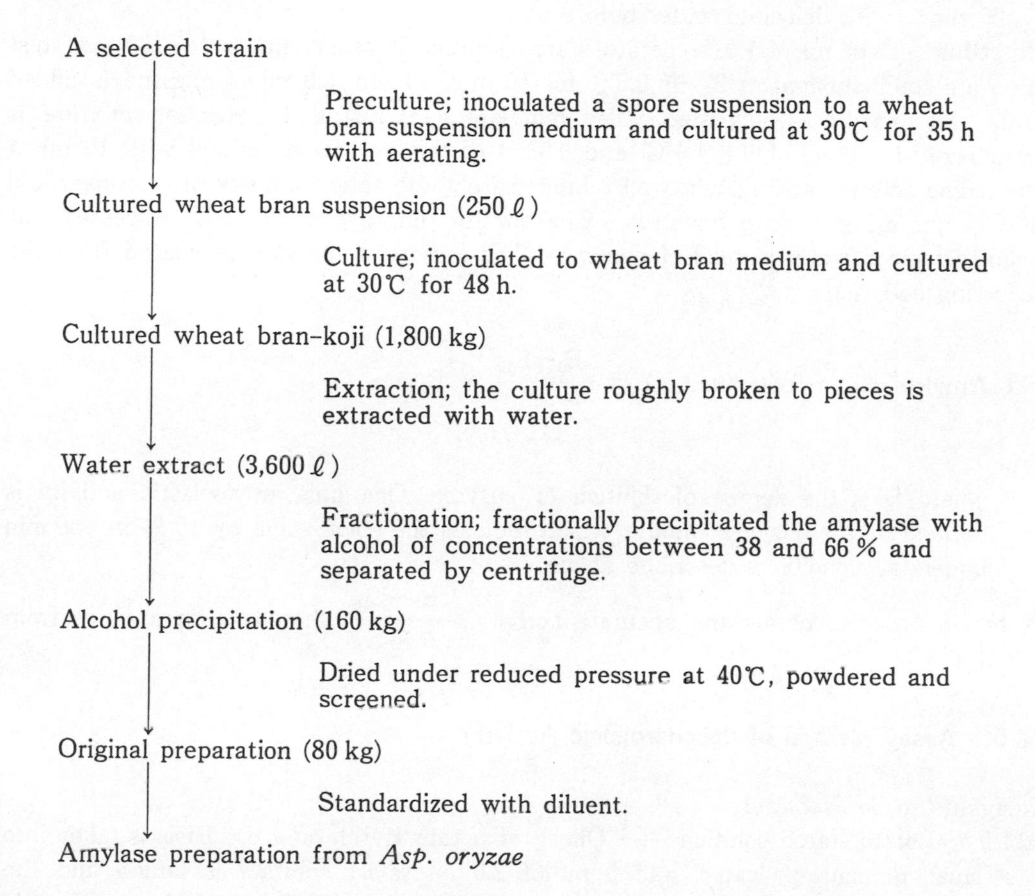

The enzyme preparation grades are based on the activity, but they are manufactured through dilution with dextrin, lactose or starch. All the preparations must be stored in a sealed container below 30℃.

VII.4. Assay Method of Amyloclastic Activity

Reagents to be Prepared;

a) 1.0 % Potato starch solution —— Two g of refined potato starch on a dry basis are placed into a 100 ml beaker and 40 ml of distilled water is added with gently shaking. Ten ml of 2.0 N NaOH is added to the starch suspension. This mixture is heated and gently boiled for 3 min. Then, 50 ml of water is added, the mixture cooled in running water, the pH adjusted to 5.0 with 2.0 N HCl, and 20 ml of 1.0 M acetate buffer (pH 5.0) added, followed by making the total volume up to 200 ml.

b) Iodine solution —— Five g of iodine and 25 g of potassium iodide are dissolved in 50 ml of deionized water, and the total volume is brought up to 500 ml with deionized water. The solution is stored in a brown bottle. This solution is diluted 200 times with deionized water before use.

Procedure ; Ten ml of 1.0 % potato starch solution is taken into a 18×180 mm test tube and preincubated at 37 ± 0.5℃ for 10 min. Then, 1.0 ml of enzyme is added and the mixture is incubated. Ten min later, 1.0 ml of the reaction mixture is transferred to 10 ml of 0.1 N HCl and 0.5 ml of this solution is mixed with 10 ml of the iodine solution and shaken well. Immediately, the optical density (E_{10}) is measured at 660 nm using 10 mm cuvettes. The control run (E_0) is similarly processed but using deionized water instead of enzyme. The enzyme activity is calculated from the following equation :

$$\text{Amyloclastic activity (unit/g)} = \frac{E_0 - E_{10}}{E_0} \times D$$

where D is the degree of dilution of enzyme. One unit amyloclastic activity is defined as the enzyme amount which reduces the color value by 10 % in one min under the conditions described above.

Note: In order to obtain the accurate activity, $\frac{E_0 - E_{10}}{E_0}$ should be in a range from 0.2 to 0.4.

Ⅶ.5. Assay Method of Saccharogenic Activity

Reagents to be Prepared ;

a) 1.0 % Potato starch solution —— One g of potato starch on a dry basis is taken into a small amount of water, and 5 ml of 2.0 N NaOH solution is added and the mixture is gently boiled for 5 min. After cooling, the pH of the solution is adjusted to pH 5.0 with 2.0 N HCl and 10 ml of 1.0 M acetate buffer (pH 5.0) is added followed by making the total volume up to 100 ml with water.

b) Fehling's reagents ——

1) Dissolve 34.64 g of cupric sulfate in water containing 0.50 ml of sulfuric acid to produce 500 ml.
2) Dissolve 176 g of sodium potassium tartrate and 77 g of sodium hydroxide in water to produce 500 ml.

Procedure ; Ten ml of 1.0 % soluble starch solution is taken into a 30×120 mm test tube, and preincubated at 37 ± 0.5℃ for 10 min. Then, 1.0 ml of enzyme is added to the starch solution. After 10 min incubation, 2 ml of Fehling's alkaline tartrate solution and 2 ml of Fehling's copper solution are added to stop the reaction, and the mixure is boiled for 15 min. After cooling in running water, 2 ml each of 30% potassium iodine solution and 25% sulfuric acid solution are added and titrated for liberated iodine with N/20 sodium thiosulfate solution (S ml). The blank value is obtained by measuring the incubation mixture but with deionized water instead of enzyme (B ml). One unit saccharogenic activity is defined as the amount of enzyme that produces 1 mg of reducing sugar as glucose per min at 37°C under above assay conditions.

Starch saccharifying activity (unit/g)

$$= (B\text{-}S) \times 1.6 \times f \times 1/10\ (\text{min}) \times D$$

where 1.6 is a constant to indicate the amount of glucose, mg, corresponding to 1.0 ml of N/20 sodium thiosulfate solution ; f, factor of N/20 sodium thiosulfate solution ; D, degree of dilution of enzyme.

(Shizuo Mihara)

Reference

1. The publication of Japan Medical Enzyme Manufacturers Association (1979)

Appendix

The commercial products of amylases and various other enzymes produced in Japan. (The name and address of the enzyme producer are indicated in the top part, and the enzymes produced, their brand names and sources, and applications are in the columns from left to right, respectively.)

Amano Pharmaceutical CO., Ltd.

1-2-7, Nishiki, Naka-ku, Nagoya 460, Japan
(Phone : 052-211-3032; Telex : 59805 BIOAMANO;
Cable : BIOAMANO NAGOYA; Fax : 052-211-3054

Enzyme	Brand name	Source	Application
Glucoamylase (amyloglucosidase)	Gluczyme	*Rhizopus* sp.	Glucose manufacturing
	Gluczyme NL (liquid)	*Asp. niger*	Glucose manufacturing
Fungal α-amylase	Biozyme	*Asp. oryzae*	Confectionery, baking and syrup manufacturing
	Biozyme L	*Asp. oryzae* (liquid)	Syrup manufacturing
Bacterial α-amylase	Amylase "Amano" A	*B. subtilis*	Starch liquefaction
β-amylase	β-amylase "Amano"	*Bacillus* sp.	Maltose manufacturing
Pullulanase	Pullulanase "Amano"	*Klebsiella* sp.	Starch debranching
Cyclodextrin glucanotransferase	CGT-ase	*Bacillus* sp.	Cyclodextrin manufacturing
α-Glucosidase	Transglucosidase "Amano"	*Asp. niger*	Isomaltose manufacturing
β-Galactosidase	Lactase F "Amano"	*Asp. oryzae*	Dairy industry
Dextranase	Dextranase "Amano"	*Chaetomium* sp.	Sugar industry
Fungal amylase	Biodiastase	*Asp. oryzae*	Digestive aid speciality

Cellulases, proteases and lipases are also available.

DAIWA KASEI K.K.

5-7-12, Uehonmachi, Tennoji-ku, Osaka 543, Japan
(Cable : ENZYMES OSAKA, Phone : 06-768-5001, Fax : 06-763-4767)

Enzyme	Brand name	Source	Application
Bacterial α-amylase	Biokleistase	*B. subtilis* var.	Textile strach size remover, additive for laundry washing mixture
Bacterial α-amylase	Kleistase	*B. subtilis* var.	Liquefaction of starch in food, brewing and starch processing industries, etc.
Bacterial α-amylase	Diasmen	*B. subtilis* var.	Pharmaceutical use as an ingredient of digestive aid

Bacterial α-amylase	Crystalline bacterial α-Amylase	*B. subtilis* var.	Laboratory use
Bacterial neutral proteinase	Protin-P	*B. subtilis* var.	Decomposition of proteinous substances in food and leather industries, etc.
Bacterial alkaline proteinase	Protin-A	*B. subtilis* var.	Decomposition of proteinous substances in food and leather industries, additive for washing compounds
Bacterial neutral proteinase	Promen	*B. subtilis* var.	Pharmaceutical use as an ingredient of digestive aid
Thermophilic-bacterial proteinase	Thermoase	*B. thermo-proteolyticus*	Decomposition of proteinous substances in food and leather industries, etc.
Thermophilic-bacterial proteinase	Thermolysin	*B. thermo-proteolyticus*	Laboratory use
Yeast cell lytic enzyme	Tunicase	*Arthrobacter* sp.	Production of yeast extract, etc.
Bacterial lactase	Biolacta	*B. circulans* var.	Decomposition of lactose

Fujisawa Pharmaceutical Co., Ltd.
4-3, Doshomachi, Higashi-ku, Osaka 541
(Phone : 06-201-4612; Fax : 06-222-6085)

Glucose oxidase	Glucose oxidase	*Asp. niger*	Oxygen remover, glucose remover
Pectin transeliminase	Zenelase	*Asp. japonicus*	Clarification of fruit juice

Godo Shusei Co., Ltd.
6-2-10, Ginza, Chuo-ku, Tokyo 104, Japan
(Phone : 03 575-2705; Fax : 03-571-5456)

Glucose isomerase	GODO-AGI	*St. griseofuscus*	High fructose syrup
Glucoamylase	Endogluzyme	*Endomyces* sp.	Food
Lactase	GODO-YNL	*Kluyveromyces lactis*	Food
Alkaline protease	GODO-BAP	*B. licheniformis*	Food, detergent
Chitinase	Chitinase -GODO	*Aeromonas hydrophila* subsp. anaerogenes	Reagent

Hankyu Kyoei Bussan Co., Ltd. (Distributor)
7-1-10, Tenjinbashi, Oyodo-ku, Osaka 531, Japan
(Phone : 06-358-1133; Fax : 06-358-5055)
Ueda Chemical Industrial Co., Ltd. (Manufacturer)
1-2-6, Takayanagi, Neyagawa-city, Osaka 572, Japan
(Phone : 0720-26-0287; Fax : 0720-26-2818)

Bacterial α-Amylase	Primase	*B. subtilis* (amyloliquefaciens)	Textile desizer
Bacterial α-Amylase	Alphamylase	*B. subtilis*	Dextrinization of starch in starch processing industry
Glucoamylase	Glutase	*Rhi. niveus*	Glucose making
β-Amylase	Hi-maltosin G	Wheat	Maltose making
Neutral protease	Orientase N	*B. subtilis*	Hydrolysis of protein
Acid protease	Orientase A	*Asp. niger*	Hydrolysis of protein
Cellulase	Cellulosin AC	*Asp. niger*	Macerating of vegetables
Hemicellulase	Cellulosin HC	*Asp. niger*	Acceleration of filtration speed of mashes
Pectinase	Orienzyme A	*Asp. niger*	Clarification of fruit juices

Hayashibara Biochemical Laboratories, Inc.
1-2-3, Shimoishii, Okayama 700
(Phone : 0862-24-4311; Fax : 0862-33-2265)

Pullulanase	Pullulanase (crystal, crude)	*Aerobacter aerogenes*	Production of maltotriose
Isoamylase	Isoamylase (crystal)	*Pseudomonas amyloderamosa*	Production of glucose Production of maltose Production of branched cyclodextrin
Cyclodextrin glucanotransferase		*B. stearothermophilus*	Production of cyclodextrin Production of coupling sugar® Production of α-glycosylstevioside

Kaken Pharmaceutical Co., Ltd.
3-4-10, Nihonbashi Honcho, Chuo-ku, Tokyo 103 (Mitsui Honcho Bldg.)
(Phone : 03-270-4351, Fax : 03-270-5305, Telex : 2222353 KAKNP J)

Streptomyces protease	Actinase E	*St. griseus*	Reagent
	Actinase AS	*St. griseus*	Food processing industry
	Actinase AF	*St. griseus*	Other industry uses

K·I Chemical Industry Co., Ltd.
328, Hamano Shioshinden, Fukude-cho, Iwata-gun, Shizuoka-ken 437-12
(Phone : 05385-8-1000; Fax : 05385-8-1263)

β-Galactosidase (Lactase)	Kumilase-LP	*P. multicolor*	Hydrolysis of lactose
β-1,3-Glucanase	Kitalase	*R. solani*	Preparation of protoplast Lysis of yeast cell wall
Chitosanase	Chitosanase K·I	*Bacillus* sp. R-4	Preparation of protoplast Hydrolysis of chitosan

Kirin Brewery Co., Ltd.
6-26-1, Jingumae, Shibuya-ku, Tokyo 150, Japan
(Phone : 03-499-6111; Telex : 242-5402 KIRINB J; Fax : 03-499-6237)

Zymolyase	Zymolyase -20T	*Arthrobacter luteus*	Lysis of yeast cell wall
	Zymolyase -100T	*Arthrobacter luteus*	Lysis of yeast cell wall (partly purified)
Diastase	Diastasum	Barley malt	Digestive Maltose syrup making

Matsutani Chemical Industry Co., Ltd.
5-3, Kitaitami, Itami City 664, Japan
(Phone : 0727-71-2010; Fax : 0727-70-0677; Telex : 5326-435 MKK J)

Amylase Mix. (α and Gluco.)	Matsulase	*B. subtilis* *Asp. oryzae* *Rhi. delemar*	Digestion of starch and cereals in brew and distilled spirits industry
Fungal protease	Protease	*Asp. oryzae*	Cereals vinegar industry

Meiji Seika Kaisha, Ltd.
2-4-16, Kyobashi, Chuo-ku, Tokyo 104
(Phone : 03-272-6511; Fax : 03-271-3528)

Fungal cellulase	Meicelase	*Tri. viride*	Digestive, Protoplast preparation
Fungal acid protease	Proctase	*Asp. niger*	Digestive, Food processing
Fungal α-amylase	Sanactase	*Asp. niger*	Digestive, Food processing
Streptomyces α-amylase	Meilase	*St. hygroscopicus*	Saccharification of starch in starch processing

Meito Sangyo Co., Ltd. Tokyo office
4-3-15 Muromachi, Nihonbashi, Chuo-ku, Tokyo 103
(Tel : 03-242-1795; Fax : 03-242-1792; Telex : 2226791 MEITO J)

Yeast lipase	Lipase-MY	*Candida cylindracea*	Enzyme digestant Milk flavor
Yeast lipase	Lipase-OF	*Candida cylindracea*	Fat-splitting
Fungal protease	Meito Rennet	*Mucor pusillus Lindt*	Cheese making

Mikuni Chemical Industries Co., Ltd.
4-1-16, Nihonbashi Muromachi 4-chome, Chuo-ku, Tokyo 103
(Phone : 03-270-8981; Fax : 03-241-5311)

Pancreatin	Pancreatin	Porcine pancreas	Digestive aid
Pancreatin	Kurapon	Porcine pancreas	Bating agent (leather industry)
Diastase	Diastase	Malt	Digestive aid
Pepsin	Pepsin	Porcine stomach mucosa	Digestive aid

Nagase Biochemicals Ltd.
1-52, Osadano-cho, Fukuchiyama, Kyoto 620
(Phone : 0773-27-5801; Fax : 0773-27-2040)

Bacterial α-amylase	Biotex	*B. subtilis*	Textile desizer
Bacterial α-amylase	Spitase	*B. subtilis*	Liquefaction of starch
Glucose isomerase	Swetase	*St. phaeochromogenes*	High fructose syrup making
Glucose oxidase	Deoxin	*P. amagasakiense*	Oxygen remover, glucose remover
Bacterial heat stable amylase	Spitase HS Biotex TS	*B. subtilis*	Starch processing
Fungal glucoamylase	Glucozyme NEO XL-128C	*Rhizopus* *Aspergillus*	Saccharification of starch Production of dextrose
Soy bean β-amylase	β-amylase #1500	Soy bean	Production of maltose Food industry
Bacterial protease	Bioprase	*B. subtilis*	Hydrolysis of proteins Marine industry Food industry, cosmetics, detergent

Bacterial catalase	Catalase "Nagase"	*Micrococcus lysodeikticus*	Oxidoreductive decomposition of H_2O_2
Fungal protease	Denazyme	*Asp. oryzae*	Food industry, improvement of taste of protein foods
Fungal acid-protease	Denapsin	*Asp. niger*	Hydrolysis of proteins
Fungal cellulase	Celluzyme	*Asp. niger*	Hydrolysis of cellulose
Fungal pectinase	Pectinase "Nagase"	*Asp. niger*	Clarification of fruit and vegetable juice
Egg white lysozyme	Egg white lysozyme	Egg white	Lysis of bacteria
Fungal lipase	Lipase (Saiken)	*Rhi. japonicus*	Hydrolysis of fat and oil Pharmaceutical uses
Plant papain	Papain	Papaya	Hydrolysis of proteins Meat tenderizer Cookie, brewing
Plant germ peroxidase	Peroxidase	Plant germ	Diagnostic reagent
Bacterial urease	Nagapsin	*Lactobacillus fermentum*	Removal of urea

Novo Industri Japan Ltd.
Sakura Bldg., 1-3-3, Uchikanda, Chiyoda-ku, Tokyo 101
(Phone : 03-295-6767; Fax : 03-233-0470)

Bacterial α-amylase	Aquazym	*B. subtilis*	Textile
	BAN	*B. subtilis*	Starch, alcohol, textile
	Ceremix	*B. subtilis*	Alcohol
	Termamyl	*B. licheniformis*	Starch, alcohol, textile, sugar
Fungal α-amylase	Fungamyl	*Asp. oryzae*	Starch, alcohol, juice, baking
Glucoamylase (amyloglucosidase)	AMG	*Asp. niger*	Starch, alcohol, juice, baking
	Dextrozyme	*Asp. niger/ Bacillus*	Starch, alcohol
Cellulase	Celluclast	*Tri. reseei (Tri. viride)*	Alcohol, wine & juice, paper, textile
	Celluzyme	*Humicola insolens*	Detergent, textile, paper
Milk coagulant	Rennilase	*M. miehei*	Dairy

Dextranase	DN	*Penicillium lilacinum*	Sugar
Galactomannanase	Gamanase	*Asp. niger*	Food processing
Glucose isomerase	Sweetzyme	*B. coagulans*	Starch processing
Glucanase	Cereflo	*B. subtilis*	Alcohol
	Ceremix	*B. subtilis*	Alcohol
	Finizym	*Asp. niger*	Alcohol
	Glucanex	*Tricoderma* sp.	Wine
	Novozym234	*Tri. harzianum*	Enzyme reagent
Lactase	Lactozym	*Kl. fragilis*	Dairy
Lipase	Lipozyme	*M. miehei*	Ester synthesis, interesterification
	Palatase M	*M. miehei*	Food processing
	Palatase A	*Asp. niger*	Dairy
Pectinase	Ultrazym	*Asp. niger*	Wine & Juice
	Pectinex Ultra SP-L	*Asp. niger*	Wine & Juice, food processing
Penicillin-V acylase	Semacylase		6-APA production
Phospholipase A-2	Lecitase	Porcine pancreas	Food processing
Bacterial proteinase	Alcalase	*B. licheniformis*	Detergent, food processing, textile, leather, dairy
	Alcamyl	*B. licheniformis*	Detergent
	Ceremix	*B. subtilis*	Alcohol
	Esperase	*Bacillus* sp.	Detergent, textile
	Neutrase	*B. subtilis*	Food processing, baking, leather, alcohol, dairy
	NUE	*Bacillus* sp.	Leather
	Novozym243	*B. licheniformis*	Enzyme reagent
	Savinase	*Bacillus* sp.	Detergent
	Subtilisin	*B. licheniformis*	Enzyme reagent
Pullulanase	Promozyme	*Bacillus* sp.	Starch processing
Trypsin, pancreatic	Crystalline Bovine Trypsin		Enzyme reagent
	Crystalline Porcine Trypsin		Enzyme reagent
	PEM		Enzyme reagent
	PTN		Food processing, leather

Nippon Shiryo Kogyo Co., Ltd.
18-1, Kanda Suda cho, Chiyoda-ku, Tokyo 101
(Phone : 03-254-1721; Fax : 03-256-7690)

Glucoamylase	Kokumilase "Touka"	*Rhi. delemar*	Partial substitute of Sake-Koji
Fungal α-amylase	Kokumilase "Ekika"	*Asp. oryzae*	Liquefaction of rice in Sake-industry
Glucoamylase + bacterial α-amylase	Kokumilase "Tsurugi"	*Rhi. delemar* + *B. subtilis*	Saccharification of rice in Sake-industry Partial substitute of Sake-Koji

Oriental Yeast Co., Ltd.
Nihonbashi Fukawa Bldg. 10-11 Kodenma-cho, Nihonbashi, Chuo-ku, Tokyo 103
(Phone : 03-663-8218; Fax : 03-663-8226, 8238)

Alcohol dehydrogenase	ADH	Yeast	OYC purified enzymes are;
Aldehyde dehydrogenase	AlDH	Yeast	
Ascorbate oxidase	ASOD	Cucumber	1) For laboratry use of biochemistry and molecular biology
Cholesterol esterase	CE	Pancreas	
Creatine kinase	CK	Rabbit muscle	2) For analysis of in vitro diagnostics and food component
Cholesterol oxidase	CO	*Nocardia* sp.	
Carboxypeptidase	CPase Y	Yeast	
Y Diaphorase	Diaphorase	*Clostridium kluyveri*	
Phosphopyruvate hydratase	Enolase	Yeast	
Formaldehyde dehydrogenase	FADH	Yeast	
β-D-Galactosidase	β-Galactosidase for EIA	*E. coli*	
Glyceraldehyde-3-phosphate dehydrogenase	GAPDH	Thermobacteria	

Glycerol dehydrogenase	GDH	Bacteria
α-Glycerophosphate dehydrogenase	α-GDH	Rabbit muscle
Glycerol kinase	GK	*E. coli*, Yeast
Glutamate dehydrogenase	GlDH	Beef liver, Yeast
α-D-Glucosidase	α-Glucosidase	Yeast
Glucose oxidase	GOD	Microorganism
Glucose-6-phosphate dehydrogenase	G-6-PDH	Yeast, *Leu. mesenteroides*
Glutathione reductase	GR	Yeast
γ-Glutamyltransferase	γ-GT	Beef Kidney
Hexokinase	HK	Yeast
3α-Hydroxysteroid dehydrogenase	3α-HSDH	Bacteria
Isocitrate dehydrogenase	iCDH	Yeast
Lactate dehydrogenase	LDH	Pig heart *Leu. mesenteroides* Rabbit muscle
Lipoxygenase	Lipoxidase	Soy bean
Malate dehydrogenase	MDH	Yeast, pig heart
Myokinase	MK	Yeast
Maltose phosphorylase	MP	Bacteria
Mutarotase	Mutarotase	Pig kidney
Phosphogluconate dehydrogenase	6-PGDH	Yeast, *Leu. mesenteroides*
Phosphoglycerate kinase	PGK	Yeast
Pyruvate kinase	PK	Pig heart, rabbit muscle
Phosphotransacetylase	PTA	*Leu. mesenteroides*
RNA Polymerase	RNA Polymerase	Yeast

Urate oxidase	Uricase	Yeast	

K.K. OSAKA SAIKIN KENKYUSHO
9-55, Taisha-cho, Nishinomiya City, Japan 662
(Phone : 0798-71-6426, Fax : 0798-71-6885)

Fungal lipase	Lipase saiken	*Rhi. Japonicus* NR400 (Exocellular type)	Pharmaceutical use Cosmetic use Milk and butter flavor making
	Olipase	*Rhi. Japonicus* NR400 (Endocellular type)	Pharmaceutical use Feeds for animal and fish
	Loose		Remove fat from raw silk

Rakuto Kasei Industrial Co., Ltd.
705 Sekinotsu-cho Tanakami Otsu City Shiga Pref.
(Phone : 0775-46-0333, Fax : 0775-46-3158)

Bacterial α-amylase	Ractase	*B. subtilis*	Textile desizer Food processing
Neutral protease	Enzylon BS	*B. subtilis*	Textile desizer Food processing
Cellulase	Enzylon C	*B. subtilis*	Textile desizer Food processing
Mannanase	Begalase	*B. subtilis*	Textile desizer Food processing

SEISHIN PHARMACEUTICAL CO., LTD.
4-13, Koamicho, Nihonbashi, Chuo-ku, Tokyo 103
(Phone : 03-669-2876, Fax : 03-669-1684)

Protease	MOLSIN	*Asp. saitoi*	Digestive enzyme
	PD	*Pen. duponti*	Food and feed industries
	IP	*Asp. sojae*	Food and feed industries
	AO	*Asp. oryzae*	Food and feed industries
Fungal α-Amylase	STALASE	*Asp. aureus*	Digestive enzyme

Pectinase	PECTOLY-ASE	*Asp. japonicus*	Fruit juice clarification
	PECTOLY-ASE Y-23	*Asp. japonicus*	Protoplast preparation from higher plants
Cellulase	CELLULASE Y-C	*Tri. viride*	Protoplast preparation from higher plants

Shin Nihon Chemical Co., Ltd.
19-10, Showa-cho, Anjo-shi, Aichi-ken 446, Japan
(Phone : 0566-76-5171; Fax : 0566-75-0010)

Acid fungal protease	Sumizyme AP	*Asp. niger*	Processing of fish solubles and whole fish, preparation of soluble and hydrolyzed proteins
β-Glucanase	Sumizyme BG	*Asp. niger*	Hydrolysis of β-glucan in brewing
Cellulase	Sumizyme C Sumizyme AC	*Tri. reesei* *Asp. niger*	Preparation of fruit juice, wine, vegetable oil and beer brewing
Fungal α-amylase	Sumizyme L (liquid and powder)	*Asp. oryzae*	Starch conversion alcohol, brewing and baking
Fungal lactase	Sumilact	*Asp. oryzae*	Hydrolysis of lactose in dairy products
Fungal protease	Sumizyme LP & LPL	*Asp. oryzae*	Baking, meat tenderizing and other protein modification and hydrolysis
Glucoamylase	Sumizyme	*Rhi. delemar*	Starch conversion, dextrose and alcoholic beverage production
Hemicellulase	Sumizyme ACH	*Asp. niger*	Fruit juice processing and baking
Pectinase	Sumizyme AP2 & MC	*Asp. niger* *Asp. oryzae*	Processing of fruits juice, vegetable juice and wine
Semi-alkaline protease	Sumizyme MP	*Asp. melleus*	Processing of fish solubles and meat
Xylanase	Sumizyme TX	*Tri. reesei*	Fruit juice processing and baking

Showa Denko K.K.
1-13-9 Shiba Daimon, Minato-ku, Tokyo 105
(Phone : 03-432-5111, Telex : J26232, Fax : 03-436-4668)

Bacterial protease	Kazusase	*Bacillus* Sp.	Detergents
Yeast protopectinase	Technase	*Trichosporon*	Pectin production

TANABE SEIYAKU CO., LTD.
3-21, Dosho-machi, Higashi-ku, Osaka 541
(Phone : 06-205-5555, Fax : 06-205-5370)

Hesperidinase	Hesperidinase "TANABE"	*Asp. niger*	Preventing clouding in syrup-preserved orange
Naringinase	Naringinase	*Asp. niger*	Removing bitter taste from citrus fruits
Pectinase	Pectinase "TANABE"	*Asp. niger*	Clarifying fruit juice and wine
Lipase	Talipase (Lipase RH)	*Rhi. delemar*	Enhancing flavor and taste of daily products
Amylase (α-amylase, glucoamylase)	Morotomin	*Asp. oryzae* *Rhi. delemar*	Liquefying and saccharifying of starch

TOYOBO CO., LTD
Biochemical Operations Div.
2-2-8 Dojima hama, Kita-ku, Osaka, 530 Japan
(Phone : 06-348-3785-8, Fax : 06-347-0839, Telex J63465 TOYOBO)

Galactose oxidase	GAO-201	*Dactylium* sp.	Useful for determination of galactose
α-Glucosidase	AGH-201	*Saccharomyces* sp.	Useful for enzymic determination of amylase
β-Glucosidase	BGH-201	Sweet almond	Useful for determination of amylase
Glucose oxidase	GLO-201	*Aspergillus* sp.	Useful for determination of amylase or glucose
Glucoamylase	GLA-111	*Rhizopus* sp.	Useful for determination of amylase

Invertase	IVH-101	*Candida* sp.	Useful for determination of saccharose
Glucose-6-phosphate dehydrogenase	G6D-311	*Leuconostoc* sp.	Useful for determination of amylase or glucosease
Fructose dehydrogenase	FCD-301	*Gluconobacter* sp.	Useful for determination of frucotse
Mannitol dehydrogenase	MND-301	*Actinobacillus* sp.	Useful for determination of mannitol
Hexokinase	HXK-301	*Saccharomyces* sp.	Useful for determination of amylase or glucose

UNITIKA Ltd.
Medical Development Department
4-68, Kitakyutaro-machi, Higashi-ku, Osaka 541
(Phone : 06-281-5021; Fax : 06-281-5256)

Acetate kinase	*B. stearothermophilus*	Analytical use (determination of acetate), ATP regeneration
Adenylate kinase	*B. stearothermophilus*	Analytical use (determination of ADP), ATP regeneration
Diaphorase [EC 1.6.99.—]	*B. stearothermophilus*	Analytical use
Diaphorase [EC 1.6.4.3]	*B. stearothermophilus*	Analytical use
Glucokinase	*B. stearothermophilus*	Analytical use (diagnostic reagent)
Glucose 6-phosphate dehydrogenase	*B. stearothermophilus*	Analytical use (diagnostic reagent)
Glutamin synthetase	*B. stearothermophilus*	Analytical use (determination of glutamine and glutamate)
Glyceraldehyde 3-phosphate dehydrogenase	*B. stearothermophilus*	Analytical use
Phosphofructokinase	*B. stearothermophilus*	Analytical use
Phosphoglucose isomerase	*B. stearothermophilus*	Analytical use

Phosphoglycerate kinase		*B. stearothermophilus*	Analytical use
Phosphotrans-acetylase		*B. stearothermophilus*	Analytical use
Polynucleotide phosphorylase		*B. stearothermophilus*	Preparation of polynucleotide
Pyruvate kinase		*B. stearothermophilus*	Analytical use (diagnostic reagent)
Superoxide dismutase		*B. stearothermophilus*	Medicine, cosmetic material, antioxidant

Yakult Honsha Co., Ltd.
1-1-19, Higashishinbashi, Minato-ku, Tokyo 105, Japan
(Phone : 03-574-6766; Fax : 03-574-7253; Telex : 02523950 YAKULT J)

Fungal cellulase	Cellulase "Onozuka" R-10,RS	*Tri. viride*	Protoplasts perparation from plants
Fungal β-1,3-glucanase	Funcelase	*Tri. viride*	Plotoplasts preparation from Eumycetes
Fungal poly-galacturonase	Macerozyme R-10, R-200	*Rhizopus* sp.	Plant cell isolation
Fungal cellulase	Cellulase "Onozuka" P-1500, FA	*Tri. viride*	Estimation of forage digestibility
Fungal cellulase	Cellulase "Onozuka"3S	*Tri. viride*	Extraction of useful components from natural products
Fungal β-1,3-glucanase	Funcelzym	*Tri. viride*	Extraction of useful components from mashrooms and yeasts
Fungal poly-galacturonase	Pectinase SS, 3S, HL	*Aspergillus* sp.	Acceleration of filtration, Juice clarification
Fungal poly-galacturonase	Macerozyme 2S	*Rhizopus* sp.	Maceration of natural products
Fungal β-galactosidase	Lactase Y-AO	*Asp. oryzae*	Hydrolysis of lactose
Fungal acid protease	Protease YP-SS	*Asp. niger*	Production of enriched extracts

Fungal neutral protease	Pancidase NP-2	*Asp. oryzae*	Ditto
Bacterial alkaline protease	Aroase AP-10	*B. subtilis*	Production of fish soluble
Fungal glucoamylase	Uniase 30	*Rhizopus* sp.	Production of glucose
Bacterial α-amylase	Uniase BM-8	*B. subtilis*	Liquefaction of starch

Subject Index

A

B

H

V

W

X

Z